AF699890

WITTGENSTEIN: MIND AND LANGUAGE

SYNTHESE LIBRARY

STUDIES IN EPISTEMOLOGY,

LOGIC, METHODOLOGY, AND PHILOSOPHY OF SCIENCE

Managing Editor:

JAAKKO HINTIKKA, *Boston University*

Editors:

DIRK VAN DALEN, *University of Utrecht, The Netherlands*
DONALD DAVIDSON, *University of California, Berkeley*
THEO A.F. KUIPERS, *University of Groningen, The Netherlands*
PATRICK SUPPES, *Stanford University, California*
JAN WOLEŃSKI, *Jagiellonian University, Kraków, Poland*

VOLUME 245

Ludwig Wittgenstein, *Bleistiftzeichnung von Drobil*, © Frau Katarina Eisenburger, reproduced here by kind permission of Frau Katarina Eisenburger.

WITTGENSTEIN: MIND AND LANGUAGE

Edited by

ROSARIA EGIDI

Department of Philosophy
Third University of Rome, Italy

KLUWER ACADEMIC PUBLISHERS
DORDRECHT / BOSTON / LONDON

Library of Congress Cataloging-in-Publication Data

Wittgenstein : mind and language / edited by Rosaria Egidi.
p. cm. -- (Synthese library ; v. 245)
"Revised version [of] the papers of the Wittgenstein conference held in November 1989 at the University of Rome La Sapienza"--Pref.
Includes indexes.
ISBN 0-7923-3171-0 (alk. paper)
1. Wittgenstein, Ludwig, 1889-1951--Contributions in philosophy of mind--Congresses. 2. Wittgenstein, Ludwig, 1889-1951--Contributions in philosophy of language--Congresses. 3. Philosophy of mind--Congresses. 4. Language and languages--Philosophy--Congresses.
I. Egidi, Rosaria. II. Series.
B3376.W564W552 1995
192--dc20 94-37362

ISBN 0-7923-3171-0

Published by Kluwer Academic Publishers,
P.O. Box 17, 3300 AA Dordrecht, The Netherlands.

Kluwer Academic Publishers incorporates the publishing
programmes of D. Reidel, Martinus Nijhoff, Dr W. Junk and MTP Press.

Sold and distributed in the U.S.A. Canada
by Kluwer Academic Publishers,
101 Philip Drive, Norwell, MA 02061, U.S.A.

In all other countries, sold and distributed
by Kluwer Academic Publishers Group,
P.O. Box 322, 3300 AH Dordrecht, The Netherlands.

All Rights Reserved.

© 1995 Kluwer Academic Publishers
No part of the material protected by this copyright notice
may be reproduced or utilized in any form or by any means, electronic
or mechanical, including photocopying, recording or by any information
storage and retrieval system, without permission from
the copyright owner.

Printed in the Netherlands

In memory of Norman Malcolm
(1911–1990)

TABLE OF CONTENTS

Preface xi

GEORG HENRIK VON WRIGHT / Wittgenstein and the Twentieth Century 1

PART I. SCIENCE, MIND, AND METAPHYSICS

DAVID PEARS / Wittgenstein on Philosophy and Science 23
ANTHONY J.P. KENNY / Wittgenstein on Mind and Metaphysics 37
JOHANNES L. BRANDL / Wittgenstein's Alleged Metaphysics of Mind 47

PART II. IN THE DOMAIN OF POSSIBILITIES: RULES, FORMAL AND INNER LANGUAGES

JACQUES BOUVERESSE / Le réel et son ombre: la théorie wittgensteinienne de la possibilité 59
CARLO CELLUCCI / Wittgenstein on the Meaning of Logical Symbols 83
AVRUM STROLL / On Following a Rule 93
DIEGO MARCONI / Fodor and Wittgenstein on Private Language 107
MASSIMO DELL'UTRI / Mentalesians and Wittgenstein's Private Language 117

PART III. INTENTIONALITY AND THE GRAMMAR OF MENTAL LIFE

ALDO G. GARGANI / Intentionality in Wittgenstein's Works 127
HERBERT HOCHBERG / Intentional Reference as a Logical Relation: A Variation on a Theme in Moore, Russell, Wittgenstein, and Bergmann 137
ROSARIA EGIDI / Wittgenstein Between Philosophical Grammar and Psychology 171
ROBERTO CASATI / Notes on Phenomenology and Visual Space 185

PART IV. BELIEF, DOUBT, AND RELATIVISM

NORMAN MALCOLM / Disentangling Moore's Paradox 195
EVA PICARDI / Malcolm on Moore's Paradox 207
WOLFGANG CARL / Wittgenstein's Refutation on Scepticism in 'On Certainty' 211
RUDOLF HALLER / Was Wittgenstein a Relativist? 223

PART V. TOPICS IN THE PHILOSOPHY OF PSYCHOLOGY

S. STEPHEN HILMY / Wittgenstein on Language, Mind and Mythology 235
JOACHIM SCHULTE / Emotion: Remarks on Wittgenstein and William James 249
GUIDO FRONGIA / Wittgenstein and Memory 263
FRITZ G. WALLNER / How Ludwig Wittgenstein Would Have Reacted to Recent Changes in Psychology 279

INDEX OF NAMES 289

INDEX OF SUBJECTS 293

PREFACE

The essays collected in this volume represent, in a revised version, the papers of the Wittgenstein Conference held in November 1989 at the University of Rome 'La Sapienza' to celebrate the centenary of his birth. They offer a systematic account of Wittgenstein's philosophy of mind and contribute to illuminate his later conception of perceptive, emotional and cognitive language. Some of the reasons why it seemed the right time to promote an ample confrontation of ideas on Wittgenstein's mature perspective are sufficiently clear as they derive from the need to sum up the state of research based on the availability of the *Nachlaß* and the publication in the last decade of a conspicuous quantity of writings dedicated to philosophical psychology; other reasons are more complex as they depend on the already noticed tendency in the recent epistemological debate to interpret Wittgenstein's provocative and controversial theses in a "perverse" way, in a way which has been used as a banner for epistemic relativism, subjectivism, and irrationalism.

The intention of this collection of essays is to construct an image of Wittgenstein's thought, which is as faithful as possible to his philosophy of mind and language from both a theoretical and exegetical point of view. The book also strives to assess the continuity and internal coherence of the theses developed throughout the different phases of his research. Attention is focused on the fundamental topics of philosophical grammar investigated by Wittgenstein starting from the Thirties: they are intentionality, analysis of belief and Moore's paradox, certainty and doubt, will, memory, sensations and emotions, the theory of aspects and of private language in its relationship with scepticism and psychologism. All these topics are part of a great plan for the treatment of non-phenomenological language to which, according to Wittgenstein's late perspective, belong both psychological, as well as mathematical concepts and whose criteria of meaning are not of an empirical nature but have the "generality of a supreme rule of the game admitting of no appeal".

The purpose of the division into thematic areas is to provide, without claiming them to be exhaustive, some important points of reference and, in a way, to create a framework for Wittgenstein's grammar of inner experience with the aim to exhibit the connections between his philosophy of mind and general

R. Egidi (ed.), Wittgenstein: Mind and Language, xi–xii.
© 1995 *Kluwer Academic Publishers. Printed in the Netherlands.*

views on metaphysics, logic, and psychology. In comparison to the recent specialized literature dedicated to the themes of philosophical psychology, some of the papers presented in this volume dedicate special attention to the historical and critical context of Wittgenstein's theses on the subject and to the examination of their roots in the European thought of his time.

I would like at this point to express my gratitude to the contributors to this volume, to the Università di Roma 'La Sapienza', to the Consiglio Nazionale delle Ricerche and to the Istituto Italiano per gli Studi Filosofici for their substantial support at the Conference. I also wish to thank my colleagues and collaborators at the Dipartimento di Filosofia e Teoria delle Scienze Umane. Last but not least, to Brian McGuinness and Rudolf Haller go my special thanks for their help and encouraging cooperation.

Rome, September 1992 ROSARIA EGIDI

GEORG HENRIK VON WRIGHT

WITTGENSTEIN AND THE TWENTIETH CENTURY

I have called my paper "Wittgenstein and the Twentieth Century". It is not my intention, however, to discuss how Wittgenstein has impressed and changed thinking in our century; nor shall I discuss the influences and impulses which he may have received from other philosophers. What I wanted to do is something which is more hazardous and indefinite – and in the opinion of many perhaps not very important either. I wanted to relate Wittgenstein to a prevailing climate of opinion or cultural situation, to something which may also be called the "moods" or *Stimmung* of a time. I have made an effort in this direction once before.[1] Here I wanted to follow it up with thoughts on how Wittgenstein's rejection of the civilization of contemporary Western society reflects a basic attitude of his to life and on how this attitude carries both his earlier and his later work in philosophy.

1. A dominant feature in the spiritual physiognomy of the twentieth century is Modernity. It has become recognized under that name largely in retrospect and in contrast to tendencies which either are critical of it or champion a new, "post-modern" mood of the time.

Modernity, thus conceived, is our legacy of the Enlightenment and the French Revolution. It is the Age of Reason matured to become an age of science and technology, of an industrial mode of production, and of democratic forms of government. In origin it was an optimistic mood. It cherished a vision of linear and unlimited perfection and progress towards a *regnum hominis* of free and equal men. The yoke of superstitious beliefs being lifted, also that of despotic government would never again be allowed to oppress man.

This original mood was partly reflected in but partly also reinforced by the ideas of evolution which were characteristic of 19th century scientific thought. They range from the historical linguistics of a Rasmus Rask and Jacob Grimm, through Charles Lyell's new biography of the earth, to Darwin's theory of the origin of species and descent of man. These scientific achievements encouraged the beliefs and sentiments, epitomized in the positivist philosophy of Auguste Comte and his followers, the evolutionist progressivism of Herbert Spencer, and also, though with more ambiguous and sophisticated overtones, in Hegel's phenomenology of the spirit.

R. Egidi (ed.), Wittgenstein: Mind and Language, 1–19.
© 1995 *Kluwer Academic Publishers. Printed in the Netherlands.*

This climate of opinion, here briefly characterized, prevailed not least in late Victorian and Edwardian England. It enjoyed an exceptionally intense "Indian Summer" of creative intellectual talent in pre-first-war Cambridge where the young Wittgenstein came to study logic with Bertrand Russell.

The First World War gave a shock to this mood but it by no means crushed it. The war could also be viewed as a great convulsion needed for breaking the fetters of unreason in which reactionary forces of the past had tried to hold back humanity on its way to Modernity. To many the revolution in Russia seemed the continuation and final breakthrough of the spiritual forces first let loose in France more than a century earlier.

The so-called modernistic movements in architecture, art and literature testify to this renewed optimism of the post-first-war period. And so do the efforts to create a global international organization to secure peace and progress for a world made – as said a slogan of the time – "safe for democracy".

This rejuvenated Modernity had one of its most consequential and serene reflections in the philosophic trend known as logical positivism. I say "consequential", thinking of the repercussions which this trend has had on philosophy throughout the century. And I say "serene", thinking of the rationalist ethos and consciousness of a message which animated many of its early protagonists. This feeling of unisonity with a tune of the times is perhaps nowhere reflected as movingly as in Carnap's Foreword to *Der logische Aufbau der Welt.* I quote from it here in English:

> We feel that there is an inner kinship between the attitude on which our philosophical work is founded and the intellectual attitude which presently manifests itself in entirely different walks of life; we feel this orientation in artistic movements, especially in architecture, and in movements which strive for meaningful forms of personal and collective life, of education, and of external organization in general. We feel all around us the same basic orientation, the same style of thinking and doing. – Our work is carried by the faith that this attitude will win the future.[2]

I vividly remember how these words moved me as a young student in Helsinki in the mid-1930s – and I think we were fortunate in my country to have received our inspiration in philosophy from a charismatic teacher, Eino Kaila, who saw himself in the vanguard of a radically new way of thinking.

The optimistic mood apart, what was new about this thinking were above all two things. One was the strong emphasis on logic as the core and center of philosophy. The other was the alignment with science, the feeling that philosophy too had, at long last, attained "den sicheren Gang der Wissenschaften".[3] "Wissenschaft" should then be understood in the traditional English sense of the word, comprising mathematics and the natural sciences, rather than in the broader German sense.

It is worth noting here that the idea of the natural sciences setting the pattern also for philosophic thinking has deeper roots in Austrian than in German or even in French or English philosophy. Schlick was successor to the chair once held by Mach in Vienna, and the Wiener Kreis could be called an inner circle of the Ernst Mach Society which introduced the *Kreis* to the world in its manifesto *Wissenschaftliche Weltauffassung*. But already long before that, Brentano had proclaimed in his *Habilitationsschrift* of 1866 that "vera philosophiae methodus nulla alia nisi scientiae naturalis est".[4] The key position of Brentano and his several pupils in the development of philosophy in our century has with time become increasingly obvious when the ties of the "new philosophy" to the phenomenalistic sensualism of Mach have gradually loosened, and the current originating as logical positivism has broadened into the mighty stream of analytic philosophy.

In spite of the many tributaries which have, in the course of the years, emptied their waters into this river, I think it is right and illuminating to call analytic philosophy the mainstream of philosophic thinking in this century. In all its heterogeneity it retains the two features which I already mentioned as typical of its origin: the emphasis on logic and the alignment with science. It is, in short, the philosophy most characteristic of a culture dominated by scientific rationality.

As is well attested, the influence of Wittgenstein's *Tractatus* on the Vienna Circle was profound. The authors – Carnap, Hahn, and Neurath – of the "manifesto" hailed Wittgenstein, together with Einstein and Russell, as the leading representative of a "scientific view of the world" or, with its more telling German name, "eine wissenschaftliche Weltauffassung".[5]

That this characterization is profoundly untrue of Wittgenstein's *Denkart* is obvious to every serious student of his philosophy. Yet it is even today quite common to label Wittgenstein, if not a "logical positivist" so at least an "analytic philosopher". There are deeper reasons for this misunderstanding than just the difficulties of getting rid of a label once attached. If Wittgenstein is not an analytic philosopher, what kind of philosopher is he then? This question certainly cannot be answered in the terms of current classifications. He is not a phenomenologist or hermeneuticist, nor an existentialist or hegelian, least of all is he a marxist.

It is, moreover, not too difficult to reinterpret Wittgenstein's thinking so as to fit the analytic mould. He said in the *Tractatus* that "Philosophy is not one of the natural sciences" (4.111). But nor would the logical positivists have claimed that to be the case. He also said that philosophy is "something which stands above or below, but not beside the natural sciences" (*ibid.*). How does this differ from Carnap's view that philosophy is the logical syntax of the language of science?[6] Is not the logic of science exactly something which stands "above" or "below" but not "beside" the sciences themselves? And Wittgenstein said many more things about what philosophy is and what it is not which can without too much ado be reconciled with the opinions both of earlier and later so-called analytical philosophers. Yet there are, I think, *profound* differences.

2. There had existed throughout the 19th century undercurrents which did not share the optimistic belief in progress through modernity and did not vest hopes for the future of man in the further development of science and technology. Kierkegaard's opposition to Hegel and system-philosophy, particularly in his *Concluding Non-Scientific Postscript*, struck an early note of discord. Dostoyevsky's *Notes from the Underground* dug deep into those dark layers of the human soul which might one day erupt in violent protest against a rationally organized, progressive society. The climax of 19th century critique of civilization is, of course, the "Umwertung aller Werte" attempted by Nietzsche. Of Nietzsche Wittgenstein once said that he had perhaps touched on "problems of the intellectual world of the West" which no other philosopher had "tackled and wrestled with" and which could only be written about "in the language of prophesy, comprehensible to the fewest".[7]

The mood of these writers is not necessarily pessimistic. But it is a sombre mood of self-reflexion and questioning of dominant currents of their time, as they saw them. And these writers, we know, were more congenial to Wittgenstein than any 19th century philosopher of the established style. From his early years he distanced himself from and condemned modernity in all its philistine manifestations.

It is but natural that the cataclysm of the first War should have nourished this mood and added to it apocalyptic overtones just – as it is also understandable that the same disasters were hailed by others as having created a *tabula rasa* for the groundwork of a brave new world. The doomsday prophet *par excellence* is, of course, Spengler.

Wittgenstein has mentioned Spengler as one of those who had influenced him. He says he took over from Spengler a line of thinking and seized on it with enthusiasm for his own work.[8] This, and the fact that Spengler's name occurs on the list – beginning with Boltzmann and ending with Sraffa – of persons, whom Wittgenstein recognized as influences, does not mean, however, that Spengler had deeply influenced the mood in which Wittgenstein viewed his times. It means, in the first place, that he had received from

Spengler's *Untergang* the germ of one of the pervasive ideas of his later philosophical thinking. This is the notion of conceptual family resemblances. It is quite another thing that Wittgenstein *also* shared the apocalyptic view of Spengler's. With the years, this even deepened to a hatred of our decaying civilization and a wish for its destruction. "Do you really think that Europe needs another great war?", I once asked him in the twilight of prewar sentiments in 1939. "Not one, but two or three", was his reply. This shook his young interlocutor deeply and seemed to him *then* unintelligible.

The apocalyptic views of Wittgenstein are best reflected in some of the "general remarks" he wrote in the late 1940s. But the best and fullest articulation of his attitude to Modernity is of much earlier date. I am thinking, in particular, of the sketches of a Foreword written in 1930 for a projected work. A version was printed with the *Philosophische Bemerkungen*, the earliest of Wittgenstein's prepared typescripts after his return to philosophy in the late 20s, published in 1964. The book for which it was written cannot, however, have been it. More likely, the Foreword was meant for the great and externally more accomplished work to which Wittgenstein's literary executors used to refer as the "Big Typescript" and which – most regrettably – to this day has not been published in its entirety. In one of the sketches he characterizes the spirit in which his work was written as follows:

> This is not . . . the spirit of the main current of European and American civilization. The spirit of this civilization makes itself manifest in the industry, architecture and music of our time, in its fascism and socialism, and it is alien and uncongenial to the author . . . Our civilization is characterized by the word 'progress'. Progress is its form . . . It is occupied with building an ever more complicated structure . . . I am not interested in constructing a building, so much as in having a perspicuous view of the foundation of possible buildings . . . So I am not aiming at the same target as the scientists and my way of thinking is different from theirs.[9]

Wittgenstein's early sketches for a Preface should be read in juxtaposition to Carnap's Preface to *Aufbau*. They afford an impressive and nice illustration to the contrast between the protagonists and the critics of the spirit of Modernity. Although we have no documentary evidence for it, the thought is close at hand that Wittgenstein wrote his words in reply to Carnap's. The way the two prefaces match seems to me too good to be a result of sheer coincidence.

3. The Hungarian philosopher and historian of ideas Christoph Nyíri sees in Wittgenstein a representative of the neo-conservative trend which rose to prominence in the 1920s, chiefly in the German-speaking world.[10] It continues an earlier conservatism which had taken a critical view of the rationalist

spirit of the Enlightenment and of the progressive democratization and secularization of 19th century society. An early exponent of this type of conservative thinking is the great Austrian poet Grillparzer. Nyíri points to the deep impression Grillparzer had made on Wittgenstein and which anybody knowledgeable in literature cannot fail to notice throughout his writings also in places where there is no mention of Grillparzer's name.

I have great respect for Nyíri's attempt to locate Wittgenstein on the cultural map. I think it in many ways illuminating of Wittgenstein's attitudes and thoughts. But I also have reservations.

The label "conservatism" does not seem to me appropriate – *pace* Nyíri's own observations on its ambiguities.[11] Wittgenstein was much more anxious to combat and distance himself from a prevailing climate of opinion than to work for the restoration of one which was already fading. He is as little nostalgic in his thinking as are Dostoyevsky or Nietzsche. Moreover, the philosopher who wrote "I destroy, I destroy, I destroy"[12] was not alien to the thought that something new could be built once the heap of rubble of a decaying culture had been cleared away. Not unlike some radicals of the left he appears to have seen something hopeful in the drastic sweeping away of an obsolete social order which had taken place in Russia. His plans of settling in the Soviet Union can be viewed in this light too. Among his Cambridge friends in the 1930s many had a pronouncedly marxist orientation. The only periodical which I have seen him reading and not frowning upon was *The New Statesman and Nation* – much more in tune with the tastes of left intellectuals than with those of apolitical conservatives.

Just as Wittgenstein's philosophy defies classification in relation to movements and trends in 20th century thought, so also is his attitude to modernism and its critics far from univocal. It may even appear to us loaded with paradox and contradiction.

4. Allan Janik, co-author of the book *Wittgenstein's Vienna* and another writer who has tried to place Wittgenstein in a broader cultural setting, has criticized Nyíri's diagnosis. He objects above all to the way in which Nyíri exploits for his purposes two key notions in Wittgenstein's later philosophy. These are the notions of rule-following and form of life. Nyíri tries to link them to political and social conservatism. This does not seem to me right. Janik's critical points against Nyíri I find well taken here.

Janik suggests that it is characteristic of Wittgenstein that "he wished to separate his philosophizing from his personal beliefs" and that, accordingly, he "took great pains to separate his cultural pessimism from, say, his view of meaning and thinking."[13] If this is right, it would be futile to look for a connection between Wittgenstein's philosophy and the mood in which he viewed his times.

One could go one step further and say that also the modernism and rationalist optimism, characteristic of many of the leading logical positivists, bears only a contingent relationship to *their* philosophy. It is true that neither Wittgenstein nor the positivists (with the sole exception of Neurath) elaborated their views of culture, history, politics, and society in the discursive form of philosophical or scientific writings. Their "Weltanschauung" shines forth only in various *obiter dicta*; in the case of the positivists the Vienna Circle manifesto or Carnap's Preface to *Aufbau*; in the case of Wittgenstein some "aphorisms" in the *Tractatus*, the prefaces to works not published in his lifetime, and a good many "general remarks" scattered throughout his writings. (Hither may also be counted the notes on Frazer's *The Golden Bough)*.

The truth of these observations notwithstanding, I disagree with Janik's suggestions above. Fichte's famous words "Was für eine Philosophie man wählt, hängt davon ab, was für ein Mensch man ist", may not be interestingly applicable to the average, mediocre, academic philosopher. But for the great ones it is, I think, profoundly true. Their philosophy reflects their personality, and *vice versa*. And if personalities differ profoundly, so will the philosophies. Therefore it is not futile to look for the way in which Wittgenstein's thought can be said to reflect his view of life.

5. A different attempt from Janik's to resolve the puzzles here is made in Stephen Hilmy's book *The Later Wittgenstein*. It is another one of those not too many writings which discuss Wittgenstein's thought in relation to intellectual currents of the time. (A virtue of Hilmy's work is that it bases its argumentation directly on Wittgenstein's *Nachlaß* and does not confine it to the printed sources only.)

Hilmy's view is that Wittgenstein's later philosophical deliberations should be seen "as a struggle against dominant intellectual trends in our modern civilization"[14] and *also* "against this trend as it manifested itself in his own earlier philosophy".[15] Thus, according to Hilmy, Wittgenstein of the *Tractatus* was himself "caught up in the scientific current of the times"[16] and had "succumbed to the scientific intellectual trends of the day".[17] Simplifying a little, one could then say that the logical positivists were right when they saw in the *Tractatus* an ally to their own philosophy, and that it is only the later thinking of Wittgenstein which reflects his *Abstandnahme* from that intellectual mood of the times to which I have here alluded with the term "modernity".

There is certainly *some* truth in Hilmy's view. At least two arguments can be given in support of it. One is so to speak "biographical". It points to changes in Wittgenstein's life. The Wittgenstein of pre-war Cambridge, a student of Frege's and Russell's logic, was in many ways a different man from the Wittgenstein who returned to Cambridge in the late 1920s. His experiences from the years of the war had deeply affected his sentiments. After return from captivity he renounced his fortune and withdrew to a life of Tolstoyan simplicity in remote countryside villages. His life later at Cambridge and in his hut in Norway is stamped by the same austere frugality. The professor of philosophy refused to let himself be integrated in the academic establishment of one of the world's most distinguished universities. This was Wittgenstein's mood when he met the philosophers of the Vienna Circle, – but it was a different mood from the one in which he had first conceived the ideas for the work which meant so much to them. The Wittgenstein who had radically changed his ways of life was now also on the road to a new philosophy which was to become less and less congenial to most of his former admirers.

The other argument in support of Hilmy's view trades on the difference between the philosophy of the *Tractatus* and that of the *Philosophische Untersuchungen*, between the philosophies sometimes referred to simply with labels "Wittgenstein I" and "Wittgenstein II". The first belongs in the atmosphere of logic and the science-oriented thinking of the positivists and later the analytic philosophers. The second is deeply infected by doubts about the influence of scientific rationalism on our thought and our lives.

What I should like to do now is the following: Against Janik I shall try to show that there is a close correspondence between Wittgenstein's thought and that which I have called his "mood". To this end we must look towards features of his personality which lay deeper than those habitually referred to as his "cultural pessimism" and which remained substantially unaffected by external changes in his way of life. Against Hilmy I shall try to show that there is a continuity in Wittgenstein's thought which links his "two philosophies" in the very way which makes them correspond to a fundamentally unchanged attitude to life. This connecting link is found, not surprisingly, in Wittgenstein's conception of philosophy and his attitude to language.

6. Wittgenstein's intellectual and moral personality must be understood against the background of the society in which he grew up. He was a son of the late Habsburg Empire. This was a very peculiar socio-political construct. The multinational and multi-lingual state was in many ways an obsolete phenomenon in a Europe *then* on its road towards democracy and industrializa-

tion in the frames of consolidated national states. It was a reactionary bulwark against progressing modernization. At the same time it appears to us today strangely modern, a forecast of what may come in a Europe now in a process of integration with national borders breaking down and a new mixing of languages and nationalities in the offing.

Of this "kaiserliche und königliche" construct – for which the author of *Der Mann ohne Eigenschaften* invented the name with ironic Greek overtones "Kakanien" – was characteristic a hypocrisy and doubleness of morality which, though certainly not unique in history, was perhaps unique in the Europe of the 19th century. Or how else shall one understand that in this atmosphere of conventional half-truth and insincerity there arose such a strong reaction against it, a violent passion for truthfulness and purity, unsparing efforts to debunk the illusions and lay bare the underground of the human soul? We witness this reaction in the puristic architectural language of Adolf Loos, in the stern atonality of Schönberg's music, in the searching cultural criticism of authors like Hermann Broch and Robert Musil, and in the apocalyptic irony of Karl Kraus. In the same circle also belongs the greatest debunker of all, Sigmund Freud.

It is against this background in Austrian *Geistesgeschichte* that we have to understand Wittgenstein: his passion for truth and sincerity and his longing for pure and simple forms of life. Wittgenstein had a rare sense for detecting or, to use a favourite word of his, "smelling" even the slightest trace of conventional lie, untruthfulness, artificiality and pretense in the people whom he met. This made relations with him strained, in some cases even unbearably difficult, for those who had the privilege of coming close to him personally. The statement holds true equally of the Wittgenstein who came to pre-war Cambridge to attend Russell's lectures and of the one who returned there in 1929 after ten years of voluntary withdrawal from the world. "Are you thinking of logic or of your sins" Russell asked him in a memorable conversation.[18] "Of both" was Wittgenstein's reply. Paraphrasing a word by Ibsen – certainly well known to Wittgenstein – one could say that philosophical thinking was to him a perpetual holding of doomsday with himself.

To the best of my knowledge, Wittgenstein is the only one among those twilight figures mentioned of Musil's "Kakanien", whose craving for honesty prompted him to radical efforts to change his (outward) life. Here the influence of Tolstoy on Wittgenstein must be rated high. Tolstoy was at home in a society at least as morally depraved as "Kakanien" but perhaps with greater resources of unconsumed vitality than the latter. Tolstoy's mature life was a relentless fight to free himself from the nets of bigotry and conventionality.

He lured, moreover, a number of men and women all over the world to try to "return" to a pure and simple life in truth. He was himself tragically incapable of this return; the last effort he made was a suicide of its kind. Also Wittgenstein's withdrawal to be a schoolmaster for peasant children was a failure. I think that both Wittgenstein and the great Russian were partly victims of an illusion about the actual existence of a country "Erehwon" uncontaminated by the moral ills of the society we live in and where we can go to start a new life. Tolstoy went to Caucasus and to the Bashkirs in search of it. Wittgenstein sought it in remote villages in lower Austria, in the wilderness of the fjords of Norway or on the shore of the Ocean in Galway. His abortive plans of settling in the Soviet Union is also a chapter in this story. But the land he sought is really the land "Nowhere".

7. In a social atmosphere of bigotry and insincerity language also tends to become corrupted. It is infiltrated by euphemisms. Things are no longer called by their blunt and simple names, but are wrapped up in evasive circumscriptions or referred to by invented technical terminology. Style becomes unperspicuous, meaning unclear. In modern society this distortion of language has assumed grotesque proportions in the jargon of public administration and the media. Austria under the late Habsburgs may have been the place where this "illness of the times", now universal, first flared up. To fight it was the set task of that generation of Austrian "purifiers" of which Wittgenstein was a member.

Fritz Mauthner's *Sprachkritik* was an early sample of this reaction to language. The fact that Wittgenstein in the *Tractatus* said (4.0031) that all philosophy is critique of language, though not in Mauthner's sense, has, I am afraid, long distracted modern philosophers from paying due attention to Mauthner's work. It belongs in the same tradition of which – not counting Wittgenstein – Karl Kraus is the supreme and most influential example.

Wittgenstein admired the work of Kraus. He counted Kraus and Loos among those who had influenced him. It is an interesting but perhaps unanswerable question, whether the Krausian moral attitude to language, which Wittgenstein shared, was perhaps a force which made him broaden his philosophy of logic of the early Cambridge years to the philosophy of language embodied in the *Tractatus*.

It is sometimes held that Wittgenstein's concerns in the *Tractatus* were the conditions which a logically ideal or perfect language has to fulfil. Thus, for example, Russell in his Introduction. This has also been disputed. I think those who dispute it are right, if the ideal language was to be a perfection of the *Begriffsschrift* first propounded by Frege and then further developed by the authors of *Principia Mathematica*. Wittgenstein's contribution to their project was in the first place the tabular theory of truth-functions and the notion of tautology. But although logic no doubt was the gate through which Wittgenstein entered philosophy, his work in the tradition of Frege and Russell was of short duration. It ended, I would say, when in the early part of the 1914–1918 war the thought of the proposition as a picture first dawned upon him and he wrote in his notebook in January 1915 "My *whole* task consists in explaining the nature of the proposition".[19] This had not been the "whole task" of Frege or of Russell.

Rather than saying that Wittgenstein in the *Tractatus* was concerned with the conditions of an ideal language we should, I think, say that he was in search of *pure* language or of language in its pure and uncorrupted form. The language which would depict the world as it really is, absolutely "true to fact". It was to this end that, at a later stage of his work on the book, he conceived of the ontology of immutable and indestructible things to which there correspond names the configurations of which in language picture the contingent configurations of things in reality. There are some indications in the notebooks that Wittgenstein's position was initially wavering on the question whether such a language could actually be found, examples given of things and corresponding names. But in the end he realized that the form and content of the "pure" language is ineffable. The language of which he was in search is, to use Max Black's happy phrase, a "never-never language".[20] It is as remote from the language we speak as is the nowhere land of pure and simple life from the contaminated societies in which we are destined to live.

Because of the way in which the pure language of the *Tractatus* is supposed to reach up to reality, one can also say that the world it is supposed to picture is a "never-never-world", a postulated ideal which is nowhere to be found. It has occurred to me that the traits of relativism and opposition to realism which some interpreters see in Wittgenstein's later work have their root in the *impasse* of the *Tractatus* approach to language. Just as there is no "pure" language there is also no "pure" reality for language to depict.

8. 17th and 18th century philosophers had entertained the fiction of "man in a state of nature". This fiction was thought useful in their search for the *raison d'être* of the state and the legitimation of societal institutions. Some fancied it to have been an existing state of affairs in the "uncorrupted" infancy of prehistorical man. Even if this was an illusion, it does not necessarily ruin the philosophic value of the fiction.

In a somewhat similar sense Wittgenstein may be said to have been in search, throughout his reflective life, of a "natural state of language". In the *Tractatus* the thought it was hidden under the disfiguring veils of ordinary speech. It existed somewhere deep under the surface of language as used. But he did not find it there. He came to the conclusion that it was not there to be found either. He had been looking for it in the wrong direction. It was not to be attained by diving under a surface but by looking at the surface itself. *There* was also to be found the uncorrupted language of which he was in search. The difficulty was only to see it. "Nothing is hidden."

It is often said that one of his early ideas which Wittgenstein never gave up is his conception of philosophy. Basically this is true, I think. But it is not a truth without modifications. It holds for his conception of philosophy as an activity and not a doctrine *(Lehre)*, and of philosophy as critique of language. But it does not hold for his specific view that philosophical problems are due to linguistic confusions, to what he later called the "bewitchment" *(Verhexung)* of our thinking by the means of language. There is no statement to this effect in the early writings. There he speaks of the philosophic problems resting on a misunderstanding of the ("true") logic of our language, on a confusion of the *grammatical* with the *logical* form of thought.[21]

The problems which arise through a "linguistic bewitchment" of our thought are not questions in search of an answer, but "bumps that the understanding has got by running its head up against the limits of language".[22] A main reason why we get these "bumps" is "that we do not *command a clear view* of the use of our words".[23] We miss the *übersichtliche Darstellung* ("perspicuous presentation") which will expose to us the undistorted use of language and "bring words back from" what Wittgenstein calls "their metaphysical to their everyday use".[24] The clarity which the perspicuous presentation gives is an absolute. It makes the philosophical problem disappear completely.[25] The disquietude is gone and the philosopher attains his aim which is "Friede in den Gedanken".[26]

Philosophy which follows these lines is strictly *descriptive.* It does not *explain* anything – for example how it is possible for signs to mean. Nor does it answer questions of essence, for example what thinking, or truth, or logical necessity are. Unlike most philosophy after Kant it is neither transcendentalist nor essentialist.[27] It puts the "how possible?"- and "what is?"-questions aside by directing our attention to the role which the problematic words play in actual communication. In this sense "the work of the philosopher" can be said to consist "in assembling reminders for a particular purpose".[28]

A philosophy which does not look for answers to questions, does not explain or theorize about the things which attract the philosopher's curiosity, and does not try to provide the foundations for our beliefs, is not a philosophy for which scientific thinking sets the pattern. It, on the contrary, fights the infiltration of this thinking into philosophy and makes it responsible for the confusions from which the philosopher tries to rid himself. It is not, need not be, hostile to science as such. But it may be said to take a critical or even hostile attitude to the influence of science outside its proper domain – and in particular on philosophic thought. In *this* it runs counter to an intellectual mainstream of the century.

9. There are a few points on which one might wish to challenge the consistency of Wittgenstein's (later) view of philosophy. On the one hand Wittgenstein says that the philosopher must "in no way interfere with the actual use of language" and that philosophy "leaves everything as it is".[29] But he also says that he is "engaged in a struggle with language".[30] In a passage in the unpublished part of the "Big Typescript" he describes this struggle as follows – I have here translated it into English:

> Human beings are deeply embedded in philosophical, i.e. grammatical confusions. Freeing them from these *presupposes* tearing them away from the enormous number of connecting links that hold them fast. A sort of rearrangement of the whole of their language is needed. (*Man muß sozusagen ihre ganze Sprache umgruppieren.*) – But of course that language has developed the way it has because some human beings felt – and still feel – inclined to think that way. So the tearing away will succeed only with those in whose life there already is an instinctive revolt against the language in question and not with those whose instinct is for the very herd which created that language as its proper expression.[31]

This sounds almost like a cry for a "cultural revolution" needed for putting the philosophic worries to rest. Because is not the "Kampf mit der Sprache" also an interference with it? Particularly if it requires a "rearrangement of the whole of language"? And a similar question can be raised about the attempt to "bring back" (in German *zurückführen*) the words from their "metaphysical" to their "everyday" use.

Wittgenstein's talk about philosophy suggests to me the following picture: There are in the large garden of language tidy plots of land where uncorrupted language-games are being played by human users. But this garden is also partly overgrown by metaphysical weeds which hide the plots from sight and blur their borders and thereby confuse those who play with language. The good philosopher should be a gardener who by clearing away the weeds displays the plots of linguistic ground in their purity and thereby helps communication to go on unimpeded by the metaphysical-philosophical confusions.

But is this picture not another vision of the pure and uncorrupted language of which Wittgenstein had been in search in the *Tractatus*? He could not find it then because he sought it under the surface of everyday language rather than on the surface itself. But if this surface is partly veiled, hidden, overgrown by metaphysical weed which must be removed before we command a perspicuous view of what things are like, may it then not be the case that this uncorrupted ordinary (use of) language too is a never-never thing and that the search for it must end in frustration? Does belief that this is not so perhaps rest on an illusory idealization of ordinary language?[32]

My impression sometimes is that Wittgenstein in the *Tractatus* had attained a position which, as it stood, was *unassailable*. As we know, he himself assailed it later. He even talked of "grave errors" in his earlier writing.[33] His thinking took what may be called a "philosophic U-turn": from the surface to the hidden depths, then from the depths back to the surface.[34] But I know from what Wittgenstein told me that there were moments when he doubted whether this was a turn in direction to final clarity or whether it too would end in an impasse of obscurity. This is the radical, "existential" doubt of one who cannot resolve the question whether all his efforts have been a failure. No outsider can resolve it for him either.

10. The bewitchment of our thinking through language happens when words take on what Wittgenstein calls a "metaphysical" use. So one can say that Wittgenstein's fight with language was a fight with metaphysics. This was precisely what the logical positivists had been engaged in. "Überwindung der Metaphysik durch logische Analyse der Sprache" was the title of one of Carnap's articles in *Erkenntnis*. So what difference, if any, was there in their respective attitudes to metaphysics?

For the positivists, questions were "metaphysical" when they could not be decided by the methods of natural science or by logical deduction from scientifically acceptable premisses. As amply testified in the documents from the heyday of positivism, metaphysics was seen by the champions of the "new philosophy" as an outgrowth and residue of religious beliefs of a pre-modern society, as a rationalized disguise of at bottom irrational attitudes. Metaphysics was philosophy which had helped reactionary forces in society to block and retard the progress of emancipated, rational, and secularized man.

Wittgenstein's fight against metaphysics was something very different. By a metaphysical use of language Wittgenstein means the "free-wheeling" of language which occurs when words get detached from their actual use in the language-games of communicative discourse and are being used for constructing what Wittgenstein calls *Luftgebäude* ("houses of cards") in the linguistic isolation of the philosopher's mind.[35] In past centuries of European history the thinking of metaphysicians was to a large extent nourished by the linguistic rituals of a religious culture. This was a culture in which language-games with words like "God", and "sin", "grace" and "doom" and "redemption" had an established everyday use. In a similar manner, the thinking Wittgenstein calls metaphysical is stamped by the linguistic patterns and thought habits of a predominantly scientific civilization.

The metaphysics which Wittgenstein is fighting is thus not one rooted in theology but one rooted in science. He is fighting the obscuring influence on thinking, not of the relicts of a dead but of the habits of a living culture. Of this he gives clear warning in the 'Blue Book' where he wrote:

> Philosophers constantly see the methods of science before their eyes, and are irresistibly tempted to answer questions in the way science does. This tendency is the real source of metaphysics, and leads the philosopher into complete darkness.[36]

He immediately gives examples: The craving for general theories, contrasted with what he calls "the contemptuous attitude to the particular case"[37]; the tendency to explain the concept of number, to reduce the infinite to the finite, mathematics to logic, intentional behaviour to bodily movement. The most vulgar examples of these tendencies we find, it seems to me, in contemporary philosophy of mind, be it in the form of the physicalists' "identification" of so-called mental states with brain processes or the eliminative materialists' rejection of our common sense psychological concepts – what they call "folk-psychology" – as a "radically false theory" eventually to be replaced by a perfected neuroscience. Farther into the jungle of metaphysics, as Wittgenstein saw it, philosophy can hardly lose itself than in these latter day phenomena of a philosophic culture gone "scientist".

Thus one can say – I am here quoting Stephen Hilmy's book on the later Wittgenstein – that Wittgenstein's "conception of metaphysics was such that it encompassed not only traditional metaphysics, but also, and especially, the dominant mode of reflection of our own epoch; and it is primarily the latter, – that constituted the intellectual current against which he was struggling".[38]

It is ironic that the metaphysics which Wittgenstein was fighting was exactly the one in the cobwebs of which the logical positivists and a good many of their followers among analytical philosophers according to him had been caught.

It is worth asking here how Wittgenstein himself viewed the great efforts of past thinkers who had tried to encompass the world intellectually in systems we call "metaphysical". The thing to say in response to the question is that he had no need to combat these systems and that he therefore also was, or would have been, able to bestow on them the respect and awe *(Ehrfurcht)* which he was disposed to feel towards *great* achievement in all spheres of human life. But he was also acutely aware that he was not *continuing* a tradition. What he did was as different from what a Leibniz or a Spinoza had been doing as is life in our times different from what it was in theirs. Yet there is also enough resemblance to make his philosophy, as he said himself, "a legitimate heir" to those intellectual pursuits of the past which traditionally go under the name "philosophy".[39]

I hope to have made it clear in which sense Wittgenstein's efforts in philosophy were a fight against a dominant climate of opinion in our century. I have called this climate that of Modernity. It was in origin coupled with a euphoric belief in progress thanks to the managerial uses of reason in industrialized democratic societies. This optimism has largely faded in the woes of two global wars and under the load which man has imposed on nature, threatening the biosphere with breakdown. But also amidst the sombre or even apocalyptic mood of scared humanity, this rake's progress continues. "Science and industry", Wittgenstein wrote a few years before he died, "might turn out to be the most enduring thing in the modern world".[40] But he also wrote that there is nothing absurd in the belief "that the age of science and technology is the beginning of the end of humanity" and that mankind in seeking to steer its course towards the future relying on scientific rationality "is falling into a trap".[41]

Wittgenstein's philosophy and also his life was a protest against these trends and an effort to set an example of "a changed mode of thought and life" which, if followed might provide a cure for what he thought of as a sick time. But no rule can, I think, be laid down for how to imitate his path.

Academy of Finland
Helsinki

NOTES

[1] von Wright (1978).

[2] Carnap (1928), p. v f.: "Wir spüren eine innere Verwandtschaft der Haltung, die unserer philosophischen Arbeit zugrunde liegt, mit der geistigen Haltung, die sich gegenwärtig auf ganz anderen Lebensgebieten auswirkt; wir spüren diese Haltung in Strömungen der Kunst, besonders der Architektur, und in den Bewegungen, die sich um eine sinnvolle Gestaltung des menschlichen Lebens bemühen: des persönlichen und gemeinschaftlichen Lebens, der Erziehung, der äusseren

Ordnungen im Grossen. Hier überall spüren wir dieselbe Grundhaltung, denselben Stil des Denkens und Schaffens. . . Der Glaube, daß dieser Gesinnung die Zukunft gehört, trägt unsere Arbeit."

[3] As early forceful expression of this view of "the new philosophy" is Russell (1914). It is interesting to read it in conjunction with the writings from the years of the flowering of the Vienna Circle.

[4] Quoted from Haller (1981).

[5] Hahn (1929), p. 54: "Einstein, Russell und Wittgenstein seien hier als diejenigen unter den führenden Denkern der Gegenwart genannt, die die wissenschaftliche Weltauffassung am wirkungsvollsten in die Öffentlichkeit vertreten und auch stärksten Einfluß auf den Wiener Kreis ausüben."

[6] Carnap (1934), pp. iii-iv: "Philosophie wird durch Wissenschaftslogik ersetzt; Wissenschaftslogik ist nichts anderes als logische Syntax der Wissenschaftssprache."

[7] See Wittgenstein (1980), p. 9.

[8] *Ibid.*, p. 19. In German: "Ich habe sie (*sc.* Gedankenbewegung) nur sogleich leidenschaftlich zu meinem Klärungswerk aufgegriffen."

[9] *Ibid.*, pp. 6–7. In German: "Dieser Geist ist . . . ein anderer als der des großen Stromes der europäischen und amerikanischen Zivilisation. Der Geist dieser Zivilisation, dessen Ausdruck die Industrie, Architektur, Musik, der Faschismus und Sozialismus unserer Zeit ist, ist dem Verfasser fremd und umsympatisch . . . Unsere Zivilisation ist durch das Wort 'Fortschritt' charakterisiert. Der Fortschritt is ihre Form . . . Ihre Tätigkeit ist es, ein immer komplizierteres Gebilde zu konstruieren . . . Es interessiert mich nicht, ein Gebäude aufzuführen, sondern die Grundlagen der möglichen Gebäude durchsichtig vor mir zu haben. Mein Ziel ist also ein anderes als das der Wissenschaftler,und meine Denkbewegung von der ihrigen verschieden."

[10] See in particular his essay 'Konservative Anthropologie: Der Sohn Wittgenstein', in Nyíri (1988). The essay opens with the words (p. 91): "Ludwig Wittgenstein begann um die Wende der zwanziger/dreissiger Jahre des 20. Jahrhunderts jenen Ideenkreis auszuarbeiten, den man heute, rückblickend, wahrscheinlich als die tiefste und umfassendste Grundlegung vom konservativen Bild des Menschen bezeichnen kann."

[11] Nyíri (1988), p. 104 ff.

[12] Wittgenstein (1980), p. 21.

[13] Janik (1985) p. 130.

[14] Hilmy (1987), p. 190.

[15] *Ibid.*, p. 192.

[16] *Ibid.*, p. 194.

[17] *Ibid.*, p. 210.

[18] Told and retold by Russell in many places. See, for example, Russell (1968), p. 99.

[19] Wittgenstein (1979), p. 39: "Meine *ganze* Aufgabe besteht darin, das Wesen des Satzes zu erklären."

[20] Black (1964), p. 11.

[21] See, for example, *Tractatus*, the Preface (Vorwort) and *ibid.*, 4.002, 4.003, and 4.0031.

[22] Wittgenstein (1953), § 119.

[23] *Ibid.*, § 122.

[24] *Ibid.*, § 116.

[25] *Ibid.*, § 133.

[26] Wittgenstein (1980), p. 43.

[27] Both attributes, in my opinion, apply to the *Tractatus*. There Wittgenstein is in search of the essence of language (the proposition) and of the trascendental a priori conditions of the possibility of language. (The resemblance with Kant has often been noted). The later philosophy of Wittgenstein is opposed to both tendencies. There is a profound change in this regard as compared with the *Tractatus*. Some attempts to "transcendentalize" Wittgenstein's later philosophy of language

therefore seem to me substantially mistaken. (Which does not prevent them from having some independent interest in themselves.)

[28] Wittgenstein (1953), § 127.

[29] *Ibid.*, § 124.

[30] Wittgenstein (1980), p. 11. "Wir kämpfen mit der Sprache. Wir stehen im Kampf mit der Sprache."

[31] MS 213, p. 423.

[32] The phrase "idealization of ordinary language" was suggested to me in conversation by Heikki Nyman.

[33] Wittgenstein (1953), Vorwort, p. x.

[34] See Wittgenstein (1953), § 108: "Die Betrachtung muß gedreht werden, aber um unser eigentliches Bedürfnis als Angelpunkt."

[35] *Ibid.*, § 118.

[36] Wittgenstein (1958), p. 18.

[37] *Ibid.*

[38] Hilmy (1987), p. 225.

[39] Wittgenstein (1958), p. 62. The fullest statement by Wittgenstein himself of his views of philosophy is the chapter "Philosophy" in TS 213 (the "Big Typescript"). This chapter has now been published (Wittgenstein 1989.) – Of great interest are the pronouncements made by Wittgenstein on his philosophy recounted by G. E. Moore in Moore (1954–1955) (particularly the third article). In the early 1930s Wittgenstein too, *somewhat* in the style of the Vienna positivists, seems to have thought of his way of doing philosophy as a great break with tradition. Moore III, p. 26: "He said that what he was doing was a 'new subject', and not merely a philosophy, a 'kink' in the 'development of human thought', comparable to that which occured when Galileo and his contemporaries invented dynamics; that a 'new method' had been discovered, as had happened when 'chemistry was developed out of alchemy'." It is perhaps significant that statements with this colouring are not found in Wittgenstein's own written work.

[40] Wittgenstein (1980), p. 63.

[41] *Ibid.*, p. 56. "Es ist z.B. nicht unsinnig, zu glauben, daß das wissenschaftliche und technische Zeitalter der Anfang vom Ende der Menschheit ist; daß die Idee vom großen Fortschritt eine Verblendung ist, wie auch von der endlichen Erkenntnis der Wahrheit; daß an der wissenschaftlichen Erkenntnis nichts Gutes oder Wünschenswertes ist und daß die Menschheit, die nach ihr strebt, in eine Falle läuft. Es ist durchaus nicht klar, daß dies nicht so ist."

REFERENCES

Black M. (1964): *A Companion to Wittgenstein's Tractatus*, Cambridge University Press, London.

Carnap R. (1928): *Der logische Aufbau der Welt,* Weltkreis Verlag, Berlin-Schlachtensee.

Hahn H. (1929): *Wissenschaftliche Weltauffassung, Der Wiener Kreis*, hrsg. vom Verein Ernst Mach, Artur Wolf Verlag, Wien.

Haller R (1981): 'Wittgenstein and Austrian Philosophy', in *Austrian Philosophy: Studies and Texts*, ed. by J. C. Nyíri, Philosophia Verlag, München.

Hilmy S. (1987): *The Later Wittgenstein*, B. Blackwell, Oxford.

Janik A. (1985): *Essays on Wittgenstein und Weininger*, Rodopi B.V., Amsterdam.

Moore G. E. (1954–55): 'Wittgenstein's Lectures in 1930–33', I-III, *Mind,* 63, pp. 1–15; 289–315; 64, pp. 1–27. Reprinted in G. E. Moore, *Philosophical Papers,* Georg Allen & Unwin, London 1959.

Nyíri J. C. (1988): *Am Rande Europas. Studien zur österreichisch-ungarischen Philosophie geschichte*, Akadémiai Kiadò, Budapest.

Russell B. (1914): *Our Knowledge of the External World as a Field for Scientific Method in Philosophy*, Allen & Unwin, London.

Russell B. (1968): *The Autobiography of Bertrand Russell*, vol. II, Allen & Unwin, London.

Wittgenstein L. (1933[2]): *Tractatus logico-philosophicus*, German text with facing translation by C. K. Ogden & F. P. Ramsey, Kegan Paul, Trench, Trubner & Co., London.

Wittgenstein L. (1953): *Philosophische Untersuchungen. Philosophical Investigations*, ed. by G. E. M. Anscombe & R. Rhees. Trans. by G. E. M. Anscombe, B. Blackwell, Oxford.

Wittgenstein L. (1958): *The Blue and Brown Books*, ed. and with a Preface by R. Rhees, B. Blackwell, Oxford.

Wittgenstein L. (1961): *Notebooks 1914–1916*, ed. by G. H. von Wright & G. E. M. Anscombe. Trans. by G. E. M. Anscombe, B. Blackwell, Oxford (1979[2]).

Wittgenstein L. (1969): *On Certainty. Über Gewissheit*, ed. by G. E. M. Anscombe & G. H. von Wright. Trans. by D. Paul & G. E. M. Anscombe, B. Blackwell, Oxford.

Wittgenstein L. (1975): *Philosophical Remarks. Philosophische Bemerkungen*, ed. by R. Rhees. Trans. by R. Hargreaves & R. White, B. Blackwell, Oxford (1964[1]).

Wittgenstein L. (1979): 'Bemerkungen über Frazers *The Golden Bough*. Remarks on Frazer's *The Golden Bough*', ed. by R. Rhees, *Synthese*, 17, pp. 233–53. Trans. by A. C. Miles, Humanities Press, Atlantic Highlands, N.J.

Wittgenstein L. (1980): *Culture and Value. Vermischte Bemerkungen*, ed. by G. H. von Wright in collaboration with H. Nyman. Trans. by P. Winch, B. Blackwell, Oxford.

Wittgenstein L. (1989): 'Philosophie', hrsg. von H. Nyman, *Revue Internationale de Philosophie*, 43, pp. 175–203.

Wright von G. H. (1978): 'Wittgenstein in Relation to His Times', in *Wittgenstein and His Impact on Contemporary Thought*, ed. by E. Leinfellner et al., Hölder-Pichler-Tempsky, Wien, pp. 73–78.

PART I

SCIENCE, MIND, AND METAPHYSICS

DAVID PEARS

WITTGENSTEIN ON PHILOSOPHY AND SCIENCE

In this paper I shall discuss the question, "How did Wittgenstein see the relation between philosophy and science?" This is not a question about his philosophy *of* science. It is not, for example, directly concerned with the brief account of scientific theories that he gave in the *Tractatus*.[1] My inquiry is a more fundamental one. I want to uncover his view of the relation between science and philosophy. If he showed scientific discourse, and, more generally, factual discourse on the same map as philosophical discourse, how would he place them? And at how many points did he see a risk of our unwittingly crossing the frontier which divides them?

It is, as usual, worth looking at the *Tractatus* first, because although it only supplies sketchy answers to my question, they do set the scene for later developments. They are, perhaps, rather disappointing answers, because Wittgenstein merely gives us a diagram in which all kinds of discourse are related to the single line dividing what can be said from what can only be shown. Everything that can be said belongs to factual discourse, all of which is sometimes counted as "scientific".[2] Everything else, including philosophy, is placed on the other side of the line. It is all done on such a small scale that many important distinctions are not clearly represented. What we are told about the differences between stating facts and explaining them seems to make science almost an offshoot of philosophy.[3] Also, not enough is said about the differences between philosophical discourse and other modes of showing. The flow of thought is always from the general to the particular and so from the theory of language to its application to various disciplines. Even important differences are treated as points to be accommodated within the theory without being given a chance to alter its structure. The result is an excessively simple demarcation of the boundary between philosophy and science.

Still, it is an interesting picture, both in itself and because it helps us to understand what came later. Philosophy is assigned the task of analyzing language in use,[4] and what it tries to say can only be shown, because it is not concerned with the mere sounds and shapes of words, but takes them with their meanings. The argument is simple: if what philosophy tries to say were true, it could not have been false and so it would not be true in the ordinary, contingent way, and, therefore, could not properly be expressed in factual

R. Egidi (ed.), Wittgenstein: Mind and Language, 23–36.
© 1995 *Kluwer Academic Publishers. Printed in the Netherlands.*

language. Its message, if it can be called one, is that, though the structure of factual language is largely optional, its fundamental features are not optional but forced on it by the nature of reality. Science[5] exploits some of the options and invents systems of description which are more perspicuous than our ordinary ones and yield factual generalizations of wider scope.[6] However, this is not a method which makes it possible to explain everything.[7]

There are points in this simple diagram which Wittgenstein elaborated after 1929, often without any changes in his basic ideas. But let me start with a development which combines continuity with radical change. In the *Tractatus* he maintains that much scientific theorizing has an *a priori* character. Instead of simply recording our discoveries, it imposes on them new systems of description which are more perspicuous and more widely applicable than our ordinary ones. This, as already remarked, brings science close to philosophy, because philosophy too is driven by the same need for a panoptic view. There are, at this stage in the development of Wittgenstein's thought, two differences between the disciplines: one is that the philosophical theory which meets this need, logical atomism, is, unlike scientific theories, total, and therefore, does not belong to factual language and can only be shown: the other is that it is a theory which is completely determined by the nature of things,[8] whereas scientific theories are under-determined by the nature of things which, in their case, leaves scope for options.

After 1929 this map undergoes a radical change. The relative positions of science and philosophy remain more or less the same, but the distance between them is increased. Philosophy still goes further than science in its search for a commanding view, but what it investigates is no longer an independent reality but only our own linguistic reactions to whatever reality there may be. The whole idea of an independent world imposing its structure on language has gone, not because Wittgenstein has become a conventionalist instead of a realist, but because he has seen that the difference between these two views vanishes at the point where theorizing becomes total. So in his later philosophy he is directly concerned with language in its place in our lives. What he looks for is no longer any kind of theory because the special features of philosophical results, which previously made him treat them as theories which can not be stated but only shown, now make him say that they are not theories at all. The commanding view that he seeks is no longer a small-scale map – the whole terrain seen from a high altitude – but a complete grasp of the relationship between the various moves made on the ground by people using language.

There are several ways of describing these developments in Wittgenstein's philosophy. One, which he himself often used, was to compare his new work with psycho-therapy. In this paper a different point of view is adopted and an attempt will be made to relate the changes to the line which divides philosophy from science.

Suppose we agree with the popular view that the task of science is to go behind the phenomena in order to explain them. Then a plausible way of assimilating philosophy to science might seem to be to try to transcend the limit of language and to theorize about possibilities[9] in the way that science theorizes about facts. That is to say, just as a scientist explains why a possibility is or is not realized as a fact, so too a philosopher might try to explain why a certain candidate for possibility does or does not achieve that status. This way of assimilating philosophy to science is rejected in the *Tractatus*, but there is only a very schematic demonstration of what is wrong with it: facts are surrounded by identifiable possibilities, but possibilities are not surrounded by a further ring of identifiable candidates for possibility. That puts in a nutshell an idea, whose range and power only become apparent in Wittgenstein's later work, where it supplies him with a strategy that can be used against many metaphysical theories.

A well developed example of this strategy is provided by his later treatment of solipsism. The solipsist uses the sentence "I do not see what you see" not in the ordinary way, to say something which sometimes happens to be true, but rather to express a universal truth about the human predicament.[10] Wittgenstein makes two moves against this position. First, he demonstrates that, if the remark, made with the solipsist's intention, has a sense, it does not have its ordinary sense. Then he argues that the solipsist is not really using existing language to express a new truth but, rather, proposing a linguistic innovation which, he hopes, will give us a better map of the possibilities than existing language. That, according to Wittgenstein, is his fundamental mistake. He is assimilating philosophy to science and modelling the relation which is supposed to make a language adequate to express the possibilities on the relation which produces true reports of the facts.

These two moves can be described with a different emphasis. The point of the first one is to convince the solipsist that his remark does not start with the sense that is required for the universal truth that he wants to express with it. Contrary to what he thinks, we do not immediately recognize the possibility, seeing what others see, which it is our misfortune never to find realized. It is important to insist on this, because the solipsist nearly always begins by assuming that he has already identified the possibility which, he claims, necessarily remains unrealized. But there he is mistaken. If he were using the words in the old way, his remark would be literally false, because according to the old way of speaking, the possibility is frequently realized – we do often see the same things as other people. So he must go back and actually do the work of linguistic innovation.

The point of the second move is that this is not the kind of work that he takes it to be. The relation between language and the world that yields sense is not at all like the relation between language and the world that yields truth. He must not assume that he can simply see the possibility that he has in mind, and that all that he needs to do is to express it in existing language, which a sympathetic audience will immediately appreciate, because they will understand its rules to have been changed in the appropriate way. But appropriate to what? We cannot know, until the possibility has actually been identified.

It is worth looking more closely at the impact of Wittgenstein's two moves against solipsism. The strategy is a general one and he uses it against other metaphysical theories, but its use against solipsism is particularly clear and fully developed. It is, for example, obvious that he does not share the positivist view that such theories are soft-headed folly. On the contrary, he believes that the tendency to assimilate philosophical problems to scientific ones is natural and inevitable, and that it can only be held in check, but never eradicated.[11] Solipsism is not a gratuitous eccentricity but the understandable consequence of reflection on two connected facts which can easily make scepticism about other people's sensations seem inescapable: I can only feel my own sensations and only I can feel them.

If we focus exclusively on Wittgenstein's first move against theories of this kind, we will be likely to find his philosophy more like Moore's or Austin's than it really is. But the first move is not intended to be a complete refutation of the solipsist. It is a preliminary skirmish designed to uncover his adversary's central position: he is really innovating, rather than using words in the old way to formulate a new truth about our predicament. The point of his innovation is that existing language pays too little heed to the facts of sensory isolation and attaches too much importance to our need to communicate, and to feel that we are communicating, with each other. The result is that the criteria adopted for everyday truth are excessively relaxed and the deeper truth about our predicament has been suppressed by a conspiracy not of silence, but of inexpressibility.

Wittgenstein's second move is to challenge the solipsist's reason for his linguistic innovation. The reason offered by him is that his new way of speaking matches the material better than existing language: but the match that he has in mind cannot be truth, because any question of truth must wait until the rules for the use of the words have been fixed, and after they have been fixed, though different ways of speaking will all aim at truth, they will not compete for it. The vaunted match can only be between the solipsist's new way of speaking and the possibilities. But what possibilities? And by what rules will this more fundamental kind of match be judged?

The solipsist thinks that he does not need to do any more work in order to answer these questions. But that is an illusion based on his unexamined assumption that the relation with the world that confers sense on language is like the relation with the world that confers truth on it. In fact, he has to do more in order to identify the possibility which, according to him, is never realized, and his further task, which is to introduce his new way of speaking, can only be carried out in existing language. The result will then necessarily be disappointing, because the suggestion of a real deprivation, originally presented in existing language, will have been cancelled by the explicit shift to the new language.

When the objection, that match can only be determined by rules, is made against solipsism, it is, of course, assumed that the target of match is identified in a general way by the function of our existing sensation-language. That is the material which the solipsist is taken to have in mind, and his question is, what language would do justice to it. So when Wittgenstein objects that the question is unanswerable, because there are no rules for judging such a match, he is criticizing a philosophical theory which relies on part of existing language to identify the target of match. The controversy then turns on what shapes the detailed construction of sensation-language after its general function has been settled.

The solipsist's claim is that, once we are committed to a language for reporting our sensations, we are forced by their nature to model it on our language for physical objects. Wittgenstein has two answers to this. One, which has not been discussed here, is the well-known anti-privacy argument, that the solipsist's claim is incoherent. The other answer, which has occupied the centre of this discussion, is that, even if it were coherent, his preferred language would not be forced on us by the material, in the way in which facts sometimes force us to acknowledge truths. We can still retain our existing language, which does justice in its own way to the facts cited by the solipsist. It certainly seems more natural to do so, and at least, it cannot be wrong.

The idea, that we can find a viewpoint from which one language can be seen to be better adapted to the possibilities than another one, is rejected in the *Tractatus*: theories about possibilities cannot be modelled on theories about facts. However, it would be an exaggeration to say that all the implications of this difference between scientific and philosophical theories were clear to

Wittgenstein when he was writing the *Tractatus*. If they had already been developed in his mind, he would have disallowed all philosophical theories from the start, even those that "can only be shown". But the *Tractatus* does offer a theory about the deep structure common to all languages which is imposed on them by the nature of reality.[12] This structure allows us certain options in the rules adopted for our languages, but it sets a limit to the variations, because, once we have made our choices, it forces their consequences on us. This view, which leaves a certain lingering similarity between philosophy and science, is clearly expressed in the *Tractatus*,[13] but its rejection is the central point of his later philosophy of language. So the idea, that theories about possibilities cannot be modelled on theories about facts, though implicit in the *Tractatus*, was not immediately taken to its logical conclusion, and the development of his critique of solipsism, described above, takes the rejection of that kind of metaphysics further than he took it in his early work.

If the idea, that our linguistic choices do not force any consequences on us, is the central point of his later philosophy of language, it will be worth looking at the way in which he developes it in his later work. For it ought to give us a clearer picture of his view of the line dividing philosophy from science. It is too brief and uninformative to say that the differences between the two disciplines struck him as so great that he disallowed all theorizing in philosophy. We need to appreciate the details and the best place to see them is in his treatment of the central problem of his later philosophy of language, rule-following.

However, before we look at that problem, there is another part of his later philosophy which needs to be examined, his views about the relation between mind and brain. These views are sometimes interpreted in a way that runs counter to his central point, that the first attachment of a language to the world never forces any consequences on us, because what we do next will always depend on what we find it natural to do and not on any fixed structure of reality.

So the remainder of this paper will be concerned with two matters, Wittgenstein's views about the relation between mind and brain, and his views about rule-following. In each case the goal will be a clearer picture of his line of demarcation between philosophy and science. The first of these two inquiries may be launched by a familiar question. If the philosophy of mind investigates mental concepts, what is its relation with neurology? Even if the further development of a language is not dictated by the metaphysical structure of the world, it might well be influenced by its physical structure. For example, there seems to be nothing in Wittgenstein's later philosophy which would lead one to expect him to reject the suggestion that the concept of memory might develope in this way, or even to reject the suggestion that memory already involves general stipulations about the physical connection between the original perception and its later representation in the subject's mind. Even if there is no metaphysical force governing the development of mental concepts, they are surely sensitive to scientific discoveries.

It is, therefore, surprising to find passages in which these natural suggestions seem to be rejected. In one of them he maintains that, when I remember someone's name, my brain may not be in a state caused by my being told his name in the past, and the *only* difference between me and someone else who had forgotten it might be that I would produce it if asked, but he would not.[14] In another passage he says that there may well be "no process in the brain connected with associating or thinking; so that it would be impossible to read off thoughts from brain-processes. I mean this: if I talk or write, there is, I assume, a system of impulses going out from my brain and correlated with my spoken or written thoughts. But why should the *system* continue further in the direction of the centre? Why should this order not proceed, so to speak, out of chaos?"[15]

It is not clear how far Wittgenstein wants to go in these passages. It is possible that he is just speculating that the central operations of the brain may be chaotic. This would hardly be a contribution to neuro-physiology, but it would show something about the concepts of "associating" and "thinking", as we now have them: their use is compatible with this speculation, and they do not, at present, have the more complex structure of physical dispositions.

But, of course, the speculation may turn out to be false, because the central operations of the brain may be discovered to be governed by causal laws. What will happen then? According to him it is not a philosopher's business to answer this question. Science constructs conceptual suburbs around the old city of ordinary language,[16] but it is science itself and not philosophy that predicts such developments. All that a philosopher can do is to ask himself which development he would find natural if a particular realized speculation turned out to be false. E.g. if Wittgenstein's neurological speculation turned out to be false, his answer might well be that he would find it natural to assimilate mental concepts to physical dispositions.

Others have found more exciting implications in Wittgenstein's remarks about brain-processes – for example, that any such conceptual fusion of psychology and philosophy is impossible. But how could he have supported such a claim? Some have taken his supporting argument to have been that, when a mental concept seemed to be including its physical basis, what would really be happening would be its replacement by a different concept. However, that would only affect the description of the change and not the change itself.

On the other hand, it must be admitted that the tone of his writing and one of the illustrations that he uses suggest that he is going beyond the point marked by my conservative interpretation. The illustration is taken from botany: according to him, the seeds of two species of tree might differ only in their origins and their later developments, and it is just a prejudice to insist that there must be an actual physical difference between them, derived from their divergent origins and the cause of their divergent developments. It is hardly necessary to observe that a geneticist would find this view entirely natural and Wittgenstein's view a prejudice. But really these speculations have little importance and the question that a philosopher must answer is, "What reaction would come most naturally to us if an underlying physical mechanism were discovered?"

This is one of those cases where Wittgenstein's thought has acquired a momentum which seems to take him further than he can legitimately go. He cannot legitimately claim that there is any metaphysical force dictating the answer to the question. The momentum which carries him perilously near to making that claim comes from his successful polemic against the postulation of mental causes to explain such achievements as understanding the meaning of a word. It looks as if the same strategy ought to be equally effective against the suggestion that the physical causes of mental performances might be discovered. This hope is encouraged by the obvious fact that no explanation extracted from physical causes could possibly yield the kind of insight that he sought as a philosopher. According to him, a typical philosophical investigation starts when people get lost in the intricacies of their own thoughts; not, of course, when they are actually thinking them, but when they pause to reflect on them. Their feeling then is that, though they experience no difficulty in thinking the thoughts, they cannot see how they are related to one another. They know their way around but they cannot see the relations between their own well-practised movements. The discovery of the physical basis of mental phenomena will obviously do nothing to dispel this kind of perplexity.

These considerations may explain the over-run of Wittgenstein's thought – or, more accurately, of its apparent implication – but they cannot justify it. Consider, first, his argument against the postulation of a mental cause to explain such achievements as understanding the meaning of a word. It is, he argues, no good postulating a picture or a formula in a person's mind to explain his grasp of a word's application. His point is not that such things never occur in the mind, or that they are no help when they do occur, but only that they cannot provide a complete explanation of the speaker's application of the word. The reason for the shortfall is simply that the picture, or the words in the formula themselves, have to be applied.[17] So it is a mistake to suppose that a mental cause of this kind could automatize a speaker's practice and eliminate all his difficulties of application. The mistake is the offering of an *explanans* which itself re-introduces the *explanandum*, and so starts an infinite regress.

However, this argument cannot be adapted for use against the suggestion that the neural basis of a speaker's grasp of the application of a word might be discovered, and even included in the mental concept. In this new area there is no point of contact between the cause and the application of the word as it is for the speaker when he makes it. In the other area there is a point of contact, because the mental picture can serve him as an aid, and that is the point of departure for Wittgenstein's argument, that such an aid could never become a complete guide for him, altogether replacing his basic discernment of the correct applications. But in this new area there never is any possibility of his using the physical cause as a guide. So there is no point of departure here for any parallel argument. To put the point in another way: a neural explanation of this kind would have no tendency to show that the speaker's proficiency was not basic. This is important, because his insistence that the speaker's reasons for his applications run out[18] is surely right.

If this interpretation of Wittgenstein's remarks on the relation between mind and brain is correct, there is no need to suppose that he put mental language under any metaphysical constraint, or to infer that he deviated at this point from his settled, later view about the line dividing philosophy from science. But if we really want to trace that view to its origin, we must seek it in his later account of rule-following. That is the next and last topic of this paper.

In *Philosophical Investigations* Wittgenstein places the examination of rule-following immediately after the critique of his early theory of meaning. There is an obvious point in this collocation: the system of the *Tractatus* is a target of the *reductio* which is the centre of the negative part of his later treatment of rule-following. It is only one of the targets, because other forms of Platonism are also under attack. For in his later book he repudiates any theory which appeals to an unspecifiable metaphysical paradigm to explain the constraint felt by rule-followers. All such theories merely project this feeling onto the world and do nothing to explain it. In fact, if his *reductio* is convincing, they offer the rule-follower a criterion of correctness which he could not possibly use.

> "Whence comes the idea that the beginning of a series is a visible section of rails laid to infinity? Well, we might imagine rails instead of a rule. And infinitely long rails correspond to the unlimited application of a rule."
>
> "All the steps are already taken" means: I no longer have any choice. The rule, once stamped with a particular meaning traces the line along which it is to be followed through the whole of logical space – But if something of this sort really were the case, how would it help?[19]

The objection is that this is a picture of guidance without any usable exemplification.

Before we ask for the argument with which he supports this objection, it is worth taking one more look at the *Tractatus*. The point was made above[20] that he there took the view that the world imposes a fixed structure on any language, so that, once we have made a linguistic choice, its consequences are forced on us. However, that was a remark made about the general structure of our language. If we ask whether he would have said the same thing about the application of a particular name to a particular object, it is not so clear what his answer would have been. Certainly, the point of the picture theory is that a name picks up from the object to which it is attached certain constraints on its subsequent use. But no account is given of the way the name is supposed to do this, nor is there any discussion of its re-applications or of the criterion of identity on which they would depend. So the later attack on Platonism, which does deal with the repeated applications of descriptive words, makes a somewhat oblique impact on the *Tractatus.*

The objection to Platonism which is developed in *Philosophical Investigations* exploits the fact that any aid in the rule-follower's mind, such as a picture or an analytical formula, will itself have to be applied, and so will raise again the same problem that was raised by the original word.[21] So if the Platonist appeals to such aids – and he must point to something in the speaker's mind which will reflect the paradigm outside his mind – he will never be able to explain the distinction between correct and incorrect applications of the word. The naturalness of certain applications is an essential part of the basis of the distinction. But the Platonist tries to dispense with it, and so in his theory the difference between obeying and disobeying a rule disappears. He will not allow people's natural tendencies to make any contribution to what counts as using the word correctly, but he cannot find any effective substitute for them.

What we have found so far in Wittgenstein's later treatment of rule-following is the rejection of an assimilation of philosophy and science which exhibits a now familiar pattern. The task of philosophy is to investigate language in use, and, when we use a language, we feel ourselves under the constraint of the rules governing its words. So we theorize about the source of this constraint, modelling our theory on science. However, Platonism makes its source inaccessible – a point made by Aristotle against Plato himself – and Wittgenstein's achievement in the negative part of his treatment of rule-following was to demonstrate that, even if the Platonic paradigm had been accessible, it would have been no help.

He also had a positive contribution to make to the investigation of rule-following, a contribution which takes his demarcation of science and philosophy one stage further. If we ask why anyone would postulate Platonic paradigms to explain the human practice of using words according to rules, we immediately open up a new approach to Wittgenstein's demarcation of philosophy and science. The philosophical investigation of rule-following is likely to start as a search for an empirically accessible criterion of correct usage which will confine itself to material available to speakers of a language. This search immediately runs into a difficulty: when people speak a language, they cannot always rely on specifiable criteria of correctness and must, in the end, trust their own competence to apply the words in its vocabulary. But this can easily seem to be a lamentable weakness in the foundations of language. So the Platonist steps in and offers to satisfy the demand for security by drawing on the resources of another world.

However, that is not the whole story. For the really fundamental mistake made in this area is to demand a single criterion of correctness in any world. For example many commentators believe that Wittgenstein himself proposed a single criterion in this world, agreement in judgements with other members of the community speaking the language. But in fact, he never goes quite so far in that direction, and the reasons why he stops short of such a theory are interesting. First, it would be a theory which took no account of the constraints imposed on the community by the world with which their language was concerned. But the correction of the mistake would not lead to a better theory about the ultimate touchstone of correct usage recognized by speakers of a language. For the only way to correct the mistake would be to go beyond the speaker's point of view and try to see what they are doing from the outside, in the wider context of their lives.

So instead of theorizing about rule-following, Wittgenstein simply describes the phenomenon. He presents language as a system of practices exhibiting regularities which cannot be tested against any single touchstone. That does not mean that any move, however bizarre, is permissible in a language game. For it is essential that the regularities be a viable part of human life in the world. That requires them to be sensitive both to the constraints imposed by the world and to the constraints imposed by other people: they must promote our activities and the satisfaction of our needs and they must make communication possible. However, these general requirements cannot be captured in any single criterion of correctness in this world or in Plato's. They preserve the regularity of linguistic practices without giving the practitioners any single point of orientation.

What this account of rule-following offers us is a way of avoiding the excessive intellectualization of rule-following which can be found in most other treatments of it. Even the *Tractatus* had offered a theory about the fixed structure of the world to explain our feeling that when we use language we are under a completely external constraint. That theory, it is true, was only concerned with the general structure of the world, but its extension to particular features, such as the system of colours, is a step that is easily taken. The idea is that the task confronting the speaker of any language is a purely intellectual one: everything that keeps his language on the rails must be picked out by him and given a place on his list of criteria. Wittgenstein's account of rule-following offers a very different picture: an essential part of the basis of the regularities exhibited by language is human nature. If that seems too subjective, we should ask ourselves why we assume that everything objective needs to be included in the speaker's picture of the world as he finds it, and nothing in his finding it so.

This is not the place to develop an interpretation of Wittgenstein's idea that language is based on a pre-existent system of perceptions and actions – the idea expressed in his remark, "In the beginning was the deed".[22] The point that needs to be made here is only that he is rejecting other philosophers' excessive intellectualization of the problem of rule-following. If they confine themselves to the point of view of people actually using language, they will never be able to give a complete explanation of the regularity of their practices.

At this point it will be objected that, if this is what Wittgenstein is doing, he must have erased the line separating philosophy from science. For this part of *Philosophical Investigations* seems merely to contain a mixture of psychology and anthropology.

It is, of course, true that his observations about the place of language in our lives are factual. However, their aim is not the construction of any theory. They are offered as an antidote to the intellectualist prejudice, that a complete account of rule-following can be given by a philosopher who confines himself to the point of view of a speaker following a rule. That, of course, was the prejudice of analytic philosophy. Wittgenstein's strategy is to suggest that we should not restrict ourselves to the rule-follower's inside story but should put it in a wider context which includes things which do not always, and, in some cases, could not ever, serve him as criteria of correct usage.

These suggestions are not a contribution to any science, but of course, it is easy to cross over into science at this point without noticing the transition. For there are many tempting empirical questions just off the narrow path trodden by Wittgenstein. Could someone solitary from birth set up a language for his own use? How much idiosyncrasy is allowed by the constraint of communication? How fast can the meaning of a word change? Interpretations of his later philosophy often stray into these scientific investigations, but it is surely possible to see how his observations could be factual but not steps towards the construction of any theory, and yet not in the least mysterious in their import and purpose. They are directed against the common philosophical tendency to over-intellectualize the practice of rule-following, and, if Wittgenstein is right, it did not take a theory to correct that mistake.

Christ Church
Oxford

NOTES

[1] See *Tractatus logico-philosophicus* (TLP), ed. by D. F. Pears & B. F. McGuinness, Routledge & Kegan Paul, London 1961, 6.3–6.372.
[2] TLP, 4.11.
[3] TLP, 6.32–6.36.
[4] TLP, 3.262 and 3.326–3.27
[5] I.e. science in the narrow sense, in which it does not include the whole of factual discourse.
[6] TLP, 6.342.
[7] TLP, 6.372.
[8] TLP, 6.124.
[9] Which are identified with the senses of sentences in TLP. See 2.202–2.221.
[10] See Wittgenstein, 'Blue Book', pp. 63ff., in *The Blue and Brown Books*, ed. and with a Preface by R. Rhees, B. Blackwell, Oxford 1958.
[11] See his remarks on the nature of philosophy in *Vermischte Bemerkungen. Culture and Value* (CV), ed. by G. H. von Wright in collaboration with H. Nyman. Trans. by P. Winch, B. Blackwell, Oxford 1980, e.g. pp. 14–17.
[12] TLP, 6.124.
[13] TLP, 3.342.
[14] *Zettel* (Z), ed. by G. E. M. Anscombe & G. H. von Wright, B. Blackwell, Oxford 1967, § 610. See *Bemerkungen über die Philosophie der Psychologie. Remarks on the Philosophy of Psychology*, Vol. I (RPP I), ed. by G. E. M. Anscombe & G. H. von Wright. Trans. by G. E. M. Anscombe, B. Blackwell, Oxford 1980, § 905.

[15] Z, § 608. See RPP I, § 903.
[16] See *Philosophische Untersuchungen. Philosophical Investigations* (PI), ed. by G. E. M. Anscombe & R. Rhees. Trans. by G. E. M. Anscombe, B. Blackwell, Oxford 1953, I, § 18.
[17] PI I, §§ 139–141.
[18] PI I, §§ 217; 485.
[19] PI I, §§ 218–219, 1st paragraph.
[20] See above, p. 24.
[21] See above, p. 32.
[22] CV, p. 31.

ANTHONY J.P. KENNY

WITTGENSTEIN ON MIND AND METAPHYSICS

Wittgenstein is often regarded as being both positivist and behaviorist: positivist in rejecting all metaphysics, and behaviorist in denying inner human life. So far as concerns philosophy of mind, this view is based on a misunderstanding of Wittgenstein's work. He did indeed attack one particular metaphysical theory of mind: the Cartesian theory. Cartesianism is metaphysical in the sense of isolating statements about mental life from any possibility of verification or falsification in the public world. But much of Wittgenstein's work in philosophy of mind is devoted to showing the importance of distinctions between different kinds of potentiality and actuality. These distinctions were one of the major concerns of the work of Aristotle which was the first book to bear the name *Metaphysics*, and were a main target of classical anti-metaphysicians. In this sense Wittgenstein himself had a metaphysics of mind; and the metaphysical sensitivity which he shared with Aristotle was what enabled him to reject Cartesianism without falling into behaviorism. In this paper I will try to illustrate different forms of metaphysics, and sketch Wittgenstein's attitude to each.

In his mature writings, Wittgenstein does not often speak of metaphysics. The word "metaphysical" occurs only twice in the *Philosophische Untersuchungen*.[1] In each case it has a pejorative sense. And Wittgenstein can sum up his method as being a corrective to metaphysics:

> Wenn die Philosophen ein Wort gebrauchen – "Wissen", "Sein", "Gegenstand", "Ich", "Satz", "Name" – und das *Wesen* des Dings zu erfassen trachten, muss man sich immer fragen: Wird denn dieses Wort in der Sprache, in der es seine Heimat hat, je tatsächlich so gebraucht? – *Wir* führen die Wörter von ihrer metaphysischen, wieder auf ihre alltägliche Verwendung zurück. (PI I, §116).

Metaphysics here seems to be identified with the search for essences. But there can be a legitimate attempt to understand essences, on which Wittgenstein himself is engaged:

> [W]ir auch in unsern Untersuchungen das Wesen der Sprache – ihre Funktion, ihren Bau – zu verstehen trachten. . . (PI I, § 92).

What is wrong is to consider the essence, not as something which lies open to view and must merely be given a perspicuous description, but something

R. Egidi (ed.), Wittgenstein: Mind and Language, 37–46.
© 1995 *Kluwer Academic Publishers. Printed in the Netherlands.*

interior and hidden: a kind of metaphysical clockwork or hardware which explains the functioning of mind and language.

The kind of metaphysics which Wittgenstein systematically attacks is the metaphysics which consists of grammar masquerading as science. One source of metaphysics is the philosopher's temptation to mimic the claims and methods of science. Metaphysics, in this sense, is a quasi physics, an imaginary physics elevated into something sublime and mysterious. The tendency to create this kind of metaphysics is well described in PI I, § 36, where Wittgenstein is talking about ostensive definition.

> Und wir tun hier, was wir in tausend ähnlichen Fällen tun: Weil wir nicht *eine* körperliche Handlung angeben können, die wir das Zeigen auf die Form (im Gegensatz z.B. zur Farbe) nennen, so sagen wir, es entspreche diesen Worten eine *geistige* Tätigkeit. Wo unsere Sprache uns einen Körper vermuten lässt, und kein Körper ist, dort, möchten wir sagen, sei ein *Geist*.

In allusion to this passage, we might call this kind of metaphysics *spiritualistic metaphysics*. The passage already cited in which Wittgenstein describes the task of the philosopher as being to bring back words from their metaphysical usage itself comes in a section where Wittgenstein has been talking about the tendency to think of the proposition as something sublime, to put a halo around thought, to think of logic as a structure of crystalline purity. (PI I, §§ 74–108) We think of the mind as a mysterious medium, different from a physical medium, where the strict laws of logic operate.

Let us look at some examples of spiritualistic metaphysics of the kind which was Wittgenstein's target. The metaphysical impulse may lead us to postulate spiritual substances, or spiritual processes. In each case we are misled by grammar: where it makes us expect a physical substance, but there is not one, we invent a metaphysical substance; where it makes us expect an empirical process, but cannot find one, we postulate an incorporeal process. (PI I, § 339).

First, metaphysical substances. One of the most bizarre, as well as the most ubiquitous, misunderstandings of the nature of the mind is the picture of mind's relation to body as that between a little man or homunculus on the one hand and a tool or instrument on the other. We smile when medieval painters represent the death of the Virgin Mary by showing a small scale model virgin emerging from her mouth: but basically the same idea can be found in the most unlikely places.

Descartes, when first he reported the occurrence of retinal images, warned us not to be misled by the resemblance between images and their objects into thinking that when we saw the object we had another pair of eyes, inside the brain, to see the images. But he himself believed that seeing was to be explained by saying that the soul encountered an image in the pineal gland. This was a particularly striking version of what has been nicknamed "the homunculus fallacy": the attempt to explain human experience and behaviour by postulating a little man within an ordinary man.

We humans are always inclined to explain things we only imperfectly understand in terms of the most advanced technology of the age we live in. As time passes and technology progresses the tool or instrument which the manikin is fancied to control gets more and more sophisticated. Thus Plato thought that the soul in its relation to the body could be compared with a sailor in a boat or a charioteer holding the reins. Many centuries later Coleridge said that what poets meant by the soul was "a being inhabiting our body and playing upon it, like a musician enclosed in an organ whose keys were placed inwards". (*Letters*, i 278). More recently the mind has been compared to a signalman pulling the signals in his signal box, or the telephone operator dealing with the incoming and outgoing calls in the brain. Most recently the boat, the chariot, the railroad and the telephone exchange have given way to the computer, so that the relation of the soul to the body is envisaged as that of the programmer who writes the software to the hardware which executes the program.

What is wrong with the homunculus fallacy? In itself there is nothing misguided in speaking of images in the brain, if one means patterns in the brain which can be mapped onto features of the sensory environment. There is nothing philosophically objectionable in the suggestion that these schematic images may be observable to the neurophysiologist investigating the brain. What is misleading is to say that these images are visible to the soul, and that seeing consists in the soul's perception of these images.

The misleading aspect is that such an account pretends to explain seeing, but the explanation reproduces exactly the puzzling features which it was supposed to explain. For it is only if we think of the relation between a soul and an image in the pineal gland as being just like the relation between a human being and pictures seen in the environment that we will think that talk of an encounter between the soul and the image has any illuminating power at all. As a metaphor, manikin talk may be no more than a harmless necessary fancy; but as an element in a theory a manikin bedevils understanding. For whatever needs explaining in the behaviour of the man turns up, grinning and unexplained, in the shape of the manikin.

From an example of a metaphysical substance, let us turn to considering metaphysical processes. Wittgenstein discusses the question: Is understanding a mental process? (PI I, § 151ff). Some philosophers have thought that understanding was a psychological process in the same sense as we might call "a psychological process" the reciting of a poem in one's head. But reflection soon shows that this is not so.

If meaning was a mental process accompanying the utterance of a sentence, it should be possible for the process of meaning to take place without the sentence being uttered at all. Can one, in fact, perform the act of meaning without uttering the sentence? If you try to do so, you are likely to find yourself reciting the sentence itself under your breath. But of course it would be absurd to suggest that simultaneously with every public utterance of a sentence there is a private one too: it would surely take great skill to ensure that the two processes were exactly synchronized with each other! And how terrible if the two came slightly out of synchrony, so that the meaning of one word got mistakenly attached to the next one!

Moreover, the question whether somebody understands a sentence, and whether she really means it, can be raised about sentences uttered in the privacy of the imagination no less than about sentences uttered before a public audience. Infuriated by a curmudgeonly relation, I may mutter to myself "I wish he would drop dead!" Luckily, I don't mean it. I hum in my mind a Russian folk-song, enchanted by the sound of the words. But I haven't the faintest idea what they mean. If understanding and meaning were processes, they would have to accompany private utterances. So if the processes involved were some kind of inner utterance, we would be set off on an endless quest for the real understanding.

Some philosophers have thought that understanding was a mental process in rather a different sense. They have conceived the mind as a hypothetical mechanism postulated to explain the observable intelligent behaviour of human beings. If one conceives the mind in this way one thinks of a mental process, not as something comparable to reciting the ABC in one's head, but as a process occurring in the special mental machinery. The process on this view is a mental process because it takes place in a medium which is not physical; the machinery operates according to its own mysterious laws, within a structure which is not material but spiritual; it is not accessible to empirical investigation, and could not be discovered, say, by opening up the skull of a thinker.

Such processes need not, on this view, be accessible either to the inner eye of introspection: the mental mechanism may operate too swiftly for us to be able to follow all its movements, like the pistons of a railway engine or the blades of a lawn mower. But we may feel that if only we could sharpen our faculty for introspection, or somehow get the mental machinery to run in slow motion, we might be able actually to observe the processes of meaning and understanding.

According to one version of the mental-mechanism doctrine, understanding the meaning of a word consists in calling up an appropriate image in connection with it. In general, of course, we have no such experience when we use a word and in the case of many words (such as "the", "if", "impossible", "one million") it is difficult even to suggest what would count as an appropriate image. But let us waive these points, allow that perhaps we can have images in our mind without noticing that we do, and consider only the kind of word for which this account sounds most plausible, such as colour words. We may examine the suggestion that in order to understand the order "Bring me a red flower" one must have a red image in mind, and that it is by comparison with this image that one ascertains which flower to bring. This cannot be right: otherwise how could one obey the order "imagine a red patch?" Whatever problems there are about identifying the redness of the flower recur with identifying the redness of the patch.

It is of course true that when we talk mental images often do pass through our minds. But it is not they which confer meanings on the words we use. It is rather the other way round: the images are like the pictures illustrating a text in a book. In general it is the text which tells us what the pictures are of, not the pictures which tell us what the words of the text mean.

In fact, understanding cannot be thought of as a process at all. Understanding is a kind of ability, and therefore is a state rather than a process (PI I, p.59). In so far as the exercise of understanding is an exercise of intelligence, we may call understanding a mental state. But it is important to guard against misunderstanding here. Understanding may be a mental state, but it is not a psychological state like pain or depression or excitement. Such states last over periods, and can be continuous or interrupted; but one cannot know uninterruptedly what a word means.

Wittgenstein's treatment of the many-faceted illusion that understanding is a mental process is an example of his critique of spiritualistic metaphysics. He attacks metaphysics not by the blunt instrument of some positivistic verification principle, but by the careful drawing of distinctions which enable him to disentangle the mixture of truism and nonsense in the metaphysician's concept of mind.

Besides spiritualistic metaphysics, there is another kind of metaphysics to which Wittgenstein was implacably opposed. This is the view that there is a fundamental branch of philosophy which underlies and underpins the rest of philosophy and the rest of the sciences. We might call this kind of metaphysics *foundationalist metaphysics*. Descartes, who was the arch-exponent of spiritualistic metaphysics, can also be taken as a spokesman for foundationalist metaphysics. "The whole of philosophy – he wrote – is like a tree, whose roots are metaphysics, whose trunk is physics, and whose branches are all the other sciences". Not only Descartes, but many other thinkers have seen philosophy as an ordered system; a system which could perhaps be most perspicuously displayed by being cast into axiomatic form, as Spinoza tried to do.

Wittgenstein's *Tractatus* has reminded many people of Spinoza; but his later philosophy was the very reverse of systematic. This does not mean that it lacked method, or rigour. It means rather that there was no part of philosophy which had primacy over any other part. One could start philosophising at any point, and leave off the treatment of one problem to take up the treatment of another. Philosophy had no foundations, and did not provide foundations for other disciplines. Philosophy was not a house, nor a tree, but a web. This is how we are to understand the famous passage PI §133:

> Die eigentliche Entdeckung ist die, die mich fähig macht, das Philosophieren abzubrechen, wann ich will. – Die die Philosophie zur Ruhe bringt, sodaß sie nicht mehr von Fragen gepeitscht wird, die *sie selbst* in Frage stellen. – Sondern es wird nun an Beispielen eine Methode gezeigt, und die Reihe dieser Beispiele kann man abbrechen. – Es werden Probleme gelöst (Schwierigkeiten beseitigt), nicht *ein* Problem.

In refusing to countenance systematic or foundationalist metaphysics, Wittgenstein was distancing himself from many of the great philosophers of the past, including Aristotle. For Aristotle there was a philosophical discipline which deserved the title "First Philosophy"; and the attempt to delineate this is a central theme in the collection of treatises which we know as "The Metaphysics of Aristotle". Sometimes first philosophy is described as the discipline which studies being qua being; sometimes as the discipline which studies being qua divine. The two formulations probably are two ways of describing a single enterprise: one accounts for everything that is the case about everything there is by appealing ultimately to the divine movers unmoved, and "the study of being qua being" describes this investigation in terms of its *explicandum*, while "the study of being qua divine" describes it in terms of its *explicans*. But however one understands Aristotle's first philosophy it is clear that Wittgenstein would have accepted neither its methodological presuppositions nor its foundational role.

However, a considerable part of Aristotle's *Metaphysics* (and of his other works which would nowadays be described as metaphysical) is devoted to a philosophical activity which resembles quite closely Wittgenstein's own method. The distinction between actuality and potentiality, and the classification of different kinds of potentiality, is universally recognized (by both friend and foe) as being one of Aristotle's most characteristic contributions to philosophy and in particular to the philosophy of mind. His distinctions were later systematised by scholastic philosophers in the middle ages. We might call the systematic study of actuality and potentiality *dynamic metaphysics*.

Thus, according to Aristotle, active powers (e.g. the power to heat) differed from passive powers (e.g. the power to be heated). Natural powers (such as water's capacity to wet) were to be distinguished from rational powers (such as a pharmacist's ability to prescribe). Natural powers needed certain preconditions for their exercise: fire will only burn wood if the wood is sufficiently dry. But if these conditions are met, then the power will infallibly be exercised. The case is not the same with rational powers. A pharmacist may have the skill to prescribe, and may have the necessary pharmacopea; but he may fail to prescribe if his patient does not have sufficient funds. Natural powers, unlike rational powers, are also *tendencies* to act in a certain manner.

The possession of rational powers, according to Aristotle, is peculiar to human beings. Among the powers of humans there are some which are innate – the senses, for instance – while others like the ability to play the flute, are acquired by practice. The liberal arts, and in general the skills which are the fruits of education, are powers of a particular kind, namely *dispositions*. Dispositions are *abilities* whose exercises are the relevant scientific, artistic and craft activities; but they are themselves *actualisations* of the capacity to learn which is presupposed by education. They can thus be called actualisations as well as potentialities.

Medieval philosophers introduced a technical terminology here: the skills were first or primary actualizations in contrast to the episodic employments of the skills which were secondary actualizations. Thus the ability to speak Greek is a first actualization, while the actual utterance of a Greek statement or command, or the understanding of a particular Greek text on hearing it, is a secondary actualization.

Wittgenstein undertook a prolonged investigation of the nature of potentiality in the 'Brown Book', where sections 58–67 are devoted to various language games with the word "can".[2] The distinctions which he draws, in writing on the philosophy of mind, between processes and states, and between different kinds of states, correspond to the Aristotelian distinctions between *kinesis*, *hexis*, and *energeia*, and the criteria by which the distinctions are made often coincide. The example which Wittgenstein discusses at length, to discuss the relationship between a power and its exercise, namely learning to read (PI I, §§ 156ff) is the same as the standard Aristotelian example of a mental *hexis*, namely, knowledge of grammar.

In addition to the Aristotelian distinction between powers, their possessors, and their exercises, we may introduce the notion of the *vehicle* of a power or ability. The vehicle of an ability is the physical ingredient or structure in virtue of which the possessor of an ability possesses the ability and is able to exercise it. The distinction between abilities and their vehicles is not something which is peculiar to human beings and their abilities. Vodka has the power to intoxicate: the vehicle of the power of vodka to intoxicate is the alcohol the vodka contains. A vehicle is something concrete, something which can be weighed and measured. An ability, on the other hand, has neither lenght nor breadth nor location. This does not mean that an ability is something ghostly: my front-door-key's ability to open my front door is not a concrete object, but it is not a spirit either.

An important instance of the distinction between possessor, ability, and vehicle is the distinction between people, their minds, and their brains. Human beings are living bodies of a certain kind, with various abilities. The mind, as we have said, is the capacity to acquire or possess intellectual abilities. The vehicle of the human mind is, very likely, the human brain. Human beings and their brains are physical objects; their minds are not, because they are capacities. Once again, to say that the mind is not a physical object is not to say that it is a ghostly spirit: denying that the mind has a length or breadth or location does not involve one in spiritualistic metaphysics.

In every age since Aristotle, philosophers have been tempted to blur the distinctions he made. In philosophy there is a perennial temptation to reduce potentialities to actualities. Some philosophers attempt to reduce powers to their exercises: thus, explicitly, David Hume, who said the distinction between a power and its exercise was frivolous. Some philosophers attempt to reduce powers to their vehicles: thus, implicitly, Descartes, who wanted to identify the powers of bodies with their geometrical properties.

Philosophical errors about capacities in general show up particularly vividly when they occur in the philosophy of mind. Applied in this area, exercise-reductionism becomes behaviourism: the attempt to identify mind with behaviour consists in treating the complex second-order capacity which is the mind as if it was identical with its particular exercises in behaviour. Applied in this area, vehicle-reductionism becomes materialism: the attempt to identify mind with brain consists in reducing my mental capacities to the parts and structures of my body in virtue of which I possess those capacities.

Materialism is a grosser philosophical error than behaviorism because the connection between a capacity and its exercise is in truth a more intimate one than the connection between a capacity and its vehicle. In the case of the mind, the connection between capacity and exercise is a conceptual connection: one could not understand what the mind was if one did not understand what kinds of thing constitute the exercise of mental capacity. The connection between capacity and vehicle, on the other hand, is a contingent one, discoverable by empirical science. Aristotle's grasp of the nature of mind will stand comparison with that of any subsequent philosopher; but he had a wildly erroneous idea of the relationship of the mind with the brain, which he believed to be an instrument to cool the blood.

Wittgenstein rejected both behaviourism and materalism: in the area of philosophy of mind he is closer to Aristotle and his scholastic successors than he is to any of the more fashionable philosophies of our scientific and scientistic age. In one of his most characteristic, and most striking remarks, he goes so far as to entertain the possibility that some of our mental capacities may lack a vehicle altogether.

> Keine Annahme scheint mir natürlicher, als daß dem Assoziieren oder Denken kein Prozess im Gehirn zugeordnet ist; so zwar, daß es also unmöglich wäre, aus Gehirnprozessen Denkprozesse abzulesen. Ich meine das so: Wenn ich rede oder schreibe, so geht, nehme ich an, ein meinem gesprochenen oder geschriebeben Gedanken zugeordnetes System von Impulsen von meinem Gehirn aus. Aber warum sollte das *System* sich weiter in zentraler Richtung fortsetzen?... Es ist also wohl möglich, daß gewisse psychologische Phänomene physiologisch nicht untersucht werden *können*, weil ihnen physiologisch nichts entspricht. . . . Warum soll es keine psychologische Gesetzmässigkeit geben, der *keine* physiologische entspricht? Wenn das unsere Begriffe von der Kausalität umstößt, dann ist es Zeit, daß sie umgestoßen werden.[3]

Here we see Wittgenstein both at his most antimetaphysical and at his most metaphysical. He is antimetaphysical in the sense that he is making a frontal attack on the scientism characteristic of our age: the assumption that there *must* be physical counterparts of mental phenomena. Yet he is here also highly metaphysical. He is not embracing spiritualistic metaphysics: what does the associating, thinking, and remembering is a human being with a body, not a spiritual substance. But in the sense of dynamic metaphysics, Wittgenstein is here very close to Aristotle. For he is envisaging as a possibility a pure Aristotelian soul, or entelechy, which operates with no material vehicle: a formal and final cause to which no mechanistic efficient cause corresponds.

We can sum up Wittgenstein's attitude thus: he was hostile to spiritualistic metaphysics and to foundationalist metaphysics; but he was one of the most consummate practitioners of the dynamic metaphysics which is one strand of the Aristotelian tradition. This can be brought out, finally, by pointing to one of the passages in which Wittgenstein most explicitly rejects spiritualistic metaphysics. In this very passage we find that, in the Aristotelian sense, he gives a surprisingly metaphysical formulation of the relationship between soul and body.

> Nur vom dem, was sich benimmt wie ein Mensch, kann man sagen das es Schmerzen *hat*. Denn man muß es von einem Körper sagen, oder, wenn du willst, von einer Seele, die ein Körper *hat*. Und wie kann ein Körper eine Seele *haben?* (PI I, § 283).

How striking, that a body's having a soul should seem more problematic than a soul's having a body!

Balliol College
Oxford

NOTES

[1] *Philosophical Investigations* (PI), ed. by G. E. M. Anscombe & R. Rhees. Trans. by G. E. M. Anscombe, B. Blackwell, Oxford 1953.

[2] *The Blue and Brown Books*, ed. and with a Preface by R. Rhees, B. Blackwell, Oxford 1958.

[3] *Zettel*, ed. by G. E. M. Anscombe & G. H. von Wright. Trans. by G. E. M. Anscombe, B. Blackwell, Oxford 1967, §§ 608–610.

JOHANNES L. BRANDL

WITTGENSTEIN'S ALLEGED METAPHYSICS OF MIND

Anthony Kenny suggests in his stimulating paper *Wittgenstein's Metaphysics of Mind* that Wittgenstein's attack on Cartesianism is driven by a metaphysical view similar to Aristotle's theory of potentiality and actuality. He recommends using the threefold distinction between *possessor*, *vehicle* and *exercise of an ability* for interpreting Wittgenstein's view of mental events. As a heuristic device, this may prove useful, but I doubt that from this proposal there will emerge anything that may legitimately be called 'Wittgenstein's Metaphysics of Mind'.

1. THREE TYPES OF METAPHYSICAL CLAIMS

What is a metaphysics of mind? That depends on what we mean by the term 'metaphysics'. I am ready to accept, with Kenny, the positivist criterion according to which a claim is metaphysical concerning some domain if it is neither verifiable nor falsifiable by observations within this domain.[1] Since the domain of possible observations can be more or less restricted, we should distinguish various *types* of metaphysical claims:

1) If a claim is neither verifiable nor falsifiable concerning *any* domain of possible observations, such a claim may be called *purely speculative*.
2) If a claim is neither verifiable nor falsifiable by observations actually available at a certain stage of investigation, such a claim may be called *investigation independent*.
3) If a claim is neither verifiable nor falsifiable given the specific conditions that make observation possible for human beings, such a claim may be called *transcendental*.

With this tripartite division in mind, let us see whether Wittgenstein says anything about the mental that could be classified in one of these categories. As concerns category (1), it hardly needs arguing that Wittgenstein rejects claims of this sort as springing from a mythological, or quasi-scientific mind-body theory. It is typical of scepticism about other minds to raise doubts about facts whose presence is beyond any conceivable evidence. This form of scepticism takes mental states as something observable *only* by the inner eye of the subject. It thus assumes from the beginning that whatever I am directly

R. Egidi (ed.), Wittgenstein: Mind and Language, 47–56.
© 1995 *Kluwer Academic Publishers. Printed in the Netherlands.*

aware of is inaccessible to anyone else. Conversely, I may have all sorts of beliefs about the minds of other persons, but I cannot *find out* whether these beliefs are correct. On the present assumption, the only way to do this would be to experience these mental states myself. But then they would no longer be the states of other minds, having become part of my own consciousness. Hence we can only *speculate* about what goes on in the mind of other people. Once the appropriate evidence for mental events has been restricted to the deliverances of an inner sense, claims about other minds turn out to be unverifiable in sense (1). The only facts which could tell in favour of their truth are facts about the way people act, and these have been excluded as the 'wrong' kind of evidence from the beginning.

There can be little doubt that Wittgenstein has spotted the flaw in this kind of reasoning. Whenever he talks of 'hidden phenomena' or 'hidden mechanisms' he warns against introducing mental entities which are accessible in *no* other way than by being aware of them immediately. For this reason he insists that instead of saying "There has just taken place in me the mental process of remembering. . ." we should rather say "I have just remembered. . .".[2] The point of this move is the following: When I talk of a mental state (or process) *in me* I am not characterizing it in any particular way, but merely am hinting at what is typical of a mental state in general. Ignoring this fact leads to confusion. Apparently my bodily states are 'in me' too; they are *my* states after all. So the question arises: what is the contrast we draw by calling some states mental and others nonmental? It seems that this contrast will only emerge if I conceive of my bodily states not as states of me, but as states *of my body* (or processes in my body). Then I may conclude by parity of reasoning that, strictly speaking, mental states are not my states either, but states *of my mind* (or processes *in my mind)*. Thus I end up with the Cartesian distinction between two substances having different kinds of properties, mental and physical.

The recommended translation should block this argument. Saying "I have just remembered. . ." should make it plain that I am talking about a mental state *of* me, not *in* me. The contrast between my bodily states and my mental states is not a contrast between states *of* my body and states *of* my mind. There is only one bearer for both kinds of states, namely *me*. The pronoun "I" in my memory report does not refer to a mental entity, but to a *person* that has both mental and bodily states. Asked for the difference between these kinds of states I may claim some sort of privileged access to my mental, but not to my bodily states. This privilege does not exclude the possibility that a mental state might make itself manifest in other ways too, besides occurring 'in my mind'. This is particularly clear in the case of feeling pains. Though this is a special form of cognitive awareness, one can know about pains also without being aware of them in this special way. If someone else cares for my pains, he knows about them, yet he does not share my feelings in pitying me.

If we observe Wittgenstein's advice not to talk of processes *in* our minds, we may hope to escape scepticism about other minds. There simply is no domain of mental facts which are accessible only to the subject experiencing them. Since all speculative metaphysics about the mind relies on such facts, there is a straightforward incompatibility between such a view and Wittgenstein's method. Whenever mental facts become the target of a sceptical argument, Wittgenstein advises us to change our way of describing these facts. Having redescribed them in the appropriate way, there is nothing left to concern a speculative metaphysician.

Turning to the idea of metaphysics as telling us what things *really* are, I think it is equally clear that Wittgenstein has no such ambitions. The need to make investigation-independent claims about the mind arises once the accessibility of mental facts is admitted both from the first person and the third person perspective. Then we find ourselves in the position of Hume, expecting the sun to rise tomorrow morning, but not having fully sufficient reasons for a rigorous proof of this claim. In the same manner we expect other people to have pains when crying and to feel pleasure when they look pleased, but for all we know their feelings might be different. This form of scepticism is strong enough to turn anything we say about facts beyond our own consciousness into metaphysical claims of the investigation-independent sort.

There is an ongoing debate over whether or not Wittgenstein endorses scepticism of a Humean variety. Unfortunately the arguments on both sides have become so complicated that it is impossible to rehearse them here. But I hope I can bring out one point on which all participants in that debate should be in agreement. This point suffices, I believe, to show that whether or not Wittgenstein was bothered by a sceptical paradox he certainly did not take refuge in some sort of metaphysics in order to solve it.

Consider the following situation. I am sitting in a bar waiting for the bill. I had two glasses of wine and I know that each of them is 2 pounds. So I expect the waiter to ask me for 4 pounds. Now a friend sitting next to me with his eye out for a quarrel raises the following question: "How in heaven's name can you know in advance what your bill will be?" "Since I have calculated it", I reply. "But that", my friend responds, "only tells you what *you* would charge yourself if you were the waiter. How can you predict the bill *he* will give you?" Already a little angry I answer: "Well, he will calculate it too, and since it is a very simple calculation it is quite improbable that he will make a mistake." At this point my friend comes up with Kripke's version of the argument about rule-following. "You expect the waiter to follow the same arithmetical rules as you do. But for all you know, and for all anybody else may know (including the waiter himself) this may not be the case. He might be a quus-person, adding in the usual way under conditions C, but doubling the result under conditions C'. On the assumption that conditions C' obtain you should get a bill of 8, not of 4 pounds. Now, tell me your reasons for thinking that the waiter is not a quus-person and that his way of doing addition is not conditional on anything you regard as irrelevant for doing arithmetic?"

As I said, it is very debatable what the next move in this game should be according to Wittgenstein. Should I join my friend in being sceptical about what the waiter's calculation will be? Or should I argue from the fact that the waiter has been trained in our community and that there is sufficient mutual agreement between us to warrant my prediction? Or should I reply that no reasons are needed for excluding such queer cases as imagined by my friend in order to justify my belief about the bill? Whatever the recommended move may be, there are other moves which are definitely wrong. It would be a clear mistake to start an argument about *understanding* what the waiter says when he asks for the bill. I could not save my prediction by adding the hypotheses that the waiter might get the words wrong, that he would not really want 8 pounds even if he said so, that he certainly *meant* he should get only 4 pounds, but mistakenly uses the word "eight" to express his demand. If that were so, I would expect him to be satisfied by getting 4 pounds because that is what he *really* asks me for. Naturally, I would also expect him to thank me for the 8 pounds I gave to him, because that is what he will call my four pounds according to the present hypothesis.

This desperate move rests on a serious mistake. It presupposes that we can always drive a wedge between what somebody *says* with the words he uses and what he really *means* by them. Of course, this strategy allows me to excuse *whatever* the waiter says or does as a purely verbal mistake, as a surface phenomenon so to speak. The correct calculation may still be there at a deeper level, as a phenomenon in the mind of the waiter. But it will be there independently of anything he does or says. This is the price one has to pay for justifying an apparently mistaken calculation by invoking a linguistic error. Of course, sometimes we do have good reasons to doubt the linguistic competence of others, but the mere fact that their calculations are wrong is not a good reason. It is the *arbitrariness* in attributing a linguistic mistake to the waiter which leads to unacceptable consequences.

If whatever seems to be wrong for a third person may be correct from the first-person perspective, it could just as well be the other way round. That is exactly how my friend in the bar wants to have it. If I allow the waiter to say "eight" and really mean "four", then he can allow the waiter to do an idiosyncratic calculation even though in the past he always came up with the expected results. That is to say, I might agree with all his previous uses of the plus-sign although in fact he always had a different calculation in mind. This difference might become apparent just in presenting me the next bill. To avoid this, we must deny that saying and meaning can diverge in an arbitrary way. An argument which opens such a gap will leave us helpless in convincing our sceptical friend that rule-following is not a matter of 'anything goes'.

So far, I think, nothing controversial has been said about the sceptical underpinnings of the rule-following problem. My only point was that the problem becomes worse as soon as we concede that rules may have been obeyed in spite of all appearances to the contrary. That is exactly what happens when we divorce mental states from their role in explaining human behaviour. If the same behaviour may be effected by just any mental state whatever, then the truth about the mental is left completely moot. Mental states are now put beyond the reach of empirical investigation. Wittgenstein wanted to block this source of metaphysical theories about the mind with his rule-following argument. In view of it, I think, there is no space left for attributing to him claims about mental phenomena whose truth is independent of any empirical investigation about their causal effects.

There is then left to consider, as a third possibility that Wittgenstein holds a transcendental metaphysics of the mind. This, again, is a big issue which should be discussed only on the basis of extensive textual evidence. But I think that something can be said in advance against this interpretation. As is well known, Wittgenstein applies his considerations about rule-following to the case of a private language, i.e. to a language which cannot be shared by several speakers. He starts this discussion by asking wheter we could *imagine* a language in which a person records his inner experiences for his private use.[3] The argument then is intended to prove that no matter how we imagine such a language, we imagine something impossible. But why should anybody want to claim that there are languages which are in principle unsuited for communication?

There are a number of prejudices about the mental which are liable to produce the myth of a private language. One such prejudice which is particularly suggestive takes every mental phenomenon to have a natural meaning, just as having pains is a natural indicator for some damage in the body. If this is the model for mental phenomena in general, then a belief or desire should be a natural indicator of the fact which obtains when the belief is true or the desire satisfied. When I learn to express my beliefs and desires in a conventional language, what happens, according to this picture, is that I learn for each sentence in the given language which mental states naturally indicate the fact described by this sentence. In doing so I will establish correlations of the following kind: Sentence S is true iff my belief that such-und-such is true. If the language has a productive vocabulary this requires an infinite list of such correlations. The only way in which I can handle such a list is by applying a procedure which generates infinitely many clauses of the form: My belief that such-and-such is true iff such-and-such. But applying such a procedure means that I already have some kind of language at my disposal. The upshot then seems to be that learning the first conventional language is only possible as a process of translating it into a more basic language, a language I somehow acquire together with my beliefs and desires even before I learn to express them verbally. This basic language must be a system of natural signs revealing their meaning simply by occurring in my mind; just as a certain feeling reveals itself to me without any need to learn in advance what it means. But now we have to observe that my pains, taken as a natural sign in my mind, reveal their meaning only *to me*. Therefore the basic language I start with must be a private one.

How does Wittgenstein respond to this reasoning? I think his answer can be interpreted in two ways. First, Wittgenstein may be taken as offering a better argument in favour of the opposite conclusion. The defender of a private language starts from the vague characterisation of mental states as natural signs and ends up with the conclusion that for any language to be possible at all every speaker *must* be equipped with a language of his own, i.e. with a system of signs that necessarily mean something only to him. The opposite view takes mental states to be necessarily linked to conventional manifestations, e.g. verbal utterances, and concludes from this that a private language is *impossible* since there are no conventional criteria for its application. The first argument amounts to a transcendental proof of the existence of a private language, the second offers an equivalent proof of the opposite. Reading Wittgenstein in this way means that he indulges in transcendental metaphysics. He does so by claiming that every mental state *has* to be linked with public criteria. But the conflict can be described in a different way too. On this second reading Wittgenstein is not aiming at any proof at all. Rather he wants to *undermine* the alleged proof either of the necessity or the impossibility of a private language: By revealing both conclusions to be too strong he argues for the falsity of their premises. As I see it, Wittgenstein was prepared to mistrust any philosopher who sets out to prove how things *must* be or how they *cannot* be. In the present case this throws doubt on the characterisation of mental states as natural signs, as well as on the idea of naming sensations by conventional signs. The philosopher's task is not to establish transcendental arguments; rather he should dismantle them whenever they are forced upon him.

Once we conceive of the private language argument not as an argument in and of itself, but as a strategic move against a philosophical prejudice, I think that turning Wittgenstein into a transcendental philosopher becomes very difficult. He induces us to mistrust the "hardness of the logical must"[4] and to question the rigidity of explanations based on mere analogy. To work out "preconceived idea[s] to which reality *must* correspond" was exactly the opposite of what he wanted to achieve. He called it the "dogmatism into which we fall so easily in doing philosophy".[5]

2. THREE GRADES OF MENTAL INVOLVEMENT

Having listed three reasons for *not* counting Wittgenstein's philosophy of mind 'metaphysical', let me turn to some considerations of a more systematic kind. One thing Wittgenstein makes us aware of, I think, is that we should be very careful in defining the vocabulary with which we attribute cognitive states. Roderick Chisholm once suggested that the terms we use in attributing *epistemic* states may all be defined in terms of the single notion of 'preferring one proposition to another'.[6] Thus we may focus on this single notion and distinguish three different ways of defining 'epistemic preference':

Df_1: *S* prefers *p* to *q* = df It is more likely that *S* will accept the proposition that *p* than the proposition that *q* (given that he considers these propositions as mutually exclusive alternatives).

Df_2: *S* prefers *p* to *q* = df It is more likely that *S* will sincerely assent to an utterance saying that *p* than to an utterance saying that *q* (given that he understands these utterances as conflicting with each other).

Df_3: *S* prefers *p* to *q* = df It is more likely that *S* will act successfully if it is the case that *p* than if it is the case that *q* given that his intentions depend on these matters of fact).

Definition Df_1 is the one that Chisholm would prefer. The idea behind it is that epistemic preference involves a decision among a given set of propositions. These are entities to which no one has a privileged access. Two persons, even when speaking different languages, may nonetheless grasp the very same proposition. That gives epistemic states an objective value. But there is an additional reason for having propositions in ones' ontological baggage. Propositions are semantically transparent entities. The only way to know a proposition is to know what it means. Linguistic entities, by contrast, may be encountered without grasping their meaning. In this respect propositions are like private experiences: They reveal to us their meaning as soon as they occur to us. But unlike private experiences they may occur to more than one individual. Therefore, defining cognitive states in terms of an awareness of propositions guards against the idea of a private language. And this may well close the door to a transcendental metaphysics of the mind.

Definition Df_2 has a different advantage. If our cognitive states are defined in terms of our readiness to assent to utterances, that makes it plain why only *persons* (and perhaps some animals) can be in such states. Neither a Cartesian ego nor a mindless body could assent to an utterance. It requires more than taking an attitude towards a proposition and also more than a bodily movement like nodding ones' head. Assent is the sincere *expression* of agreement, which involves both mental and bodily states. Thus Df_2 guards against theories which turn the domain of other minds into a domain of speculative metaphysics.

Both of these advantages are shared by definition Df_3. A Cartesian ego may have mental states, but only a person with a body is able to perform actions. Since Df_3 rests on this latter notion it secures the link between the mental and the physical. The other advantage, namely the objectivity of mental states, can be guaranteed not only by introducing abstract propositions, but also by relating a person's actions to what is actually the case. When two people act successfully in the same way under the same conditions, we have reason to assume that their preferences are the same as well.[7] Thus Df_3 prevents mental states from being locked up in a private sphere.

Besides sharing the advantages of the other definitions, Df_3 has an advantage of its own. It shows that there is no more to the cognitive states of a person than what is manifested in her actions. This is not so clear from the other definitions. They require us to know which propositions a person considers as the relevant alternatives or to translate the utterances to which the person assents into our own language. In each case there may be more than one way to fulfill this requirement. Yet, judging from what the person *does*, there may be no difference between choosing one translation over the other or one set of propositions rather than another. According to definitions Df_1 and Df_2 this indicates a difference in the cognitive states of the person. Any difference in the objects of the attitudes implies a difference in the mental state itself. In drawing such fine-grained distinctions we find that the definitions Df_1 and Df_2 lead us to investigation-independent claims about our cognitive states. Definition Df_3 avoids this danger by reporting mental states without relating a person to any specific object. The definition only relies on what a person *S* achieves given that it is the case that *p* or the case that *q*. Here the variables "p" and "q" may be replaced by non-synonymous sentences without implying that a different mental state is attributed to *S*. In this way we are prevented from a conception of mental states that we cannot verify or falsify by even the best investigative procedures available.[8]

3. MENTAL AND MODAL INVOLVEMENT

Let me close this note with a comparison between the three definitions of epistemic preference and Quine's objections to modal logic. Quine recommends taking "necessity" as a semantic predicate to be attached to names of sentences.[9] This corresponds to the meta-linguistic definition Df_2. He objects to using "necessarily" as an operator on open sentences because this amounts to introducing essential properties. A similar criticism may be urged against definition Df_1. According to this definition there ought to be propositions that a person prefers to any other alternative. Such a commitment to 'epistemically perfect' propositions is just as objectionable as a commitment to essentialism.

But there is a middle way in the case of modal logic and also in the case of defining epistemic preferences. The third option is to take "necessity" as an operator on closed sentences only. This is still objectionable from Quine's point of view because it sacrifices extensionality. But, as Arthur Prior once remarked, it would be mere stubbornness to deny that the sentence *"S thinks that p"* is a function of the sentence *"p"* only because the function has to be intensional.[10] I think it would be equally stubborn to insist on a definition of mental states in terms of assent or dissent, only to avoid the full blown mentalistic idiom of grasping propositions. There is a modest form of getting involved both in mental and in modal affairs. It was Wittgenstein's advice, I think, to find the right degree of involvement in the first case. That was his way to avoid dubious metaphysical claims about the mind.[11]

Universität Salzburg

NOTES

[1] Kenny mentions this criterion also in his book *The Metaphysics of Mind*, Oxford University Press, Oxford 1989, p. ix.

[2] L. Wittgenstein, *Philosophische Untersuchungen. Philosophical Investigations*, ed. by G. E. M. Anscombe & R.Rhees. Trans. by G. E. M. Anscombe, B. Blackwell, Oxford 1953, § 306.

[3] *Ibid.*, § 243.

[4] *Ibid.*, § 437.

[5] *Ibid.*, § 131.

[6] Roderick M. Chisholm & R. Keim, 'A System of Epistemic Logic', *Ratio*, 14, 1972, pp. 95–110.

[7] Charles Travis pointed out to me that this inference cannot be drawn simply from the fact that two persons achieve the same results under the same conditions, because that may be a matter of pure luck. The demand that they must 'act successfully in the same way' should rule this out, but it is not clear to me how to explicate this phrase appropriately.

[8] Most of the literature on individuating mental states focusses on the opposite problem, namely, that the criteria of individuation offered (for instance) by possible-worlds-semantics are too *coarse grained*. But there is pressure from *both* sides here to draw just as many distinctions as there are in psychic reality.

[9] W.V.O. Quine, 'Three Grades of Modal Involvement', in: *The Ways of Paradox and other Essays*, Harvard University Press, Cambridge Mass., 1966, pp. 158–176.

[10] Arthur Prior, *Objects of Thought*, Clarendon Press, Oxford 1971, p. 48.

[11] I am indebted to Tony Anderson, Barry Smith and Charles Travis for their help with this paper.

PART II

IN THE DOMAIN OF POSSIBILITIES: RULES, FORMAL AND INNER LANGUAGES

JACQUES BOUVERESSE

LE RÉEL ET SON OMBRE: LA THÉORIE WITTGENSTEINIENNE DE LA POSSIBILITÉ

ABSTRACT

Wittgenstein repeatedly criticised the idea of certain "shadowy beings" which we believe to be necessary to relate two radically different things which apparently cannot be put directly in contact with each other, like thought and what it represents, sentence and what verifies it, desire and what realizes it, expectation and what satisfies it, etc. Wittgenstein tried to show that we are brought no nearer to the goal by the interposition of intermediary entities of that kind. In the case of a declarative sentence, the shadowy being must be a possibility which is expressed by the sentence and can be or not be realized. If the sentence is true, the possibility is supplanted by the corresponding fact; and if it is false, the only reality we have is so to speak the shadow, the possibility itself. One of the most fundamental theses of Wittgenstein's philosophy is that all possibility (at least "logical" or "grammatical" possibility, as he calls it) must be contained in language, and not between language and reality, in and intermediary realm of possibilities, and that it is all of the same kind and on the same level. To say that something is possible is simply to say that an expression has sense; and to say that it is impossible is not to say that the expression represents an impossible sense, as if we had first to consider a possible sense and then to exclude it as unreal or absurd, but simply that it has no use and does not belong to language. An important consequence of that is that, for Wittgenstein, there are no degrees of impossibility, no impossibilities that could be deeper and more radical than others. We cannot distinguish between a nonsense which makes (or at least could make) sense and a nonsense which does not. As Wittgenstein says, "what we exclude has no semblance of sense", it is not thinkable and does not have to be thought in some way in order to be excluded. What we exclude is always a use and not, so to speak, a possibility which turns out to be impossible.

I

Wittgenstein a consacré une part importante de ses réflexions à la critique d' "êtres fantômatiques" *(shadowy beings)*, qui sont supposés servir

R. Egidi (ed.), Wittgenstein: Mind and Language, 59–81.
© 1995 *Kluwer Academic Publishers. Printed in the Netherlands.*

d'intermédiaires obligés entre deux choses qui ne paraissent pas pouvoir être mises en relation l'une avec l'autre: la pensée et ce qu'elle représente, la phrase et ce qui la vérifie, le désir et ce qui le réalise, l'attente et ce qui la satisfait, etc. Puisque la chose que nous attendons peut ne pas se réaliser, "ce que nous attendons n'est pas le fait, mais une ombre de ce fait; pour ainsi dire, la chose qui est presque le fait".(*)[1] (BLB, p. 36) De même, puisque le fait qui rendrait notre pensée vraie, s'il existait, peut très bien ne pas exister, nous sommes tentés de dire que l'objet de notre pensée n'est pas le fait lui-même, mais plutôt une ombre de ce fait. "Il y a, remarque Wittgenstein, différents noms pour cette ombre, par exemple 'proposition', 'sens de la phrase'." (*ibid.*, p. 32) Puisqu'une pensée peut être fausse, elle doit contenir, à défaut du fait lui-même, qui, en l'occurrence, n'existe pas, du moins la possibilité considérée comme une sorte de substitut fantômatique ou de double annonciateur du fait qui la vérifierait. Wittgenstein s'efforce de montrer que nous pourrions tout aussi bien faire l'économie de ces entités intermédiaires qui peuplent le royaume des ombres, parce qu'elles ne font que repousser un peu plus loin le problème qu'elles étaient supposées résoudre:

> Nous n'avons rien gagné en admettant qu'une ombre devait intervenir entre l'expression de notre pensée et la réalité à laquelle se rapporte notre pensée. Nous avons dit que, si nous voulions une image de la réalité, la phrase elle-même *est* une image de ce genre (bien que ce ne soit pas une image par ressemblance). (*ibid.*, p. 41)

Le problème est que nous demandons aux ombres que nous évoquons, de réaliser une chose tout à fait impossible. Nous disons qu'un état intentionnel comme le désir doit contenir une ombre de la chose désirée en voulant dire par là qu'il doit en quelque sorte contenir la chose elle-même, mais à l'existence près. C'est ce que nous pouvons dire de moins absurde, si nous ne voulons pas être obligés de proférer une chose encore plus absurde, à savoir que la chose désirée elle-même doit être présente dans le désir. Mais, objecte Wittgenstein:

> Comment pouvons-nous désirer que *ceci précisément* arrive, si ceci précisément n'est pas présent dans notre désir? Il est tout à fait vrai de dire: la simple ombre ne ferait pas l'affaire; car elle s'arrête immédiatement avant l'objet; et nous voulons que le désir contienne l'objet lui-même. (*ibid.*, p. 37)

Dans ses leçons des années 1930–1932, Wittgenstein souligne ce point de la façon suivante:

> Si j'attends que Mr Smith entre, mon attente contient quelque chose de semblable à Mr Smith; si j'attends une tache rouge, mon attente ne peut pas contenir la tache rouge elle-même, mais seulement quelque chose qui lui est semblable. Mais attendre quelque chose de 'semblable'

> est déjà attendre quelque chose de différent. L'élément 'semblable', l'ombre, dans l'attente, est différente de la réalisation; vous n'êtes pas *plus près* de la réalisation.[2]

En d'autres termes:

> Interpoler une ombre ou une chose du même genre entre un symbole ou une proposition et un fait ne résoudra rien à moins que nous n'ayons une autre ombre pour servir d'intermédiaire entre l'ombre et le fait; nous avons alors une régression à l'infini, nous ne sommes pas arrivés plus près. Une similitude n'explique pas l'ombre, elle n'explique pas non plus la relation au fait. (WL1930–32, p. 32)

Dans le cas d'une proposition déclarative, l'ombre médiatrice est la possibilité du fait concerné, c'est à dire le fait précis qui devrait être réalisé pour que la proposition soit rendue vraie, moins la réalisation. Et l'objection est évidemment la même que précédemment. Pas plus que la réalisation du désir, l'actualisation de la possibilité ne peut être comprise en termes de concordance d'un fait réel avec une image ressemblante. Ou plus exactement, si l'expression du désir et la description du fait ne sont pas capables d'aller elles-mêmes jusqu'à la chose concernée, aucune image plus ressemblante ne le fera davantage. Dire qu'une proposition exprime une possibilité n'est donc pas dire que la proposition contient une image suffisamment ressemblante pour être l'image de ce fait à l'exclusion de tout autre, mais simplement qu'elle a un sens; et dire qu'une proposition a un sens, ne veut pas dire qu'il lui correspond, dans un univers qui n'est pas celui des faits réels ni celui des expressions, une entité qu'on appelle son sens.

Nous ne pouvons donc pas essayer d'expliquer ce qu'on veut dire en disant qu'une proposition p a un sens en remplaçant "p a un sens" par "p représente une possibilité" et "p n'a pas de sens" par "p est une proposition à laquelle ne correspond aucune possibilité" ou, ce qui est pire encore, "à laquelle correspond une impossibilité". "p a un sens" et "p exprime un fait possible" doivent avoir exactement le même sens; et, si l'on considère les choses dans la perspective du *Tractatus*,[3] l'une et l'autre sont des propositions impropres, qui essaient de dire quelque chose qui peut seulement se montrer. Il va sans dire que l'on n'améliorerait pas les choses en disant que "(La phrase) p a un sens" veut dire que "La proposition exprimée par p est possible", alors qu'une phrase dénuée de sens exprime ou tente d'exprimer une proposition impossible. "p a un sens", "p exprime une proposition" et "p exprime une proposition possible" disent en fait rigoureusement la même chose. Dans un passage dont la rédaction (qui n'est pas due à Wittgenstein lui-même) est assez confuse, mais qui est néanmoins suffisamment clair sur le point essentiel, Wittgenstein dit:

> Nous ne pouvons pas dire 'Une proposition (*proposition*) p est possible'. Si p a un sens, cela en lui-même montre qu'elle est possible. Si p n'était pas possible, ce ne serait pas une proposition du tout. Tout ce que nous pourrions vouloir dire par 'p est possible' serait 'p a un sens'; il ne pourrait y avoir aucune indication concernant la question de savoir si ce que p asserte est arrivé ou pourrait arriver. De ce fait, 'p est possible' n'est pas une proposition légitime. (Possibilité veut dire ici possibilité logique). (WL1930–32, p. 9)

C'est une des thèses fondamentales de la philosophie de Wittgenstein que toute la possibilité (du moins la possibilité de l'espèce qu'il appelle "logique" ou "grammaticale") est contenue dans le langage. C'est le système de représentation que nous adoptons qui détermine les possibilités accordées à la réalité concernée. Wittgenstein dit que "la possibilité est exprimée (contenue) dans le langage lui-même". (WL1930–32, p. 8) Et il faut comprendre cela comme signifiant en particulier que la possibilité n'est pas contenue dans un univers de possibilités qui occuperait une position intermédiaire entre le langage et la réalité. Plus précisément, la possibilité est contenue dans la grammaire du langage, puisque c'est la grammaire qui détermine ce qui a un sens et ce qui n'en a pas, c'est à dire ce qui est possible et ce qui ne l'est pas. "La grammaire, écrit Wittgenstein, n'est pas l'expression de ce qui est le cas, mais de ce qui est possible; il y a donc un sens auquel la possibilité est la forme logique". (WL1930–32, p. 10)

Il résulte de cela que ce qui est possible et ce qui ne l'est pas est arbitraire dans un certain sens, le sens auquel, comme le dit et le répète Wittgenstein, la grammaire est arbitraire: "Ce qui est appelé possible et ce qui n'est pas appelé possible est en un certain sens arbitraire. Nous disons que, bien qu'aucun homme ne soit assis sur cette chaise, quelqu'un le pourrait". Cela signifie en gros que:

> La phrase 'Quelqu'un est assis sur cette chaise' a un sens, c'est à dire, il y a une possibilité logique que quelqu'un y soit assis. Il est théoriquement possible, c'est à dire possible dans une théorie, qu'un certain hydrogène ait six fois la valence normale. Cela pourrait être possible dans certaines théories, mais peu pratique. Les systèmes peu pratiques sont rejetés, et le rejet est traité comme si ce qui est rejeté était quelque chose de faux. Le rejet d'un système grammatical est comme le rejet d'un étalon de longueur, et l'acceptation d'une grammaire, d'un symbolisme, est comme l'acceptation d'un étalon de longueur.[4]

On pourrait donc dire que "p est possible" veut dire "p a un sens" et "p est possible chimiquement" veut dire "p a un sens en chimie", c'est à dire dans le cadre d'une certaine théorie chimique admise. Le choix d'un système de représentation détermine un espace de possibilités pour les faits de la catégorie concernée; et inversement, ce qui est possible est toujours ce qui peut se dire de façon douée de sens dans un certain système de représentation. Wittgenstein compare, du reste, à l'occasion, le cas de la possibilité logique à celui de la possibilité chimique. Dans les *Recherches Philosophiques*[5] on trouve la suggestion suivante:

> Comparez 'logiquement possible' avec 'chimiquement possible'. On pourrait appeler par exemple chimiquement possible une combinaison pour laquelle il y a une formule de structure avec les bonnes valences (par exemple H-O-O-O-H). Une telle combinaison n'existe évidemment pas forcément; mais même à une formule HO_2, il ne peut correspondre moins dans la réalité qu'aucune combinaison. (PI I, § 521)

Vue du point de vue de la réalité, l'impossible devrait, semble-t-il, représenter quelque chose de moins que la simple non-existence, une forme aggravée de la non-existence, tout comme la nécessité ressemble à une forme accentuée de l'existence. Pourtant, à une formule qui représente une combinaison impossible correspond dans la réalité exactement la même chose, à savoir simplement aucune combinaison. La différence entre les deux est faite uniquement par le système de représentation adopté.

Quand Wittgenstein dit que la grammaire est arbitraire, il veut dire qu'il n'existe pas de faits auxquels nous pourrions essayer de nous référer pour établir que les possibilités et les impossibilités qui ont été décrétées par la grammaire étaient bien des possibilités et des impossibilités réelles. L'argument essentiel qu'il utilise sous des formes diverses au début des années trente pour justifier cette impossibilité de soumettre la grammaire au verdict d'une réalité quelconque revient toujours à dire que la description de possibilités et d'impossibilités qui sont supposées précéder la grammaire et rendre ses règles obligatoires, présupposerait elle-même une grammaire et donc une distinction préalable, qui ne peut être justifiée, entre le sens et le non-sens. De toute façon, comment pourrions-nous justifier les exclusions qui ont été opérées par la grammaire sans être contraints d'utiliser des expressions auxquelles la grammaire que nous cherchons à justifier ne donne précisément aucun sens?

> Pouvons-nous donner une description qui justifiera les règles de la grammaire? Pouvons-nous dire pourquoi nous devons utiliser ces règles? Notre justification ne pourrait prendre que la forme consistant à dire: 'Etant donné que la réalité est telle ou telle, les règles doivent être telles ou telles'. Mais pour décrire une réalité dans laquelle la grammaire serait autrement, je devrais utiliser les combinaisons précises que la grammaire défend. Les règles de la grammaire distinguent le sens et le non-sens et, si j'utilise les combinaisons défendues, je profère du non-sens. (WL1930–32, p. 47)

Dans la *Grammaire Philosophique*,[6] Wittgenstein formule sa position ainsi: "J'aimerais dire: 'Je dois *commencer* avec la distinction du sens et du non-sens. Avant elle, rien n'est possible. Je ne peux pas la justifier'". (PG, § 81)

La difficulté qui se présente ici tient au fait que, lorsque nous disons d'une chose qu'elle est impossible, il semble que nous ayons dû d'abord en quelque sorte penser une possibilité et plus précisément nous la représenter, pour pouvoir ensuite l'exclure. A l'époque du *Tractatus*, Wittgenstein avait été aux prises avec la question de savoir ce qui peut bien correspondre dans la réalité à la proposition "non-p", lorsque celle-ci est vraie. Si l'on exclut, comme il le fait, que le fait négatif puisse être conçu comme résultant du fait positif par l'adjonction d'un constituant supplémentaire qui correspondrait à la négation (un fait négatif est la "non-existence d'états de choses" (2.06) et non l'existence d'états de choses négatifs), ce qui correspond en pareil cas à "non-p" ne peut être que la non-réalisation du fait décrit par "p". Dans les leçons des années 1930–1932, Wittgenstein formule les choses ainsi:

> On demande parfois: comment pouvons-nous nier une proposition? Car si non-p est vrai, il n'y a rien pour lui correspondre; il n'y a quelque chose pour correspondre à p que lorsque p est vrai. Qu'est-ce qui correspond à 'La porte n'est pas ouverte' quand elle *est* ouverte? Mais il y a ici une fausse analogie: p ne *correspond* pas à *quelque chose*. Et ce qui correspond à non-p est *que p n'est pas le cas*. (WL1930–32, p. 52)

Mais cela suppose évidemment que le fait que p, dont la propostion négative asserte la non-réalisation, soit représenté – et cela signifie nécessairement, représenté comme une possibilité – dans l'expression de la proposition.

Que se passe-t-il à présent, si nous essayons de nier une proposition qui dit, non pas que quelque chose est le cas, mais que quelque chose est possible, c'est à dire si nous essayons d'asserter que quelque chose n'est pas possible? La possibilité de la chose, qui est exprimée par la proposition positive, n'est-elle pas du même coup nécessairement présupposée? Ou encore, pour pouvoir dire que p n'a pas de sens, ne faut-il pas avoir donné d'abord un sens à p ou en tout cas envisagé une possibilité de sens pour p? Wittgenstein reconnaît qu'il y a quelque chose de paradoxal dans le fait que nous puissions dire que quelque chose est impossible, puisque cela implique, semble-t-il, que nous puissions dire ce qu'est la chose dont nous sommes en train d'exclure la possibilité:

> Il est étrange que nous disions ce que c'est, qui est impossible, par exemple que le dessus de cheminée ne peut pas être jaune et vert en même temps. En parlant de ce qui est impossible, il nous semble que nous concevions l'inconcevable. Lorsque nous disons qu'une chose ne peut pas être verte et jaune au même moment, nous excluons quelque chose, mais quoi? Si nous devions découvrir quelque chose que nous décririons comme vert et jaune, nous dirions immédiatement que ce n'était pas un cas exclu. Nous n'avons pas exclu du tout un cas quel qu'il soit, mais plutôt l'usage d'une expression. Et ce que nous excluons n'a pas de semblance de sens. La plupart d'entre nous croient qu'il y a un non-sens qui a du sens, et un non-sens qui n'en n'a pas – que

> c'est un non-sens qui n'est pas du même genre que de dire 'Ceci est vert et jaune au même moment' et de dire 'Ab sur Ah'. Mais ce sont des non-sens dans le même sens, la seule différence étant dans la façon dont sonnent les mots. (WL1932–35, p. 64)

Ce qui nous induit en erreur dans des cas de ce genre est que nous avons l'impression que nous faisons un effort désespéré pour réussir à penser ou à imaginer quelque chose et qu'une proposition comme "Aucun objet ne peut être vert et jaune au même moment", par exemple, enregistre simplement l'échec de cette tentative. Si j'essaie de soulever un homme avec une seule main, je peux m'apercevoir que j'en suis incapable et dire qu'il m'est impossible de soulever un homme avec une seule main. Mais si la possibilité dont il s'agit est la possibilité logique, le point important est que ce que je ne peux imaginer est également quelque chose que je ne peux pas essayer d'imaginer, parce que je ne sais tout simplement pas ce qu'il pourrait être question d'essayer d'imaginer. Rien, en effet, n'a été décrit ou même simplement suggéré par l'expression dont on entend exclure l'usage: "Nous pourrions demander: à quoi ressemble une structure impossible? Peut-elle être décrite? Lorsque nous appelons une chose quelconque impossible, il n'y a rien que nous appelions sa description".[7] (LFM, p. 148)

La situation est évidemment un peu plus délicate lorsqu'une expression qui donnait l'impression d'avoir un sens apparaît comme étant en réalité dépourvue de sens au terme d'un processus de démonstration plus ou moins compliqué. Wittgenstein est revenu fréquemment sur la question de savoir comment nous pouvons, en mathématiques, chercher une chose dont nous découvrons finalement qu'elle est impossible, comme par exemple la trisection de l'angle ou la construction de l'heptagone régulier avec la règle et le compas, ce qui donne l'impression que la chose que l'on cherchait est une chose que l'on ne pouvait pas non plus réellement chercher. Dans ses leçons sur les fondements des mathématiques de 1939, Wittgenstein dit que:

> Tout ce que la démonstration mathématique du fait qu'un heptagone régulier ne peut être construit avec la règle et le compas effectue, est de nous donner de bonnes raisons d'exclure l'expression 'construction de l'heptagone' de notre notation. Par conséquent, 'Smith a tracé la construction de l'heptagone' n'est pas faux, mais dénué de sens. Il utilise une expression qui n'est pas seulement une expression à laquelle on n'a pas donné de signification, mais qui a été exclue. Elle a été exclue pour des raisons d'expérience, bien que l'énoncé selon lequel c'est impossible ne soit pas un énoncé d'expérience. (LFM, p. 47)

Wittgenstein ne va pas jusqu'à nier que l'on puisse faire une différence entre le non-sens immédiat et le non-sens que l'on pourrait dire différé, si l'on entend par là que certaines expressions peuvent être exclues immédiatement comme dénuées de sens, alors que d'autres ne le peuvent qu'au terme d'une suite d'opérations dont le résultat peut être, au moins du point de vue psychologique, tout à fait inattendu. Mais le non-sens est toujours le non-sens d'une expression exclue, et non d'un sens impossible, d'un sens qui, en quelque sorte, n'a pas de sens. Wittgenstein soutient qu'il n'y a pas de degrés dans l'impossibilité ou le non-sens, que tout ce qui est impossible ou dénué de sens l'est au même degré et de la même façon, même s'il ne l'est pas immédiatement:

> Les règles pour l'usage des mots peuvent exclure certaines combinaisons, et cela de deux façons: (1) lorsque ce qui est exclu est reconnu comme non-sens dès qu'il est entendu, (2) là où des opérations sont requises pour nous permettre de le reconnaître comme non-sens. Le fait que la négation d'une tautologie complexe soit une contradiction est découvert par les mêmes moyens par lesquels on trouve que $x^2 + 6x + 7$ *a deux racines entières* est vrai – à l'aide d'opérations. Nous pourrions croire, par exemple, que 'S a x paires de chaussures, avec $x^2 = 2$' a un sens, parce que nous aurions la sensation que nous pourrions extraire un sens de cette phrase en résolvant l'équation quadratique $x^2 = 2$. Le fait que nous ne voyions pas ce qu'est le résultat est une raison de croire que nous pourrions appeler la phrase une espèce de non-sens différente de 'tables, chaises, chaussures'. Le mot 'non-sens' est utilisé pour exclure certaines choses, et pour des raisons différentes. Mais il ne peut pas être vrai qu'une expression est exclue et cependant pas tout à fait exclue – exclue parce qu'elle désigne l'impossible, et pas tout à fait exclue parce qu'en l'excluant nous devons penser l'impossible. Nous excluons des phrases comme 'C'est à la fois vert et jaune' parce que nous ne voulons pas les utiliser. Bien entendu, nous pourrions donner un sens à ces phrases. J'ai dit antérieurement que ce qui est possible ou impossible est une question arbitraire. Nous pourrions instaurer comme règle, par exemple, que 'Vert et jaune peuvent être à la même place au même moment' devra avoir un sens. (WL1932–35, p. 64)

Qu'une expression soit exclue directement ou que, comme c'est le cas en mathématiques, elle le soit à la suite d'un calcul ou d'une démonstration, l'exclusion est donc dans tous les cas aussi complète et radicale et ne laisse rien subsister de ce qui pouvait ressembler au premier abord à une possibilité ou une potentialité de signification.

Wittgenstein, comme on l'a vu, refuse de considérer la possibilité comme une forme affaiblie ou amoindrie de l'existence, et donc comme capable de préexister en un sens quelconque à notre décision d'adopter une certaine norme d'usage pour les expressions concernées. Le "réel du possible", comme on disait à l'époque de Leibniz, ne peut résider ailleurs que dans le langage lui-même:

> Si vous voulez comprendre ce qu'est la possibilité logique, une bonne analogie est la possibilité chimique. Prenez une formule de structure en chimie:

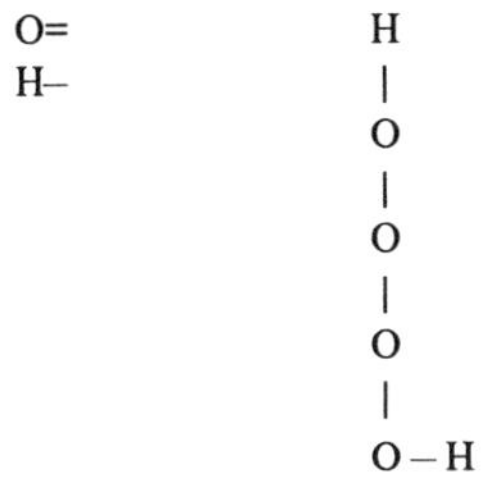

H_2O_4

'H_2O_4 est chimiquement possible bien qu'il n'existe pas'. Cela étant, qu'est-ce qui est possible à propos de H_2O_4, s'il n'existe pas? Est-il dans ce cas un peu plus *réel* par le fait d'être possible? Est-il plus proche de l'existence que HO_2? Vous pourriez dire: 'Y a-t-il des stades de l'existence?' (cf 'Cela n'est pas arrivé, mais c'était tout à fait possible, ce qui est déjà quelque chose!). Quelle est l'utilité de cette idée que c'est une possibilité chimique? Y a-t-il ici aussi une réalité qui est de la nature d'une ombre? Et où est l'ombre? – Les chimistes ne pourraient-ils pas croire qu'il y a un monde dans lequel il y a un H_2O_4, bien qu'il ne soit pas passé à la réalité?

N'y a-t-il pas ici une réalité faite d'ombre – que vous avez réellement dessinée? Votre *langage* n'est-il pas la réalité faite d'ombre?

Lorsque vous dites que 'H_2O_4 est possible', vous voulez dire simplement que c'est un signe dans votre système. Ce système de valences n'a pas été choisi au hasard, mais parce qu'il s'accordait bien avec les faits. Mais une fois qu'il a été choisi, ce qui est possible est ce dont il y a une image dans le langage des valences. Nous avons adopté un langage dans lequel cela *a un sens* de dire 'H_2O_4. . .' – ce n'est pas vrai, mais cela a un sens.

Exactement la même chose est vraie de la possibilité géométrique. Dire qu'une ligne droite peut être tracée, c'est dire que cela a un sens de parler de tracer une ligne droite. (LFM, p. 146)

La métaphore de l'ombre interposée entre la réalité et nous rend justice à notre idée que le possible précède et annonce dans certains cas le réel. Nous pouvons par exemple, dans certaines conditions, voir l'ombre d'un homme qui s'apprête à entrer avant de voir l'homme lui-même. Mais si le possible n'est pas réel, c'est l'ombre qui devient en quelque sorte la réalité, la réalité demeurée à l'état de possible. Aussi déterminantes que puissent être les raisons que nous avons d'instaurer une possibilité ou une impossibilité grammaticales en adoptant un mode d'expression déterminé, Wittgenstein soutient que nous sommes radicalement empêchés de les concevoir et de les présenter comme équivalant à la reconnaissance directe ou indirecte d'une possibilité ou d'une impossibilité qui existaient en dehors du langage et attendaient d'y être enregistrées. Puisque les propositions mathématiques sont des propositions grammaticales, on ne peut évidemment pas non plus dire d'elles qu'elles se contentent de décrire fidèlement des possibilités et des impossibilités objectives qui sont en l'occurrence reconnues essentiellement par la procédure de la démonstration. Comme toutes les propositions grammaticales, elles déterminent des possibilités et des impossibilités qu'elles donnent l'impression de découvrir dans une réalité qu'elles sont supposées décrire, en l'occurrence la réalité mathématique, l'exemple le plus parfait de ce que pourrait être une réalité purement grammaticale, si cela avait un sens de parler de quelque chose de ce genre.

Pour Wittgenstein, les propositions de mathématiques pures opèrent au niveau grammatical de la détermination de sens (*Sinnbestimmung*), et non au niveau descriptif de l'utilisation de sens (*Sinnverwendung*). Wittgenstein dit de la proposition mathématique qu'elle "doit en vérité nous montrer ce que cela a un SENS de dire"[8] et de la démonstration qu'elle "doit nous déterminer à reconnaître ceci comme sens et à ne pas reconnaître cela comme sens." (*ibid.*) Il en résulte immédiatement que la possibilité en mathématiques n'est pas et n'a pas besoin d'être le reflet d'une existence qu'elle révèle et qui la fonde. Wittgenstein évoque à ce propos, avec le mélange de déférence extrême et de distanciation radicale qu'il manifeste généralement en pareil cas, le point de vue tout à fait caractéristique de Frege:

> Frege, qui était un grand penseur, a dit que, bien qu'il soit dit dans Euclide qu'une ligne droite *peut* être tracée entre deux points quelconques, en fait la ligne existe déjà, même si personne ne l'a tracée. L'idée est qu'il y a un univers de la géométrie, dans lequel les entités géométriques existent. Ce que dans le monde ordinaire nous appellerions une possibilité est dans le monde géométrique une réalité. Dans le ciel euclidien, deux points sont déjà reliés l'un à l'autre. C'est une idée très importante: l'idée de la possibilité comme étant une espèce différente de réalité; et nous pourrions l'appeler une ombre de la réalité. (LFM, pp. 144–145)

Wittgenstein fait allusion ici à un passage de *Über die Zahlen des Herrn Schubert* dans lequel Frege écrit que:

> [Les postulats d'Euclide] ne sont à proprement parler rien d'autre que des axiomes qui énoncent qu'il y a des constructions – points, lignes, surfaces – constituées d'une certaine manière. C'est ainsi que la proposition qui demande que l'on trace une ligne droite d'un point quelconque à un autre dit qu'il y a, pour deux points pris de façon quelconque, une ligne droite qui les joint. Lorsque nous traçons une ligne nous dirigeons notre attention sur elle, qui est à proprement parler déjà là. La possibilité objective de tracer une ligne est à proprement parler la même chose que l'existence objective de cette ligne.[9]

Selon une conception courante que l'on a de la distinction qui existe, dans le système d'Euclide, entre les axiomes et les postulats, on est enclin à dire que les axiomes énoncent que certaines choses sont ou ne sont pas le cas, alors que les postulats garantissent certaines possibilités, essentiellement la possibilité d'effectuer des constructions d'un certain type. Frege considère, bien entendu, cette distinction comme illusoire. Pour lui, asserter la possibilité de tracer une certaine figure veut dire asserter l'existence de cette figure dans l'univers des entités géométriques. L'image qu'il y a derrière cette conception est, comme le remarque Wittgenstein (*ibid.*), qu'une droite est déjà tracée entre deux points quelconques, mais une droite extrêmement mince et à peine perceptible, sur laquelle nous en retraçons en quelque sorte une plus épaisse et plus grossière (nous la "matérialisons" simplement). On peut dire que Wittgenstein propose, comme Frege, d'ignorer la distinction qui vient d'être évoquée; mais, chez lui, la réunification s'effectue en sens inverse: les mathématiques construisent des possibilités et en éliminent d'autres, sans jamais se référer à une réalité déterminée et immuable qui serait présente à l'arrière-plan.

Wittgenstein a toujours considéré comme une source de confusion désastreuse l'idée que les mathématiques peuvent distinguer entre leur possibilité et leur réalité et traiter en quelque sorte séparément de l'une et de l'autre. Dans les *Philosophische Bemerkungen*,[10] il observe que: "Sans cesse on se laisse égarer par l'idée: 'Mais peut-il y avoir une possibilité sans qu'il y ait une factualité qui lui corresponde?'" (p. 164, note). Dans le cas des mathématiques, la situation est la suivante:

> Le sentiment que l'on a est: en mathématiques, il ne peut pas y avoir réalité et possibilité. Tout est sur un *seul* plan. Et c'est à dire, en un certain sens, réel.
>
> Et c'est exact. Car ce que les mathématiques expriment avec leurs signes est tout sur un *seul* plan; c'est à dire: elles ne parlent pas tantôt de leur réalité et tantôt de leur possibilité. Elles ne doivent, au contraire, pas du tout essayer de parler de leur possibilité. Mais il y a bien dans leurs signes une possibilité, à savoir la même qui réside dans les propositions proprement dites, dans lesquelles les mathématiques sont appliquées.
>
> Et lorsqu'elles essayent (comme dans la théorie des ensembles), d'*énoncer* (aussprechen) leur possibilité, c'est à dire lorsqu'elles la confondent avec leur réalité, alors on a le droit de les renvoyer à l'intérieur de leurs frontières. (PB, pp. 164–165)

La différence que les mathématiques cherchent à faire entre leur possibilité et leur réalité se manifeste évidemment en priorité dans la distinction que l'on croit devoir instaurer entre l'infini potentiel et l'infini actuel. Cantor souligne que c'est à peine si l'infini potentiel peut encore être appelé "infini", puisque ce que l'on désigne sous le nom d'infini potentiel est en réalité simplement du "fini variable". En revanche, l'infini proprement dit, l'infini actuel, est quelque chose de tout à fait déterminé: "Le concept ω, par exemple, ne contient *rien de fluctuant, rien de variable, rien de potentiel*, il n'est pas un ἄπειρον, mais un ἀφωρισμένον, et la même chose est vraie de tous les autres nombres transfinis".[11] Si l'on se demande quelle est la relation exacte qui existe entre ces deux concepts si différents de l'infini, la réponse de Cantor est tout à fait claire: "... Au vrai, l'infini potentiel n'a qu'une réalité empruntée, dans la mesure où il renvoie toujours à un infini actuel, qui seul le rend possible. D'où l'épithète accolée de façon pertinente par les scholastiques à l'infini potentiel: συνκατηγορηματικως". (*ibid.*, p. 404)

Wittgenstein plaide, au contraire, pour ce qu'on pourrait appeler l'autonomie radicale de la possibilité. Il soutient que l'on ne peut pas la décrire en la transformant, comme le fait la théorie des ensembles, en une réalité infinie qui échappe par essence à toute possibilité de symbolisation directe:

> La théorie des ensembles est fausse pour la raison qu'elle présuppose apparemment un symbolisme qu'il n'y a pas, au lieu de celui qu'il y a (qui est seul possible). Elle construit sur un symbolisme fictif, donc sur le non-sens. (PB, p. 211)

Wittgenstein soutient que l'infini ne peut apparaître dans les mathématiques que sous la forme de la possibilité infinie, qui est reflétée, et non pas décrite, dans les possibilités infinies du symbolisme. Un axiome qui asserte l'existence de l'infini ne pourrait donc qu'être un non-sens:

> L'*axiom of infinity* est déjà un non-sens pour la raison que la possibilité de l'énoncer présupposerait un nombre infini de choses – donc ce qu'il veut affirmer. Des concepts logiques, par exemple de l'infinité, on peut dire que leur essence démontre leur existence. (*ibid.*, p. 124)

Dire qu'il n'y a pas de nombre entier naturel plus grand que tous les autres revient à dire que cela n'a pas de sens de parler d'un nombre de ce genre. La théorie des ensembles cherche à transformer une détermination grammaticale en une énonciation descriptive portant sur une extension infinie. La raison ultime de l'hostilité de Wittgenstein à la théorie des ensembles est, de toute évidence, que le mode de pensée ensembliste contribue plus que n'importe quel autre à masquer le caractère de propositions grammaticales que possèdent les propositions mathématiques et à rapprocher beaucoup plus qu'il ne le faudrait le statut des mathématiques de celui d'une science naturelle. L'origine et le fondement des possibilités et des impossibilités que les mathématiciens découvrent sont supposées, en effet, résider dans des collections infinies données, qui ne peuvent être explorées que par des moyens indirects. On ne sait pas actuellement s'il est possible ou non de trouver quatre entiers positifs x, y, z et n tels que $n > 2$ et $x^n + y^n = z^n$. Mais cela signifie que l'on ne sait pas s'il y a ou non quatre nombres de ce genre dans l'ensemble des entiers positifs. Ils y sont ou ils n'y sont pas. Mais, pour l'instant, seul un être omniscient comme Dieu, qui connaît les ensembles infinis dans leur totalité, sait laquelle de ces deux choses est vraie.

Dans le *Tractatus* Wittgenstein remarquait déjà que:

> La théorie des classes est en mathématiques tout à fait superflue.
> Cela se rattache au fait que l'universalité dont nous avons besoin en mathématiques n'est pas l'universalité *accidentelle*. (TLP, 6.031)

Dans les *Philosophische Bemerkungen*, Wittgenstein souligne que la généralité qui intéresse les mathématiques ne peut pas être la généralité amorphe, celle d'une proposition comme "Toutes les pommes sont mûres", mais uniquement la généralité dominée, structurée et finalement constituée par la démonstration. Si les nombres naturels constituaient une totalité donnée en extension, il serait concevable qu'ils possèdent tous une certaine propriété de façon accidentelle, c'est à dire d'une manière telle que l'on pourrait seulement constater qu'ils la possèdent en les examinant tous l'un après l'autre, sans avoir les moyens de démontrer que cette propriété qu'ils possèdent tous est une propriété qu'ils doivent nécessairement posséder. (Gödel s'oppose directement à Wittgenstein lorsqu'il soutient que nous pourrions très bien nous trouver dans une situation dans laquelle il est possible de conjecturer à la fois que l'on sera en mesure de vérifier, pour n'importe quel nombre donné, qu'il possède une certaine propriété, et qu'il n'existe cependant pas de démonstration universelle de ce fait.) Wittgenstein écrit:

> Il n'y a pas tous les nombres, justement parce qu'il y en a une quantité infinie. Et parce qu'il ne s'agit pas ici du 'tous' amorphe, comme dans la phrase 'Toutes les pommes sont mûres', où l'ensemble est donné par une description externe, mais de la totalité de structures qui doivent précisément être données en tant que telles.
> Cela ne concerne pour ainsi dire pas la logique de savoir combien il y a de pommes, lorsqu'on parle de *toutes* les pommes. En revanche, il en va autrement dans le cas des nombres, de ceux-ci elle est responsable individuellement. (p. 148)

Le mode de pensée extensionnel de la théorie des ensembles encourage l'idée désastreuse que, même lorsqu'on a affaire à une collection infinie, il y a encore deux façons d'établir une proposition universelle: l'une qui consisterait à passer successivement en revue tous les éléments et que des êtres finis comme nous le sommes ne peuvent utiliser, et l'autre qui consiste à aboutir au résultat par un moyen indirect ou en prenant un raccourci dont on ne peut pas être certain *a priori* qu'il existe, ce raccourci étant constitué en l'occurrence par la démonstration. Wittgenstein soutient que, s'agissant de l'infini, une telle distinction n'a justement aucun sens. De la théorie des ensembles, il dit qu'elle raisonne comme si tous les éléments d'une collection infinie étaient en quelque sorte contenus dans une caisse que l'on peut transporter avec soi. Si l'on emporte la caisse, on est sûr d'emporter tous les éléments. Mais si l'on veut s'assurer qu'ils possèdent tous une certaine propriété, il faut ouvrir la caisse et trouver un moyen quelconque de les en sortir tous:

La description le saisirait [l'infini] à peu près comme on porte emballées dans une caisse une quantité de choses que l'on ne peut pas tenir dans les mains. Elles nous sont alors invisibles, et pourtant nous savons que nous les portons (pour ainsi dire de façon indirecte). La théorie des agrégats vend en quelque sorte chat en poche. L'infini n'a qu'à s'arranger comme il veut dans cette caisse. (PB, p. 206)

II

L'aspect le plus déconcertant de la conception que Wittgenstein développe au début des années trente à propos de la possibilité et de l'impossibilité (logiques) est certainement son idée que toutes les impossibilités correspondent à des interdits grammaticaux et que toutes les choses que la grammaire exclut sont exclues de la même façon et dans le même sens. Dans son compte rendu des leçons que Wittgenstein a données au cours des années 1930–1933, Moore avoue qu'il avait été passablement surpris par l'usage que Wittgenstein faisait de l'expression "règle de grammaire", et qu'il avait tenté en vain de lui faire admettre qu'en tout cas il n'utilisait pas cette expression dans son sens ordinaire. L'idée de Moore était que l'on devrait distinguer au moins deux sens auxquels on peut dire qu'une règle de grammaire a été violée: (1) le sens usuel, qui fait référence à une règle grammaticale reconnue et explicite comme celle qui interdit de dire, par exemple, "Trois hommes va en bateau"; (2) le sens philosophique wittgensteinien, en vertu duquel on peut qualifier de faute contre la grammaire l'affirmation que deux couleurs différentes figurent au même endroit en même temps dans un champ visuel. Moore ne consent pas à mettre sur le même plan une simple incorrection grammaticale, au sens usuel, comme celle de la phrase "Trois hommes va en bateau" et une chose que la grammaire exclut comme inimaginable, inconcevable, impensable ou impossible (au sens logique), par exemple la co-occurrence de deux couleurs différentes au même endroit du champ visuel. La réponse de Wittgenstein est la suivante:

L'expression correcte est 'Cela n'a pas de sens de dire . . .'; mais d'ordinaire nous utilisons un mode d'expression défectueux qui consiste à parler d'une règle de grammaire. Ainsi, cela n'a pas de sens de dire 'Cette table est aussi identique que l'autre'. Voir ce qui se passe si l'on utilise le même échiquier et les mêmes pièces que nous le faisons pour les échecs, mais en faisant des mouvements qui ne sont pas prévus par les règles.

Nous avons le sentiment que le premier mésusage auquel Moore fait référence est anodin, mais le second, vicieux. Mais en fait les deux sortes de règles sont des règles dans le même sens. Ce qui se passe est simplement que certaines d'entre elles ont fait l'objet d'une discussion philosophique et d'autres non. Si nous discutons une règle, nous devons l'énoncer.

Ces difficultés proviennent d'une fausse analogie. Ainsi le puzzle selon lequel nous ne 'pouvons pas' mesurer le temps est dû à l'analogie du temps physique. Nous sommes enclins à dire que nous *ne pouvons pas* imaginer ou penser quelque chose, et nous sous-entendons que nous *pourrions* l'exprimer correctement si nous en avions l'expérience. Dire que quelque chose est 'logiquement impossible' ressemble à une proposition. Ainsi, si nous disons que nous ne pouvons pas penser rouge et bleu ensemble dans le même espace visuel, nous avons l'impression d'*essayer* de le faire, comme si nous *parlions* du monde physique; nous nous dupons d'une certaine façon nous-mêmes et croyons que cela *peut* être fait.

Les règles grammaticales sont toutes de la même espèce, mais ce n'est pas la même erreur si un homme viole l'une d'entre elles que s'il en viole une autre. S'il utilise 'va' au lieu de 'vont', cela ne provoque pas de confusion; mais dans l'autre exemple, l'analogie avec l'espace physique (cf deux personnes sur la *même* chaise) entraîne une confusion. Lorsque nous disons que nous ne pouvons pas penser à deux couleurs à la même place, nous commettons l'erreur de croire que c'est une proposition, bien que cela n'en soit pas une; et nous n'essaierions jamais de dire cela si nous n'étions pas induits en erreur par une analogie. Il est trompeur d'utiliser les mots 'ne peut pas' parce que cela suggère une mauvaise analogie. Nous devrions dire: 'Cela n'a pas de sens de dire . . .'.

La règle concernant rouge et bleu ((2) ci-dessus) est une règle concernant l'usage du mot 'et'; et nous ne dirions que 'va' ((1) ci-dessus) n'a pas de sens que si quelqu'un disait que cela pose un problème philosophique. (WL1930–32, pp. 97–98)

La raison pour laquelle la première impossibilité ne pose pas de problème philosophique, alors que la deuxième en pose un, est que, dans le premier cas, nous avons affaire, semble-t-il, à une convention purement grammaticale, dont la violation n'empêche pas la phrase de rester compréhensible et de continuer à représenter un fait possible, alors que, dans le deuxième cas, la règle grammaticale, si l'on consent à l'appeler ainsi, donne l'impression de refléter une impossibilité objective et d'énoncer un fait important qui résulte de la nature même de la couleur. Une incorrection grammaticale de l'espèce ordinaire est une chose que l'auditeur peut corriger implicitement et qui n'affecte généralement pas ou pas beaucoup la compréhensibilité de la phrase; mais la violation d'une règle grammaticale au deuxième sens, ne laisse apparemment subsister aucune possibilité pour la phrase de signifier encore quelque chose.

L'idée que Wittgenstein critique est celle qui consiste à croire que, dans le cas d'une phrase grammaticalement bien formée et représentant néanmoins une impossibilité logique ou un non-sens comme "A cet endroit du champ visuel figurent en ce moment deux couleurs différentes", nous essayons bel et bien de faire quelque chose que nous ne parvenons pas à faire, à savoir combiner entre elles des significations de mots qui, à première vue, se prêtent tout à fait à ce genre d'opération et qui pourtant se révèlent être telles que le sens propositionnel obtenu n'en est pas un. Un sens qui avait l'air possible se trouve être, en dépit des apparences, un sens impossible, un sens qui n'a pas de sens. Wittgenstein soutient qu'un non-sens n'a rien à voir avec un sens qui n'aurait pas de sens. Lorsqu'on dit que "Cela n'a pas de sens de dire que p" on veut dans tous les cas simplement dire que la phrase 'p' est, pour une raison ou pour une autre, retirée de la circulation, et non qu'elle a en quelque sorte un sens qui n'en est pas un. Wittgenstein souligne que:

> Lorsqu'on dit qu'une phrase est dénuée de sens, ce n'est pas pour ainsi dire son sens qui est dénué de sens. En réalité, une combinaison de mots est exclue de la langue, retirée de la circulation. (PI I, § 500)

Les mots qui indiquent un état de choses impensable

> sont éliminés de notre langage, comme par exemple n'importe quel bruit arbitraire, et la raison de leur élimination *expresse* ne peut consister que dans *la tentation où nous sommes* de les confondre avec une phrase de notre langage. (PG, p. 130)

La raison pour laquelle un non-sens syntaxique caractérisé comme "ah est à" semble beaucoup plus radical et ne nous pose normalement aucun problème philosophique sérieux, alors qu'un non-sens sémantique comme "César est un nombre premier" peut nous en poser un, est donc uniquement qu'une analogie trompeuse, qui est tentante dans le deuxième cas, ne peut pas nous tenter et nous tromper dans le premier.

La position de Wittgenstein sur ce point est évidemment liée de façon tout à fait directe à la critique qu'il formule au début des années trente contre l'idée de ce qu'il appelle le corps de signification *(Bedeutungskörper)* qu'un mot a derrière lui et dont il n'est que la surface apparente. Il écrit:

> Considérez, l'analogie suivante: entre un cube ou une pyramide avec une surface peinte, derrière laquelle il y a un corps invisible, et un mot et la signification derrière lui. Toute position dans laquelle cette surface pourrait être placée dépendra de la position du corps solide qui est derrière elle. Nous sommes tentés de croire que, si nous savons qu'un cube est derrière la surface peinte, nous pouvons connaître les règles qui gouvernent l'arrangement de cette surface avec d'autres surfaces. Mais ce n'est pas vrai. On ne peut pas déduire la géométrie du cube de la contemplation d'un cube. Les règles ne suivent pas d'un acte de compréhension. (WL1932–35, p. 50)

L'analogie du corps de signification est discutée également dans la *Grammaire Philosophique* (p. 54) et évoquée en passant dans les *Recherches Philosophiques*:

> On aimerait parler par exemple de la fonction du mot dans *cette* phrase. Comme si la phrase était un mécanisme dans lequel le mot a une fonction déterminée. Mais en quoi consiste cette fonction? Comment se manifeste-t-elle? Car il n'y a en vérité rien de caché, nous voyons bel et bien la phrase dans sa totalité! La fonction doit se montrer dans le cours du calcul (Corps de signification.) (PI I, § 559)

Ce que Wittgenstein conteste est justement l'idée que les possibilités et les impossibilités grammaticales qui s'expriment dans les règles que nous adoptons pour l'usage d'un mot soient le reflet de possibilités et d'impossibilités cachées qui concernent le corps de signification que le mot a derrière lui. L'imagerie du corps de signification suggère que si nous connaissons les possibilités de combinaison des solides dont les mots constituent les surfaces peintes apparentes, nous connaissons *ipso facto* les possibilités d'agencement de ces surfaces, c'est à dire les combinaisons de mots qui sont possibles et utilisables et celles qui ne le sont pas. Lorsque deux ou plusieurs expressions ne peuvent être combinées en un tout signifiant, c'est parce que cette combinaison est exclue par la géométrie de leurs corps de signification respectifs. Une combinaison de mots est dénuée de sens lorsque les sens qu'il y a derrière les mots sont de nature telle qu'ils ne se prêtent pas à un arrangement spatial susceptible de constituer à nouveau un sens.

Le principe de compositionnalité du sens dit que le sens d'une phrase est une fonction compositionnelle du sens des expressions qui la composent ou encore que le sens d'une partie de la phrase est une partie du sens de la phrase. Comme le constatait déjà Frege, si la dénotation de la phrase est également une fonction de la dénotation de ses parties, ce n'est cependant pas une fonction compositionnelle, puisque la dénotation d'une partie n'est généralement pas une partie reconnaissable de la dénotation du tout. Mais aussi longtemps que l'on en reste au niveau du sens, cette façon de voir les choses semble parfaitement en ordre, et elle est très tentante. Lorsqu'on se demande pourquoi la phrase 'Socrate est identique' n'a pas de sens, on a l'impression d'essayer de combiner entre eux deux sens partiels, ou, dans le langage de Wittgenstein, deux corps de signification et de s'apercevoir que c'est impossible. On pense que, si une phrase de ce genre n'a pas d'usage, c'est parce qu'elle ne peut pas en avoir; et si elle ne peut pas en avoir, c'est parce qu'elle représente une tentative pour signifier quelque chose, que la nature des significations partielles utilisées empêche d'aboutir.

Wittgenstein voudrait, semble-t-il, que nous nous en tenions autant que possible à l'idée que la phrase en question n'a simplement pas d'usage, qu'elle a été, avec une quantité d'autres, et pour une multitude de raisons qui peuvent différer considérablement d'un cas à l'autre, mise en quelque sorte sur une voie de garage, sans essayer de justifier les exclusions de ce genre en termes de signification et de possibilités ou d'impossibilités de signifier. Le rejet de l'idée du corps de signification est dirigé justement contre l'illusion que les règles dont nous constatons la présence dans le langage découlent de significations préexistantes et ont été imposées par elles. Wittgenstein dit dans la *Grammaire Philosophique* qu'"on ne peut pas pénétrer derrière les règles, parce qu'il n'y a pas de derrière." (p. 224) Les possibilités et les impossibilités de signifier auxquelles nous sommes tentés de faire appel pour expliquer et justifier les règles sont en réalité le résultat immédiat de l'existence des règles elles-mêmes et l'expression directe du fait que le jeu qui se joue est défini par les règles en question. La description la plus sobre et la plus appropriée de la situation serait d'ailleurs, selon Wittgenstein, celle qui consiste à dire non pas "On ne peut pas. . .", ce qui, en raison de la tendance à croire que toute possibilité doit reposer en fin de compte sur une factualité corrélative, peut aisément nous induire en erreur, mais plutôt simplement "Dans ce jeu, il n'y a pas. . .".[12]

Dans la mesure où Wittgenstein estime que toutes les prohibitions de la grammaire sont finalement du même type et qu'il n'y a pas lieu de considérer certaines d'entre elles comme reflétant des impossibilités plus fondamentales et plus radicales que d'autres, il choisit de négliger ouvertement, du point de vue philosophique, des distinctions qui ont généralement été considérées comme essentielles. Husserl, par exemple, soutient qu'il ne faut pas confondre la véritable absence de signification, celle des expressions du type "vert est ou", qui, à proprement parler, ne sont pas des expressions, avec ce qu'il appelle "l'*impossibilité* a priori *d'un sens remplissant*"[13], qui caractérise des expressions contradictoires comme *cercle carré*. Une expression a une signification dans le deuxième sens "quand un remplissement possible, en d'autres termes la possibilité d'une illustration intuitive, formant une unité, correspond à son intention." (*ibid.*) Elle n'en a pas lorsqu'au contraire "nous appréhendons l'impossibilité idéale de ce remplissement de la signification en vertu du vécu de l'incompatibilité des significations partielles dans l'unité de remplissement intentionnée."(*ibid.*) Husserl conteste la tendance que nous avons à qualifier de dépourvues de signification au premier sens, des expressions contradictoires ou comportant des incompatibilités évidentes. Il remarque que, si nous procédons de cette façon, nous devons également considérer comme dépourvues de signification toutes les expressions dont les mathématiciens établissent par des démonstrations quelquefois très compliquées qu' "elles sont *a priori* sans objet" (*ibid.*, p. 64) et être prêts à nier, par exemple, qu'un concept comme celui de *décaèdre régulier* soit simplement un concept. Ce qui se passe en pareil cas, n'est pas que l'expression apparaît au terme de la démonstration comme étant en réalité dénuée de sens, mais plutôt que son sens apparaît comme ayant été construit d'une manière telle qu'il est *a priori* exclu qu'un objet puisse lui correspondre.

En d'autres termes, il faut faire une distinction entre ce qui n'a pas de sens, le non-sens proprement dit *(Das Unsinnige)* et l'absurdité ou le contresens *(Das Widersinnige)*, qui n'est pas une absence de sens:

> La combinaison un *carré rond* donne lieu vraiment à une signification unitaire, qui a son mode d'être dans le monde des significations idéales, mais c'est une évidence apodictique qu'à la signification existante ne peut correspondre aucun objet existant. Si par contre nous disons *un rond ou; un homme et est*, etc., il n'existe même pas là de significations qui correspondent à ces assemblages comme leur sens exprimé. Les mots qu'on a rassemblés éveillent sans doute en nous la représentation indirecte d'une *certaine* signification unitaire exprimée par eux, mais nous avons en même temps l'évidence apodictique qu'une telle signification ne peut exister, que des parties de signification de cette nature et combinées de cette manière sont incompatibles dans une signification unitaire. (*Recherche* IV, p. 121)

Dans le premier cas, c'est simplement l'objet visé qui est impossible; dans le deuxième, c'est la signification elle-même. Dans le premier, "la signification elle-même existe" (*ibid.*, p.122), sans que l'objet puisse exister; dans le deuxième, c'est la signification elle-même qui ne peut exister.

Ce que Husserl appelle la "morphologie pure des significations" a pour but d'exclure le non-sens, et non le contre-sens, c'est à dire de séparer les significations qui ne peuvent exister de celles qui existent, de façon complètement indépendante de la question de savoir si ces dernières sont "objectives" ou "sans objet" (cf. p.129). La "grammaire pure logique", comme l'appelle encore Husserl, traite des lois du sens, au sens strict, ou des formes de signification possibles en général; et les lois qu'elle formule, qui correspondent à ce qui, dans la grammaire, est "purement" grammatical, c'est-à-dire apriorique, sont universelles. Elles peuvent l'être dans la mesure où les entités sur lesquelles elles portent, les significations, existent ou n'existent pas, sont possibles ou ne le sont pas d'une façon qui est indépendante de la question de savoir si elles sont ou non exprimées dans une langue quelconque et de la façon particulière dont elles le sont, si elles le sont. Par *a priori*, Husserl entend ici évidemment quelque chose de beaucoup plus fort qu'une simple universalité à la Chomsky. Ce qu'il désigne par là n'est pas simplement "ce qui est universellement humain au sens empirique" (pp. 132–133), mais un a priori *pur*, c'est à dire purifié de tout élément anthropocentrique et de toute relativisation implicite de ce genre.

Dans la sixième *Recherche*, Husserl distingue entre des significations possibles (réelles) et des significations impossibles (imaginaires). Mais il entend cette fois par possibilité (réalité) d'une signification le fait qu'il lui corresponde un sens remplissant ou "une intuition complète dont la matière est identique à la sienne" (p. 129):

> Quand, par exemple, nous réalisons, sur la base de l'intuition, la signification de l'expression *surface blanche*, nous avons l'expérience vécue de la réalité de ce concept, le phénomène intuitif représente véritablement quelque chose de blanc et une surface, et cela précisément en tant que surface blanche; et c'est ainsi que l'intuition remplissante ne représente pas seulement, en général, une surface blanche, mais fait de celle-ci, au moyen de son contenu, une donnée intuitive, et cela d'une manière aussi complète que l'exige l'intention de signification. (p. 131)

Quand à l'impossibilité, elle ne doit pas seulement être définie comme négation de la possibilité, mais "réalisée au moyen d'un fait phénoménologique propre." (*ibid.*) Ce fait phénoménologique propre doit être cherché dans la sphère du conflit ou de l'inconciliabilité entre contenus.

Le corrélat de la conciliabilité des contenus est la "possibilité" des *significations* complexes. Par exemple, la conciliabilité des espèces *rougeur* et *rondeur* fait que l'expression "rouge et rond" exprime une signification possible, même si les deux espèces ne se trouvent jamais réunies empiriquement. En revanche, l'incompatibilité du rouge et du vert rend impossible ou imaginaire la signification de l'expression "rouge et vert" (au même moment et sur toute sa surface). Husserl prend soin de préciser que:

> . . . *Quand on parle de conciliabilité en ce qui concerne les significations* (les 'concepts'), il ne s'agit pas là seulement, en général, de la possibilité qu'elles se concilient en un tout, serait-ce même en un tout de significations – ce qui serait bien plutôt la conciliabilité *grammaticale pure logique* au sens de la IVème *Recherche* – mais (. . .) de la conciliabilité de la signification en une signification *possible*, c'est à dire en une signification qui est conciliable avec une intuition correspondante en l'unité d'une connaissance objectivement adéquate. (p. 133)

A propos de l'exemple sur lequel Wittgenstein est revenu sans cesse, l'incompatibilité du vert et du rouge, Husserl note que le conflit produit une "séparation" entre rouge et vert, mais contribue en même temps à produire une nouvelle unité concernant les éléments: *conflit, rouge, vert*. En revanche, il y a désunion entre les éléments: *conflit, rouge, rond*.

> La synthèse de la connaissance était conscience d'une certaine concordance *(Übereinstimmung)*. Or, à la concordance correspond, à titre de possibilité corrélative, la 'non-concordance', le 'conflit' *(Widerstreit)*. L'intuition ne 'concorde' pas avec l'intention de signification, elle 'est en conflit' avec elle. Un conflit sépare, mais le vécu du conflit met en relation et pose une unité, il est une forme de la *synthèse*. Si la synthèse précédente était du genre de l'identification, celle-ci est du genre de la *distinction* (nous ne disposons malheureusement pas ici d'un autre nom positif). (*ibid.*, p. 58)

Si l'on considère les choses à la façon de Husserl, on peut dire que, si une expression comme "un objet rouge et vert" n'a pas de sens, c'est parce qu'elle tente d'exprimer une signification qui est en réalité impossible ou imaginaire.

En d'autres termes, l'absence de sens tire son origine d'une impossibilité; et l'impossibilité repose en dernière analyse sur un certain type de factualité ou de vérité concernant en l'occurrence les significations elles-mêmes. Celles-ci, dans l'exemple considéré, bien que conciliables en un tout au sens de la grammaire pure logique, se révèlent inconciliables avec une intuition correspondante dans une unité de type cognitif. Comme on l'a vu, toute la conception de Wittgenstein cherche au contraire à discréditer l'idée que les possibilités et les impossibilités de surface (le fait que certaines combinaisons de mots sont autorisées et d'autres non) reflètent des possibilités et des impossibilités réelles concernant des choses qui se passent à l'arrière-plan du langage. Ce qui rend très tentante une vision des choses comme celle de Husserl est justement le fait que, pour quelqu'un qui connaît le langage, les mots n'ont pas seulement une surface, mais également une profondeur, dans laquelle se passent apparemment les choses essentielles. Wittgenstein est revenu fréquemment sur cette idée ou cette impression, qu'il décrit de la façon suivante dans le § 594 des *Recherches Philosophiques*:

> Mais les mots, prononcés d'une façon qui a un sens *(sinnvoll)* n'ont tout de même pas seulement une surface, mais également une dimension en profondeur: 'Quelque chose d'autre a bien lieu, justement, lorsqu'ils sont prononcés avec un sens, que ce qui a lieu lorsqu'ils sont seulement prononcés.' – La manière dont j'exprime cela n'est pas la chose importante. Que je dise qu'ils ont dans le premier cas une profondeur, ou que quelque chose se passe à ce moment-là en moi, dans mon intériorité, ou qu'ils ont une atmosphère – cela revient toujours au même.
>
> 'Si maintenant nous sommes tous d'accord là-dessus, est-ce que cela ne va pas être vrai?'
>
> (Je ne peux pas accepter le témoignage d' 'autrui', parce que ce n'est pas un *témoignage*. Cela me dit seulement ce qu'il est enclin à dire.)

Il y a bien des façons d'exprimer le fait que les mots que nous prononçons en voulant dire quelque chose par eux contiennent quelque chose de plus que ce que l'on voit et entend, quelque chose qui est justement l'essentiel. Mais Wittgenstein semble estimer que tout ce que l'on peut être tenté de dire sur ce point correspond davantage à une inclination pour une certaine forme de pensée ou un certain mode d'expression qu'à la solution d'un problème. Et comme il le dit: "En philosophie, on doit distinguer entre des propositions qui expriment notre inclination de pensée et celles qui *résolvent* le problème".[14]

Il n'est pas contestable que la prononciation et l'audition de mots d'une langue qui nous est familière sont accompagnées de sensations, d'impressions et d'expériences caractéristiques qui manquent totalement lorsque nous prononçons ou entendons des mots dans une langue que nous ne comprenons pas:

> Celui qui lit une phrase dans une langue courante pour lui ressent les mots des différentes catégories verbales de façon tout à fait différente. (L'image du corps de signification). Nous

> oublions totalement que les images phoniques ou graphiques 'non' et 'table' et 'vert' sont des choses homogènes et ne voyons clairement l'uniformité des mots que dans une langue qui nous est étrangère. (Cf William James sur les sensations qui correspondent à des mots comme 'non', 'mais', etc.). (PG, p. 58)

La réalité des expériences mentales qui sont ici évoquées n'est évidemment pas en question. Ce qui l'est, aux yeux de Wittgenstein, c'est notre tendance à expliquer leur occurrence par l'action des corps de signification très différents que les mots des différentes catégories ont derrière eux, alors que celui qui ignore le langage n'est sensible qu'à l'uniformité de surface. Il lui manque justement la dimension essentielle, à savoir la profondeur.

Wittgenstein manifeste la plus grande méfiance à l'égard de l'idée que la signification d'une phrase douée de sens est quelque chose comme un tout composé des significations de ses parties:

> La signification n'est pas l'expérience vécue que l'on a lors de l'audition ou de la prononciation du mot, et le sens de la phrase n'est pas le complexe de ces vécus. – (Comment le sens de la phrase 'Je ne l'ai toujours pas vu' se compose-t-il à partir des significations de ses mots?) La phrase est composée des mots, et cela suffit. (PI, pp. 500–501)

Une fois rendue suspecte cette idée que la signification de la phrase est le complexe des vécus de signification associés à ses mots, il est évidemment difficile de continuer à dire que l'expérience du non-sens est quelque chose comme l'expérience de l'impossibilité de combiner des significations partielles en un tout cohérent:

> 'Le concept S. n'est pas un S.' Est-ce un non-sens? Eh bien, je ne sais pas ce que quelqu'un qui dit cela veut dire par là: c'est-à-dire de quelle façon il a l'intention d'utiliser cette phrase. Je peux m'imaginer mainte utilisation facile à concevoir pour elle, – 'Mais tu ne *peux* justement pas l'utiliser ou même simplement la penser d'une manière telle que par les mots 'Le concept S.' et par le deuxième 'S.' on *veuille dire* la même chose que tu veux dire *habituellement* par ces mots'. C'est là que réside l'erreur. On pense ici comme si on avait devant les yeux *cette* comparaison: les mots dans la phrase vont ensemble, c'est à dire on peut écrire la suite de mots dénuée de sens; mais la signification de chaque mot est un *corps* invisible, et ces corps de signification *ne* vont *pas* ensemble. ('Le vouloir dire donne à la phrase une dimension supplémentaire'.).
>
> D'où l'idée que l'on ne peut pas penser la phrase, car dans la pensée je devrais maintenant *composer* les significations des mots en un sens, et cela ne va pas. (Puzzle).
>
> Mais la contradiction n'est-elle pas défendue par la loi de contradiction? – 'non (p & non-p)' ne défend en tout cas rien. C'est une tautologie. Mais si nous interdisons une contradiction, alors nous excluons les formes contradictoires de notre langage. Nous éliminons ces formes.[15]

Université de Paris I
Panthéon-Sorbonne

NOTES

(*) Page numbers of the quotations of the German or English texts here translated into French refer to the Basil Blackwell edition of Wittgenstein's works.

[1] *The Blue and Brown Books* (BLB and BRB), ed. and with a Preface by R. Rhees, B. Blackwell, Oxford 1958.

[2] Cf. *Wittgenstein's Lectures. Cambridge, 1930–1932* (WL1930–32). From the Notes of J. King & D. Lee, ed. by D. Lee, B. Blackwell, Oxford 1980, p. 30.

[3] Cf. *Tractatus Logico-Philosophicus* (TLP), ed. by D. F. Pears & B. F. McGuinness, Routledge & Kegan Paul, London 1961.

[4] See *Wittgenstein's Lectures. Cambridge, 1932–1935* (WL1932–35). From the Notes of A. Ambrose & M. Macdonald, ed. by A. Ambrose, B. Blackwell, Oxford 1979, p. 69.

[5] Cf. *Philosophische Untersuchungen. Philosophical Investigations* (PI), ed. by G. E. M. Anscombe & R. Rhees. Trans. by G. E. M. Anscombe, B. Blackwell, Oxford 1953.

[6] Cf. *Philosophische Grammatik* (PG), hrsg. von R. Rhees, B. Blackwell, Oxford 1969.

[7] Cf. *Wittgenstein's Lectures on the Foundations of Mathematics. Cambridge, 1939.* (LFM). Based on Notes by R. G. Bosanquet, N. Malcolm, R. Rhees & Y. Smyhies, ed. by C. Diamond, The Harvester Press, Hassocks, Sussex 1976.

[8] *Bemerkungen über die Grundlagen der Mathematik. Remarks on the Foundations of Mathematics*, ed. by G. H. von Wright, R. Rhees & G. E. M. Anscombe. Trans. by G. E. M. Anscombe, B. Blackwell, Oxford 1964, p. 164.

[9] Cf. Gottlob G. Frege, *Kleine Schriften*, hrsg. von I. Angelelli, G. Olms, Hildesheim 1967, p. 253.

[10] (PB). Aus dem Nachlaß hrsg. von R. Rhees, B. Blackwell, Oxford 1964.

[11] Cf. Georg Cantor, 'Mitteilungen zur Lehre vom Transfiniten', in *Gesammelte Abhandlungen, mathematischen und philosophischen Inhalts*, hrsg. von E. Zermelo, G. Olms, Hildesheim 1966, p. 390.

[12] Cf. *Zettel*, ed. by G. E. M. Anscombe & G. H. von Wright. Trans. by G. E. M. Anscombe, B. Blackwell, Oxford 1967, § 134.

[13] Cf. Edmund Husserl, *Recherches Logiques*, Tome I, traduit par Hubert Elie, Tome II et III, traduits par Hubert Elie avec la collaboration de Lothar Kelkel et René Schérer, P.U.F., Paris 1959–1962.

[14] L. Wittgenstein, *Letzte Schriften über die Philosophie der Psychologie. Last Writings on the Philosophy of Psychology*, Vol. I, ed. by G. H. von Wright & H. Nyman. Trans. by C. G. Luckhardt & M. A. E. Aue, B. Blackwell, Oxford 1982, § 109.

[15] *Bemerkungen über die Philosophie der Psychologie. Remarks on the Philosophy of Psychology*, Vol. I, ed. by G. E. M. Anscombe & G. H. von Wright. Trans. by G. E. M. Anscombe, B. Blackwell, Oxford 1980, §§ 42–44.

CARLO CELLUCCI

WITTGENSTEIN ON THE MEANING OF LOGICAL SYMBOLS

1. The nature of meaning was a major concern of Wittgenstein throughout his work. In this note I shall focus on a special aspect of the matter, the question of the meaning of the universal quantifier and more generally the question of the meaning of logical symbols.

2. Wittgenstein maintains that the meaning of logical symbols is given by their inference rules:

> We can conceive the rules of inference as giving the signs their meaning because they are rules for the use of these signs ([W4] VII 30.)

It seems indisputable that the inference rules meant here by Wittgenstein are formal rules, i.e. mechanical rules:

> Every rule you give for the use of a symbol could be given in the form of a mechanism [. . .]. *Any* rule can be imagined to be a description of a mechanism ([W5] XXX p. 282.)

Of course, there are several kinds of inference rules. Which rules give the meaning of logical symbols? Wittgenstein provides an answer to this question in the case of the universal quantifier:

> One learns the meaning of 'all' by learning that *fa* follows from $(x) \,.\, fx$ ([W4] I 10.)
> The meaning of $(x) \,.\, fx$ is made clear by our insisting on *fa*'s following from it ([W4] I 11.)
> The interesting case of generality is this: we often have a means of ascertaining the general proposition before considering particular cases: and we then use the general method to judge the particular case ([W4] V 50.)

What Wittgenstein is claiming here is that the meaning of the universal quantifier is given by the rule of universal elimination:

$$(\forall E) \quad \frac{\forall x \varphi(x)}{\varphi(t)}$$

Such a view was conceived very early by Wittgenstein, as it appears from a letter to Russell of 1.7.1912:

> Will you think that I have gone mad if I make the following suggestion?: The sign $(x) \,.\, \varphi x$ is not a complete symbol but has meaning only in an inference of the kind: from $\vdash \varphi x \supset_x \psi x \,.\, \varphi(a)$ follows ψa. Or more generally: from $\vdash (x) \,.\, \varphi x \,.\, \varepsilon_0(a)$ follows $\varphi(a)$. I am – of course –most uncertain about the matter but something of the sort might really be true ([W1] p. 12)

R. Egidi (ed.), Wittgenstein: Mind and Language, 83–91.
© 1995 *Kluwer Academic Publishers. Printed in the Netherlands.*

This should be seen against the background of a 'consequentialist' theory of meaning according to which the meaning of a proposition is to be given in terms of its consequences:

> We must be able to understand a proposition without knowing if it is true or false. What we know when we understand a proposition is this: We know what is the case if the proposition is true, and what is the case if it is false. But we do not know (necessarily) whether it is true or false ([W2] p. 98).

3. Wittgenstein's view was a very isolated one. At about the time when he elaborated it, an alternative view was put forward by the German logician Gerhard Gentzen. According to him, the meaning of logical symbols is given by the introduction rules rather than by the elimination rules:

> The introductions represent, as it were, the 'definitions' of the symbols concerned, and the eliminations are no more, in the final analysis, than the consequences of these definitions. This fact may be expressed as follows: In eliminating a symbol, we may use the formula with whose terminal symbol we are dealing only 'in the sense afforded it by the introduction of that symbol' ([G] p. 80.)

Thus, in particular, the meaning of the universal quantifier is given by the rule of universal introduction:

$$(\forall \mathrm{I})\ \frac{\varphi(a)}{\forall x \varphi(x)}$$

Gentzen's view has become popular through the writings of Dag Prawitz and Michael Dummett who have been building upon it (see [Su] for references). On the other hand, Wittgenstein's view has long been overlooked in the literature. As far as I know, it was first considered by Prawitz [P], without explicitly crediting it to Wittgenstein:

> Since the introductions and eliminations are inverses of each other, Gentzen's idea to justify the eliminations by the meaning given to the constants by the introductions may be reversed. Instead of interpreting the constants [. . .] as asserting the existence of certain constructions that build up formulas with these constants, we may interpret them as stating the performability of certain operations ([P] p. 289.)

Later on the possibility of such a view was also considered by Dummett:

> Would it not be equally feasible to adopt the reverse procedure and take the meaning of a sentence as given in terms of the consequence of uttering it? [. . .] It is a matter of taking the meanings of the logical constants as given by the elimination rules governing them and formulating a procedure for justifying arbitrary forms of argument, including introduction rules, by reference to them ([Du] p. 17.)

4. Prawitz, however, considers the view that the meaning of logical symbols is given by the elimination rules as generally untenable: in his opinion such a view can be maintained only for a restricted class of formulas. He draws this conclusion by pointing out the difficulties involved in the notion of *validity based on the elimination rules*. Such a notion is introduced by Prawitz as follows.

An *atomic system* **S** is a set of inference rules whose premises and conclusion are atomic.

If **S** is an atomic system, then **M(S)** is the natural deduction system whose inference rules are the rules of minimal logic **M** plus the rules of **S**.

A derivation $\mathscr{D}$ in **M(S)** is *closed* if it has no open assumptions and no open parameters; $\mathscr{D}$ is *open* if it is not closed.

Let $\mathscr{D}$ be a closed derivation of a sentence φ in **M(S)**. The notion: $\mathscr{D}$ is *valid w.r.t.* **S**, is defined inductively as follows:

(i) if φ is atomic, then $\mathscr{D}$ is valid w.r.t. **S** if $\mathscr{D}$ is a derivation in **S**;

(ii) if $\varphi \equiv \varphi_1 \wedge \varphi_2$, then $\mathscr{D}$ is valid w.r.t. **S** if the derivation

$$(\wedge\text{E}) \quad \frac{\begin{array}{c}\mathscr{D}\\ \varphi_1 \wedge \varphi_2\end{array}}{\varphi_i}$$

is valid w.r.t. **S** for $i = 1,2$;

(iii) if $\varphi \equiv \varphi_1 \rightarrow \varphi_2$, then $\mathscr{D}$ is valid w.r.t. **S** if, for each extension **S′** of **S** and for each closed derivation $\mathscr{D}_1$ of φ_1 that is valid w.r.t. **S′**, the derivation

$$(\rightarrow\text{E}) \quad \frac{\begin{array}{c}\mathscr{D}_1\\ \varphi_1\end{array} \quad \begin{array}{c}\mathscr{D}\\ \varphi_1 \rightarrow \varphi_2\end{array}}{\varphi_2}$$

is valid w.r.t. **S′**;

(iv) if $\varphi \equiv \forall x \psi(x)$, then $\mathscr{D}$ is valid w.r.t. **S** if, for each name t, the derivation

$$(\forall\text{E}) \quad \frac{\begin{array}{c}\mathscr{D}\\ \forall x \psi(x)\end{array}}{\psi(t)}$$

is valid w.r.t. **S**.

An open derivation $\mathscr{D}$ in **M(S)** is *valid w.r.t.* $\mathscr{D}$ in **M(S)** if, for each extension **S′** of **S** and for each closed derivation $\mathscr{D}'$ obtained from $\mathscr{D}$ by replacing (1) each open assumption ψ by a closed derivation of ψ that is valid w.r.t. **S′** and (2) each open parameter by a name, $\mathscr{D}'$ is valid w.r.t. **S**.

This validity notion seems to apply only to derivations of formulas containing no disjunction or existential quantification:

> Disjunctions and existential quantifications seem impossible to handle in this way since the induction over the complexity of the end formulas used in the definition of validity breaks down in these cases ([P] p. 290.)

Indeed the clause for disjunction would be of the following form:

(v) if $\varphi \equiv \varphi_1 \vee \varphi_2$, then $\mathscr{D}$ is valid w.r.t. **S** if, for each open derivation

$$\begin{array}{c}[\varphi_1]\\ \mathscr{D}_1\\ \psi\end{array} \quad \text{and} \quad \begin{array}{c}[\varphi_2]\\ \mathscr{D}_2\\ \psi\end{array}$$

that is valid w.r.t. **S**, the derivation

$$(\vee E)\ \frac{\begin{array}{c}\mathscr{D}\\ \varphi_1 \vee \varphi_2\end{array} \quad \begin{array}{c}[\varphi_1]\\ \mathscr{D}_1\\ \psi\end{array} \quad \begin{array}{c}[\varphi_2]\\ \mathscr{D}_2\\ \psi\end{array}}{\psi}$$

is valid w.r.t. **S**.

Since ψ is a formula whose complexity can be higher than that of $\varphi_1 \vee \varphi_2$, here the induction on the complexity of φ breaks down. Similarly in the case of existential quantification.

5. Prawitz's claim, however, seems to depend on the assumption that the rules of disjunction elimination ($\vee$E) and existential elimination ($\exists$E) can be formulated only in the form of Gentzen [G]. We know now that we can introduce systems of natural deduction – the so-called *sequent natural deduction systems* (see [Bo], [C1], [C2], [C3]) – which are free of most limitations of Gentzen's systems. While the original formulation of such systems still contained an existential instantiation rule of a kind similar to Gentzen's, this is by no means necessary, as shown by the systems of [C1], [C2].

In the latter systems the rules of disjunction elimination and existential elimination take the form:

$$(\vee E)\ \frac{\Delta, \varphi_1 \vee \varphi_2}{\Delta, \varphi_1, \varphi_2}$$

$$(\vee E)\ \frac{\Delta, \exists x \varphi(x)}{\Delta, \varphi(\varepsilon)}\ .$$

The notion of validity based on the elimination rules can be formulated in terms of these rules in such a way that induction no longer breaks down.

An alternative approach is provided by the so-called *multiple conclusion natural deduction systems* (see [K], [KK], [Si], [SS]), which however are of a different character.

6. The view that the meaning of logical symbols is given by the elimination rules seems to provide an approach to the theory of meaning of logical symbols that is at least as viable as the more popular Gentzen-Prawitz approach. I am not claiming that it is an adequate approach, but only that it is at least as adequate as the Gentzen-Prawitz approach.

Indeed, in certain respects it may even be more adequate. For example, consider the rule of induction,

$$(\mathrm{IND})\quad \frac{\varphi(0) \qquad \forall x(\varphi(x)\rightarrow\varphi(x'))}{\forall x\varphi(x)}\ .$$

Let **PRA** be primitive recursive arithmetic, formulated as an atomic system. Then (IND) is valid w.r.t. **PRA** in the following sense: for each closed derivation $\mathscr{D}_1$ of $\varphi(0)$ and $\mathscr{D}_2$ of $\forall x(\varphi(x)\rightarrow\varphi(x'))$ that is valid w.r.t. **PRA** and for each numeral n, the derivation

$$(\mathrm{IND})\quad \cfrac{\cfrac{\begin{matrix}\mathscr{D}_1\\ \varphi(0)\end{matrix} \qquad \begin{matrix}\mathscr{D}_2\\ \forall x(\varphi(x)\rightarrow\varphi(x'))\end{matrix}}{\forall x\varphi(x)}}{\varphi(n)}$$

can be transformed into a derivation $\mathscr{D}$ that is valid w.r.t. **PRA**. Just let $\mathscr{D}$ be the following derivation:

$$(\rightarrow\mathrm{E})\quad \cfrac{\begin{matrix}\cfrac{\begin{matrix}\mathscr{D}_1\\ \varphi(0)\end{matrix} \qquad (\forall\mathrm{E})\ \cfrac{\begin{matrix}\mathscr{D}_2\\ \forall x(\varphi(x)\rightarrow\varphi(x'))\end{matrix}}{\varphi(0)\rightarrow\varphi(1)}}{\varphi(1)}\ (\rightarrow\mathrm{E})\\ \vdots\\ \varphi(n-1)\end{matrix} \qquad (\forall\mathrm{E})\ \cfrac{\begin{matrix}\mathscr{D}_2\\ \forall x(\varphi(x)\rightarrow\varphi(x'))\end{matrix}}{\varphi(n-1)\rightarrow\varphi(n)}}{\varphi(n)}$$

Therefore (IND) is valid w.r.t. **PRA** in a sense that is perfectly in line with Wittgenstein's view about (IND):

> 'The proposition $(n).fn$ follows from the induction' only means that every proposition of the form $f(n)$ follows from the induction ([W3] VI 31.)

Such a view of induction is already expressed by Wittgenstein in his conversations with Waismann (cf. [Wa1] IV) and is similar to the one reported by Waismann (cf. [Wa2], end of chapter 9).

On the other hand (IND) is not valid w.r.t. **PRA** in the sense of Gentzen-Prawitz, i.e. in the sense of the notion of validity based on the introduction rules. For, generally two closed derivations, $\mathscr{D}_1$ of $\varphi(0)$ and $\mathscr{D}_2$ of $\forall x(\varphi(x)\rightarrow\varphi(x'))$, that are valid w.r.t. **PRA** are not sufficient to obtain an open derivation of $\varphi(a)$ that is valid w.r.t. **PRA**, for a parameter a. However such a derivation of $\varphi(a)$ is required in order to obtain a closed derivation of $\forall x\varphi(x)$ that is valid w.r.t. **PRA** in the sense of Gentzen-Prawitz. Therefore the Gentzen-Prawitz view of the meaning of the universal quantifier is essentially more restrictive than Wittgenstein's view.

Dummett and Prawitz do not seem to question Wittgenstein's principle that meaning is fully manifested in use. On the other hand, admittedly the Gentzen-Prawitz meaning of the implication and the universal quantifier, which is constructive in a strong sense, is very seldom used in mathematics.

> It must be admitted that such a strong meaning of $\supset$ and $\forall$ is seldom used [. . .] this notion of validity of a formula in a system does not have natural mathematical applications. For instance, the induction axioms are not generally valid ([P] p. 288.)

On that account one might conclude that, if meaning is to be fully manifested in use, then Wittgenstein's view of the meaning of the universal quantifier is more adequate than the Gentzen-Prawitz view.

7. Prawitz's notion of validity based on the elimination rules is not the only possible implementation of the view that the meaning of logical symbols is given by the elimination rules. An alternative implementation is provided by tableaux. It was already suggested by Beth that

> the [tableau] rules [. . .] exactly reflect the *meaning* of the words '*all*', '*some*', '*not*', '*and*', '*if* . . . *then*', '*or*', '*if and only if*', and of the symbols by which they have been replaced ([Be] p. 21.)

This view has been recently taken over by Mondadori (see [M1], [M3]) who maintains, again without special reference to Wittgenstein, that tableaux are a way of implementing the view that the meaning of logical symbols is given by the elimination rules:

> Although tableau proofs really are (quasi-mechanical) tests for validity [. . .], their structure somehow reflects the meaning of the logical operators involved in them [. . .]. In fact, tableau rules are just one way of making precise what can be called the *consequentialist* theory of meaning according to which the meaning of a logical operator * is connected with the consequences of propositions of the form *(*P*1, . . . , *Pn*), in which * is the main logical constant. This conception of meaning implies that elimination rules – as opposed to introduction rules – are to be taken as primitive meaning rules ([M3] p. 4)

Mondadori (see [M1], [M2], [M3], [Da]) also proposes an alternative implementation of this view, in terms of a calculus **KE** consisting of elimination rules plus (a version of) the principle of excluded middle:

(EM) $\varphi \quad \neg\varphi$ (branching)

He considers (EM) to be a structural rule reflecting the assumption of a classical notion of truth.

This proposal, however, does not sound very convincing. The main reason for considering (EM) as a structural rule seems to be that it is strictly related to Gentzen's cut rule:

$$(\mathrm{Cut}) \quad \frac{\Gamma \Rightarrow \Delta, \varphi \quad \varphi, \Gamma \Rightarrow \Delta}{\Gamma \Rightarrow \Delta}$$

Indeed, reading (Cut) upside-down, what the rule says is that, in all circumstances and for all formulas φ, either φ is true or φ is false (and hence $\neg\varphi$ is true).

There is, however, a basic difference between the role of (Cut) in Gentzen's calculus of sequents **LK** and the role of (EM) in Mondadori's elimination calculus **KE**. By Gentzen's *Hauptsatz*, (Cut) is eliminable in **LK**, or equivalently (EM) is eliminable in the system of tableaux (see [Sm] pp. 110–115 or [BM] pp. 31–33, 79–83). On the other hand (EM) is not eliminable in **KE** because eliminating (EM) from **KE** would yield an incomplete system (see [M1]).

This means that, while (Cut) does not contribute to the meaning of logical symbols in **LK**, (EM) contributes to such a meaning in **KE** in an essential way. Thus, while (Cut) plays a role of a proper structural rule in **LK**, (EM) does not play such a role in **KE**: it is rather an additional logical rule, distinct from the elimination rules.

The situation is somewhat similar to that of Gentzen's natural deduction system **NK** where, as stressed by Gentzen (see [G] p. 82), the law of the excluded middle $\varphi \vee \neg \varphi$ plays a role of a logical rule distinct from both the introduction rules and the elimination rules. It is just because of such a special role that Gentzen's view that the meaning of logical symbols is given by the introduction rules does not extend to classical logic.

Since (EM) plays a role of a logical rule distinct from the elimination rules, thus contributing to the meaning of logical symbols in an essential way, **KE** does not seem to provide a viable approach to a consequentialist theory of meaning.

Università di Roma 'La Sapienza'

REFERENCES

[Be] Beth E.W. (1969): 'Semantic entailment and formal derivability', in *The philosophy of mathematics*, ed. by J. Hintikka, Oxford University Press, Oxford, pp. 9–41.

[BM] Bell J. and Machover M. (1977): *A course in mathematical logic*, North-Holland, Amsterdam.

[Bo] Boričić B.R. (1985): 'On sequence-conclusion natural deduction systems', *Journal of Philosophical Logic* **14**, pp. 359–377.

[C1] Cellucci C. (1988): 'Efficient natural deduction', in *Temi e prospettive della logica e della filosofia della scienza contemporanee*, vol. I, ed. by C. Cellucci and G. Sambin, Clueb, Bologna, pp. 29–57.

[C2] Cellucci C. (1991): 'Sequent natural deduction and intuitionistic logic', in *Nuovi problemi della logica e della filosofia della scienza*, vol. II, ed. by G. Corsi and G. Sambin, Clueb, Bologna, pp. 259–266.

[C3] Cellucci C. (1992): 'Existential instantiation and normalization in sequent natural deduction systems', *Annals of Pure and Applied Logic* **58**, pp.111–148.

[Da] D'Agostino M. (1990): *Investigations into the complexity of some propositional calculi*, Oxford University Computing Laboratory, Technical Monograph PRG-88, Oxford.

[Du] Dummett M. (1980): 'Comments on Prof. Prawitz's paper', in *Logic and philosophy*, ed. by G.H. von Wright, Nijhoff, Dordrecht, pp. 11–18.

[G] Gentzen G. (1969): 'Investigations into logical deduction', in *The collected papers of Gerhard Gentzen*, ed. by M.E. Szabo, North-Holland, Amsterdam, pp. 68–131.

[K] Kneale W. (1956): 'The province of logic', in *Contemporary British Philosophy*, 3rd series, ed. by H.D. Lewis, Allen & Unwin, London, pp. 237–261.

[KK] Kneale W. and Kneale M. (1962): *The development of logic*, Oxford University Press, Oxford.

[M1] Mondadori M. (1988): 'Classical analytical deduction', *Annali dell'Università di Ferrara*, n,s, section III, Discussion Paper No. 1.

[M2] Mondadori M. (1989): 'Classical analytical deduction, Part II', *Annali dell'Università di Ferrara*, n,s, section III, Discussion Paper No. 5.

[M3] Mondadori M. (1989): 'An improvement of Jeffrey's deductive trees', *Annali dell'Università di Ferrara*, n,s, section III, Discussion Paper No. 7.

[P] Prawitz D. (1971): 'Ideas and results in proof theory', in *Proceedings of the Second Scandinavian Logic Symposium*, ed. by J.E. Fenstad, North-Holland, Amsterdam, pp. 235–307.

[Si] Šikić Z. (1990): 'Premiss tree proofs and logic of contradiction', *Zeitschrift für mathematische Logik und Grundlagen der Mathematik* **36**, pp. 273–280.

[Sm] Smullyan R.M. (1968): *First-order logic*, Springer-Verlag, Berlin.

[Su] Sundholm G. (1986): 'Proof theory and meaning', in *Handbook of philosophical logic*, vol. III, ed. by D. Gabbay and F. Guenthner, Reidel, Dordrecht, pp. 471–506.

[SS] Shoesmith D.J. and Smiley T.J. (1980): *Multiple-conclusion logic*, Cambridge University Press, Cambridge.

[Wa1] Waismann F. (1967): *Wittgenstein und der Wiener Kreis; Gespräche, aufgezeichnet von F. Waismann*, ed. by B.F. McGuinness, Blackwell, Oxford.

[Wa2] Waismann F. (1959): *Introduction to mathematical thinking*, Harper, New York.

[W1] Wittgenstein L. (1974): *Letters to Russell, Keynes and Moore*, ed. by G.H. von Wright, Blackwell, Oxford.

[W2] Wittgenstein L. (1979): 'Notes on logic', in L. Wittgenstein, *Notebooks 1914–1916*, 2nd edition, ed. by G.H. von Wright and G.E.M. Anscombe, Blackwell, Oxford, pp. 93–107.

[W3] Wittgenstein L. (1974): *Philosophical Grammar*, Engl. transl. by A.J.P. Kenny, Blackwell, Oxford.

[W4] Wittgenstein L. (1978): *Remarks on the Foundations of Mathematics*, Engl. transl. by G.E.M. Anscombe, Blackwell, Oxford.

[W5] Wittgenstein L. (1976): *Lectures on the Foundations of Mathematics, Cambridge 1939*, ed. by C. Diamond, The Harvester Press, Hassocks.

AVRUM STROLL

ON FOLLOWING A RULE

The question of what it is to follow a rule, much debated in the current literature on Wittgenstein, is more complicated than most of its discussants seem to have realized. In general, their writings exhibit a tendency to overlook or blur important distinctions, lapses that make it difficult or even impossible to determine exactly what is at issue. Baker and Hacker are certainly right when they say that unless it is understood what rules are "confusions about rules ramify into muddles about following rules".[1]

In addition, Baker and Hacker are to be commended for their own discussion of rules, a discussion that is extensive and illuminating. Among the examples of rules they give are rules of games, legal rules, the traffic code, and rules of etiquette; and they point out that the range of sentence types that may be used to state rules is very wide. It includes deontic sentences, imperatives, and ordinary declarative sentences *inter alia*, and they go on to emphasize that such sentence-types can be used for altogether different purposes as well (p. 39). They also bring out certain essential features of rules, for instance, i) that any rule can be expressed, ii) that it must be possible to follow (or violate) a rule, iii) that rules are creatures of the will, iv) that they are standards of correctness and guides to action, and v) that they must be more or less transparent to participants in a rule-governed practice (pp. 62–63).

Yet as broad-ranging as is their coverage of the topic there are things they tend to underemphasize and more importantly, there are things they overlook. My intention in this paper is to carry the subject forward and in ways that are philosophically significant. Their significance can best be appreciated by comparing what follows with Kripke's approach to the issue of rule following. In my view Kripke's interpretation of Wittgenstein, as deep and interesting as it is, is marred by focusing on too narrow a set of examples, in particular those involving the expansion of mathematical formulas. A considerable portion of this paper will thus be devoted to Kripke's treatment of this problem.

There is one important issue, and the subtle distinctions connected with it, that I will not pursue here. This concerns the relationship between rule following and the notion of a private language, including the question of whether rule following is simply a matter of regularity (so that in principle someone

R. Egidi (ed.), Wittgenstein: Mind and Language, 93–105.
© 1995 *Kluwer Academic Publishers. Printed in the Netherlands.*

can be said to be following a rule if his or her behavior exhibits such regularity) or whether rule following presupposes regularity and also a human community. The question is directly relevant to such conundrums as what could be meant by a private language and whether someone could have or use such a language. In their discussions of rule following Baker and Hacker in effect contend that regularity is a sufficient condition for rule following. Their interpretation thus comes dangerously close to asserting what Wittgenstein is generally thought to be denying; and even if this is not precisely so, the interpretation seems to me wrong on its own merits. On this point, Kripke seems to me to be right as does Norman Malcolm. Here is what Malcolm says in a recent paper, 'Wittgenstein on Language and Rules':

> To speak a language is to participate in a way of living in which many people are engaged. The language I speak gets its meaning from the common ways of acting and responding of many people. I *take part* in a language in the sense in which I *take part* in a game – which is surely one reason why Wittgenstein compares languages to games. Another reason for this comparison is that in both languages and games there are *rules*. To follow the rules for the use of an expression is nothing other than to use the expression as it is ordinarily used – which is to say, as it is used by those many people who take part in the activities in which the expression is embedded. Thus the meaning of the expression is *independent* of me, or of any particular person; and this is why I can use the expression correctly or incorrectly. It has a meaning independent of my use of it. And this is why there is no sense in the supposition that a forever-solitary person could know a language any more than he could buy and sell (*Inquiry*, 32, 1989, p. 22).

However, as I have indicated, this issue is too complicated to be explored here; so I shall set it aside. But there are at least three other, overlooked, strands in this complex web that must be discriminated if certain liabilities are to be avoided and if the issue, or at least certain permutations on it, can be made susceptible to resolution and it is these upon which I shall concentrate.

First, there is the question of how to describe *what it is* that is being followed. To my knowledge no commentator has distinguished from one another such things as edicts, decrees, rescripts, orders, ordinances, statutes, maxims, resolutions, regulations, judicial decisions, canons, precepts, formulas, regimens, rituals, recipes, habits, customs, standards, criteria, principles, and certain sorts of practices. Yet some or all of these can, in certain contexts, be referred to as "rules", whereas in other contexts some of them may be distinguished from or contrasted with rules. Most of them are the sorts of things that can be said to be followed, but some of them are not, such as criteria. We do not follow criteria but apply them. In addition, it is important to distinguish between things that express rules – such as road signs –

and the rules that are expressed by these things, a point that Baker and Hacker do not emphasize. And there are more complex cases.

Take "formula" for instance. A formula is defined, in one of senses of the term, as a set of words that indicate a procedure which is to be followed in order to bring about some particular result. There are both mathematical and non-mathematical formulas. One who follows the injunctions laid down in a mathematical formula – for example, by extending a series in a prescribed way – may be said to be following the rule for the expansion of that series. Such a rule does not permit improvisation or flexibility.

Among the non-mathematical formulas are recipes; these consist of sets of words which give directions for preparing foods, medicines etc., typically by describing the proportions which go into such preparations. But cooks would not generally describe recipes as rules. Nor are they formulae that have an obligatory force. A mathematical formula must be be followed exactly and no doubt that is true of certain recipes; but it is assuredly not true of others. Recipes are often construed by chefs or cooks as guidelines or suggestions that allow for considerable variation and creativity, and not as injunctions to be followed slavishly. A cook who is following a recipe is not following an injunction nor is a cook who changes or modifies a recipe violating an injunction. It is thus possible to assert that a chef who deviates from a recipe is nonetheless following it and also without inconsistency to assert that a student who deviates from a mathematical formula is not following it. Clearly then what counts as a rule will depend on the kind of thing that is being adhered to or deviated from.

Furthermore, we can distinguish various degrees of "strength" in the set of terms I have listed above. Consider, for instance, the differences between regulations and precepts, both of which can be described as types of rules. Regulations tend to be rules of maximum strength and obligatory force. They are generally issued by some person or by some group that has the power or authority to do so, such as a governor or a city council. Accordingly, they often have the force of law behind them and their violations frequently carry sanctions. Because regulations tend to be detailed and specific in what they mandate, such sanctions also tend to be specific and detailed.

In many of these respects we can contrast regulations with precepts. Both, of course, being rules are things that can be followed or not followed. But precepts are, as it were, "softer" in character than regulations. They tend to be personal rather than institutional, are generally not "issued" by authorities, but are adopted voluntarily, and usually after some reflection. They are on the whole advisory rather than obligatory and are not typically susceptible to penalties, let alone to specific penalties. One who fails to act according to his or her percepts may suffer, not from the imposition of an external sanction but rather from guilt or shame.

Both regulations and precepts are principles that may directly affect the ways that ordinary people live, as distinct say from mathematical formulas which do not, but their impacts on people's lives differ. Regulations tend to be devised and to be laid down by others and are often regarded by those they affect as irksome or even as unfair. But precepts are hardly ever regarded as unfair, even if they are demanding or even if one finds himself unable to live up to them. In such cases, a person will not infrequently blame himself and not his precepts, since these are taken to express ideals towards which he ought to aspire. Both regulations and precepts are subject to change and modification, but precepts are less subject to outright replacement and are more susceptible to development, broadening and growth than regulations. They are normally general in scope rather than detailed and often have a pronounced moral character instead of being neutral, as when it is said that a person's life is being guided by certain precepts. In many circumstances of use, it is patent that we could not, while preserving meaning, replace the term "precept" by the term "regulation". What will count as rule following will therefore vary depending upon whether one is referring to regulations or to precepts in using the term "rule" and similar remarks to apply to most of the other items I have listed above.

Furthermore, it is important to stress that such rules differ in their functions and therefore are subject to differing kinds of assessments, such as what (if anything) counts as a violation or an infringement of a rule, or what sorts of inaptnesses or infelicities in its formulation or application it is heir to. For instance, in a game such as contract bridge, there are what might be called "constative" rules. These are formative rules that define the game of contract bridge. One who does not play in total compliance to these rules is not playing contract bridge. They include the provision that the deck used in the game must consist of exactly 52 cards, divided equally between four suits and the rule that each player must be dealt exactly 13 cards if the course of play that follows is to be bona fide. What is called a "misdeal" would be a violation of the latter rule and the hand would be annulled.

But there are also rules of strategy in bridge and these differ from constative rules. One who fails to take a finesse is still playing bridge if he is in compliance with the set of constative rules, but he may not be optimizing his chances of winning by such a lapse. But the hand is not nullified for that reason. Each of these types of rules is thus subject to different infelicity conditions, rules of strategy being connected with winning, whereas constative rules are not; and it may be that ambiguities can arise about rule following depending on which sorts of felicity conditions are observed or not observed. Rules also differ with respect to how, and whether, questions concerning them are decidable, and by whom. In contract bridge, questions about whether a certain constative rule has been followed can in principle always be decided, for example by counting the cards, but whether one should have taken a finesse may be the subject of disagreement even among experts and even after post-game analysis.

In legal contests, for instance, even where there is no doubt about what a particular rule means there can be disagreement among lawyers about whether the rule has been violated in a particular case; and even different levels of appellate courts may disagree. The issue of what it is to follow a "rule" may thus vary depending on such factors as whether a constative or strategic rule is being referred to and upon various contextual factors.

In the *Investigations*[2] Wittgenstein is comparatively selective in the number (five in all, according to my count) of the items he identifies as rules. He speaks of rules themselves, of course, and also of formulae, orders, criteria, and practices. Most of those who write about Wittgenstein or about the problem of rule following are much more myopic. They tend to focus, as Kripke does, on one or two examples, generally mathematical expressions, where the issue posed is how to extend a particular formula; but it may be quite misleading to generalize or to draw presumably important philosophical conclusions – as say Kripke does – from such special or restricted cases. I shall say more about this later.

Second, one must distinguish the diverse roles that the concepts of understanding, meaning, and explanation play when the issue of rule following is broached. And these in turn must be distinguished from the concept of applying a rule, and that from operating in accordance with a rule. It is often said that understanding what a term means is a sufficient condition for one's correctly applying it. And then by analogy it is claimed that if one can understand what a rule means it follows that one can correctly apply it. But one of Wittgenstein's main points is to reject the supposed tie between the concepts of meaning, understanding, and that of correct application. Consider this passage. Wittgenstein writes:

> I say "There is a chair". What if I go up to it, meaning to fetch it, and it suddenly disappears from sight? – "So it wasn't a chair, but some kind of illusion." – But in a few moments we see it again and are able to touch it and so on. – "So the chair was there after all and its disappearance was some kind of illusion". – But suppose that after a time it disappears again – or seems to disappear. What are we to say now? Have you rules ready for such cases – rules saying whether one may

> use the word "chair" to include this kind of thing? But do we miss them when we use the word "chair"; and are we to say that we do not really attach any meaning to this word, because we are not equipped with rules for every possible application of it? (PI I, § 80)

In the entry that follows, Wittgenstein extends the point to operating in accordance with a rule.

> All this, however, can only appear in the right light when one has attained greater clarity about the concepts of understanding, meaning, and thinking. For it will then also become clear what can lead us (and did lead me) to think that if anyone utters a sentence and *means* or *understands* it he is operating a calculus according to definite rules. (PI I, § 81)

Wittgenstein's analysis thus indicates that one can drive a wedge between the concepts of understanding (or meaning) and that of rule following and/or of operating in accordance with rules. Wittgenstein thus allows for the possibility that one might understand a rule without knowing how to apply it or without acting in accordance with it. The latter point is especially obvious since nonconformists are persons who understand rules they refuse to obey.

Third, we must distinguish three different perspectives on rule following from one another: a first person from a third person perspective (a contrast that is also frequently expressed in a plural form, in which "we" is contrasted with "they"), and a neutral point of view that is different from either.

We find Wittgenstein's discussion in the *Investigations* moving in exceedingly complex ways between these three different positions. He will frequently begin a query from a first person aspect and then without emphasis shift to one, or not infrequently, to both of the other aspects even within the same sentence. Here are some examples, beginning with a first person case. As the quotation will illustrate, the issue is posed in terms of how can *I* know that *I* am developing the series correctly.

> ". . . When *I* say I understand the rule of a series, I am surely not saying so because I have *found out* that up to now I have applied the algebraic formula in such-and-such a way! In my own case at all events I surely know that I mean such-and-such a series; it doesn't matter how far I have actually developed it." (PI, § 147)

But then we have a shift to a third person perspective, as in the following. Wittgenstein asks:

> How can he *know* how he is to continue a pattern by himself – whatever instruction you give him? – (PI I, § 211)

And here is a mixed example:

> So how am I to determine the rule according to which he is playing? (PI I, § 82)

Even more common are instances in which the point of view is neutral or in which the issue is raised from other than a point of view. In the quotation I am about to read no mention of a point of view occurs.

> To carry it out correctly! How is it decided what is the right step to take at any particular stage? (PI I, § 186)

Or again,

> What is the criterion for the way the formula is meant? (PI I, § 190)

In contrast, Saul Kripke's emphasis is invariably on the first person. His characteristic question is: "How do I know that I am following a rule?" Kripke imagines a speaker saying to himself, "I follow directions that determine how I should go on, but what are these directions since I never explicitly gave them to myself?" Note the emphasis upon the word "I" in the following quotation.

> The basic point is this: Ordinarily, I suppose that in computing '68 + 57' as I do, I do not simply make an unjustified leap in the dark. I follow directions I previously gave myself that uniquely determine that in this new instance I should say '125'. What are these directions? By hypothesis, I never explicitly told myself that I should say '125' in this very instance. Nor can I say that I should simply 'do the same thing I always did', if this means 'compute according to the rule exhibited by my previous examples'. That rule could just as well have been the rule for quaddition (the quus function) as for addition. The idea that in fact quaddition *is* what I meant, that in a sudden frenzy I have changed my previous usage, dramatizes the problem.[3]

I believe that Kripke's concentration upon the first person point of view and his apparent unawareness of the frequent shift in perspective that we find in the text of the *Investigations* is evidence of a failure to appreciate the complex problem that Wittgenstein is considering. We can bring this out, beginning with a simple case where the issue of rule following is raised from a first person perspective. Let us see how the matter of whether a rule has been followed can be decided in that case and then compare it with what Kripke says. We can then consider more difficult cases where there are shifts in perspective to see how they might be resolved. Our purpose in examining this spectrum of cases is to emphasize the complexities that are involved both in stating the putative problem of rule following and also of deciding how, if at all, it might be resolved.

Our first example, then, will be a simple case, where the perspective is first person. We shall see that the question is both easily stated and easily answered. Let us assume that what is in question is whether the constative rule of contract bridge that each person must be dealt exactly thirteen cards has been violated. In an important respect this is like the question of whether a mathematical formula has been followed correctly, for the question is clear and there is a determinate, correct solution to it.

If I am the dealer how do I find out that I have violated the rule? This can occur in a number of ways: I may note that I have a card left over when I finish dealing; or another player may tell me that he has counted and found either more or fewer than thirteen cards in his hand. If in doubt about the accuracy of what he has said, I can also count his cards. There seems to be no special problem about resolving this first person case.

But Kripke sets up his problem in such a way that it is not clear what the problem is supposed to be or how it would bc solved. Suppose I am the person adding two numbers according to the usual rule for the addition of the positive integers. As Kripke describes the situation the problem is that I do not know whether in calculating the sum '68 + 57' I am using '+' (i.e., 'plus') in the same way as I did in the past. Have I forgotten what '+' means? That would be a natural way of interpreting the situation. If that were so, there are various steps to find out whether I am now using '+' in the way I formerly did.

It is, of course, frequently possible to find out that one has forgotten something, but once that has been discovered, one can then go on. Yet Kripke explicitly denies that the problem arises because I have forgotten what '+' means. But then we have some puzzles. How is it that I know what '68' and '57' mean but not what '+' means? What has happened to prevent me from knowing what '+' means? Why should I suppose that I am following the rule for quaddition instead of the rule for addition?

Kripke does not answer any of these questions. But he should because they are crucial. For example, if I, the rule follower, do not know whether I am following the rule for addition or the rule for quaddition then why should I suppose that I have violated the rule for addition? And how can I go about arranging for my future actions to conform to it? We can contrast Kripke's formulation of Wittgenstein's supposed problem with that we have just presented about whether a constative rule of bridge has been followed. In the latter case we not only know what it would like be to follow or to violate the rule, but also what procedure or procedures can be invoked in order to decide the matter of which of these we have done.

But now let us expand the range of examples and consider a situation where there is both a first person and a third person point of view at issue. We can ask with respect to the same constative rule of bridge: "How do *I* know that *he* (where he is now the dealer) has violated the rule?" Once again, the answer is readily at hand. I, a non-dealer, count my cards and if I do not have exactly thirteen then I know that the dealer has violated the rule; or if I find I have thirteen, then I can ask the other players if they do so as well. If some of them do not, then we know that a misdeal has taken place; both the question and the mode of resolution are clear. And finally, suppose that the question is posed from "the view from nowhere", to use Tom Nagel's felicitous phrase, namely in these words: "How can *it* be decided that the rule has been violated?" This question does not invoke first, second or third person perspectives. And yet it is easily settled. For the cards can be counted and in this way the question can be settled. If there is some unease felt by my putting the point in the passive voice, we can simply say that anyone can count the cards and determine whether a violation has occurred. All of this is perfectly straightforward.

It is instructive to contrast this simple case, where a constative rule has been under consideration, with a case, that of trespass, where the issues are of a higher order of difficulty; but in so doing, we should recognize that there are, say, legal cases such as that involving the Miranda Rule or state regulations concerning abortion that are even more complex.

The first difficulty one faces with respect to trespass arises with respect to the question of what the concept itself means. There are both legal and non-legal uses for this notion and all of them raise conceptual problems. In some broad sense, the concept of trespass involves an unwarranted passage by someone or some group into a domain whose proprietor is another person or group. But how shall we describe a case of trespass – was it a simple crossing of boundaries, an intrusion, an infringement, an invasion? And what is meant by "domain", and "proprietor" in this definition?

The notion of a *domain*, for instance, is normally interpreted broadly. It might include certain activities, such as those of a busy executive whose time is his domain and which can be trespassed upon by another. The notion of "being a proprietor" can be expanded to include such an executive; so one can be a proprietor of something susceptible to trespass other than land or real property. Shall we say that cases of trespass are always cases of intrusions? And what does "unwarranted" mean? Let's briefly explore some of the complexities these questions raise.

The notion of an intrusion covers a broad spectrum of activities, ranging from simply stepping into or walking across (say) the property of another to an actual invasion of that property. We can distinguish within this range such differing cases as encroaching, entrenching, infringing upon, and invading the property of another.

Each of these concepts will have different felicity conditions. One who, say, while hunting, and without realizing that he has entered restricted boundaries, walks across a field owned by a farmer is not to be described as *invading* that piece of property though, depending on the circumstances, he can be said to be intruding upon it. And depending on the circumstances, as we shall see, such intrusion may be characterized as a case of trespass.

In contrast, invasion presupposes a specific intent to enter the domain of another whereas this is not necessarily true of intrusion. Nor is our hunter encroaching upon the owner's property, since "encroach" conveys the sense of doing something gradually or perhaps stealthily – the term would more suitably describe the actions of the white settlers in America who gradually displaced the native Indians, for instance. The hunter is simply passing across that piece of land without authorization. But even unauthorized entry into a piece of property is not to be identified with trespass. Trespass requires that the owner would specifically forbid certain classes of persons from entering his property without his permission. But in this case, if the owner does not object to strangers hunting upon his property, then we do not have a case of trespass. So in addition to unauthorized entry, the owner's attitudes about such entry will make a difference with respect to how the crossing of that piece of land is to be characterized; and in particular, to the determination of whether a rule has been violated or not.

Furthermore, what does "unwarranted" mean? Here again we have a vast array of possible answers, ranging from deliberate violations of explicit regulations (such as those committed by conscientious objectors who breach a promulgated government regulation by entering a fenced area at a military establishment) to simply entering some facility without being invited by the host and where no laws are involved at all. The concept of "unauthorized entry", for instance, is one type of unwarranted entry yet cases of unauthorized entry are susceptible to differing assessments ranging from those not having the express approval of say, the owner of a piece of property, to those which violate specific ordinances. A tradesman intent upon soliciting business may enter a property without authorization, annoying the owner by so doing, and in that sense his entry is unwarranted; but is it trespass? Simply on the basis of what has been said so far, it is doubtful that we would describe it as such. Chutzpah it certainly is, but trespass – well it is hard to say without a more detailed scenario. And we should distinguish that case from the case of a poacher, who with a wholly different intent and purpose, may enter an area without authorization. Assuming the satisfaction of certain other conditions, it raises no doubts: the poacher is violating a specific ordinance, and is guilty of actions that are punishable by law, whereas those of the aggressive salesman are not.

One could explore these notions at greater length if time permitted, but since it does not, let us see how even these distinctions affect questions about rule following when they have perspectival implications of the sort mentioned above.

Let's make the case as simple as possible. Suppose I am a hunter intent upon shooting some birds and while following a flock I cross a farmer's fields. How do I know that I have violated the rule of trespass in this case? The answer is much more complicated than it would be in the case of the constative rule of bridge where I am the dealer. It will depend on all sorts of circumstances. Has the farmer posted signs stating that anyone entering his property is guilty of trespass, or signs stating that shooting is not allowed? Suppose he hasn't – then how do I know I have entered forbidden terrain? If there are no signs I may not know that I am in unauthorized territory. Well perhaps there are no signs, but it may be that the field has been fenced with wire. Then if I have cut the wire, or crossed over or under it in order to enter the field, I would at least know that I have entered a territory which is enclosed.

But the fact that the field is enclosed does not entail that one who enters it is committing trespass; that, as I noted earlier, requires the satisfaction of other conditions as well. But suppose the field has not been fenced. Then how do I know that I have trespassed? One possible answer, given the present scenario, is that I may not know that I have, and indeed I may never know that I have. So the situation is different from the case where I have misdealt some cards or from the case of the correct expansion of a mathematical formula. The question of whether I have or have not followed the rule in these latter cases does not typically depend on the sorts of contextual factors I have been describing. Such queries differ from those about trespass in that they are in principle completely resolvable. And as we have seen in the case of constative rules, violations can be easily decided no matter what the point of view or perspective that is assumed. But this is not true of trespass. Here the point of view may make a substantive difference.

Let us vary the hunting scenario to see why. There I am, within a fenced enclosure, gun in hand aiming at some birds. The farmer, going about his daily activities, suddenly sees me and orders me off his property even using the word "trespass". Have I trespassed? Of course I have. Do I know I have – without a doubt. The question has been decided for me by the farmer.

Now let us alter the point of view. Suppose the question is: How do *I* know that *another* has trespassed? Suppose the "I" in this case is the farmer. Then I know that the hunter has entered the enclosed field without my authorization. I may also have posted signs that the hunter has not seen. There is no doubt in my mind that the hunter has entered the field without permission; the prohibition against trespass has been violated. So once again the farmer has resolved the issue – but this time it is because I am the farmer. But these two cases are not generalizable, and if we change the scenario again, even slightly, we can see why.

We shall ask the same question: "How do I know that another has trespassed?" But now suppose that the person who asks is the hunter. Further, suppose a friend and I are hunting together and that we have become separated. Suddenly I see my friend in an adjacent field. How do I know that he has trespassed? Well, I may not: I may not have seen any signs forbidding shooting, I may not from my position have seen any fences, and I may not have seen my friend crawl under the wire. Now as I look at him, I just see him walking in a nearby field. I don't believe he has trespassed and I certainly do not *know* that he has. My situation is thus different from that of the farmer. He knows things that I do not know – and the fact that he does makes a difference with respect to my answering the question: "How do I know that he has violated a rule?" Clearly then the circumstances make an enormous difference with respect to a correct answer.

My point in this essay has been that the question of rule following that Wittgenstein poses in the *Investigations* is more complex than most commentators have realized. I am suggesting here that before we can state and then hope to answer this question the three factors I have described in this paper must be taken account of. First, we must broaden our range of examples, and in particular not allow mathematical formulae to become paradigms of what rules are; second, we must distinguish what is involved in understanding a rule from knowing how to apply it, and finally, we must distinguish from one another various perspectival questions about rule following. If we follow these prescriptions – they are surely not rules – then perhaps some "muddles" about rule following can be avoided.

University of California,
San Diego

NOTES

[1] G. P. Baker & P. M. S. Hacker, *An Analytical Commentary on the 'Philosophical Investigations'*, vol. 2, *Wittgenstein: Rules, Grammar, and Necessity*, B. Blackwell, Oxford 1985, p. xi.
[2] *Philosophical Investigations. Philosophische Untersuchungen* (PI), ed. by G. E. M. Anscombe & R. Rhees. Trans. by G. E. M. Anscombe, B. Blackwell, Oxford 1953.
[3] Quoted in *The Philosophy and Language*, ed. by A. P. Martinich, Oxford University Press, New York, p.478.

DIEGO MARCONI

FODOR AND WITTGENSTEIN ON PRIVATE LANGUAGE

In his book *The Language of Thought* (= LoT),[1] Fodor makes, among many claims, the following two:

(a) Wittgenstein's private language argument (= PLA) "isn't really any good" (p. 70);

(b) Wittgenstein's argument is anyway not really against Fodor's own theory of the language of thought (pp. 68, 70).

Concerning the first claim, I shall try to show that Fodor's refutation of Wittgenstein's argument fails. On the second, Fodor is right but only trivially: Wittgenstein's is not an argument against the possibility of a "language of thought" in Fodor's sense because the language of thought is not a private language in Wittgenstein's sense. Indeed, it is not a *language* in Wittgenstein's sense. I do not mean that it is not a language for it is a *private* language, and a private language can't be a language for Wittgenstein. I think it can be shown not to be a language (for Wittgenstein, and perhaps for most of us) without appealing to its privacy.

On the other hand, I shall say nothing against (or for) Fodor's main views in the book: I will not be denying (or asserting) that we have something like an innate system of internal representations which plays a crucial role in the learning of our first natural language. I believe that such views are independent of (a) and (b).

1. According to Fodor, the reason why the PLA isn't any good is that it is a *verificationist* argument. Suppose we grant to Wittgenstein that "unless there are public procedures for *telling* whether a term is coherently applied, there will be no way of *knowing* whether it is coherently applied" (p. 70).[2] To go from this to the further conclusion that, therefore, the term cannot *be* coherently applied is to assume verificationism – i.e. to assume that it makes no sense to assert that a term is (or isn't) coherently applied unless one has some way of telling. But verificationism is surely wrong, so Wittgenstein does not prove that terms of a private language cannot *be* coherently applied – even if we grant that in the case of a private language we could not know that they are so applied.

But this is not Wittgenstein's argument. The trouble with a private language

R. Egidi (ed.), Wittgenstein: Mind and Language, 107–115.
© 1995 *Kluwer Academic Publishers. Printed in the Netherlands.*

is not that we have no way of assuring ourselves that a term belonging to it is being used correctly (or coherently); the trouble is that we can't be *mistaken* about whether it is being used correctly, for "whatever is going to seem right to me is right".[3] The private language is not a game at which we can't win, it is a game at which we can't *lose*.[4] If Wittgenstein were a verificationist in the PLA, his argument ought to be the following:

(1) It makes sense for me to say that 'S' is being used correctly (or coherently) (by me) just in case I know how to ascertain that 'S' is being so used;

(2) But I don't know (and can't possibly know) how to ascertain whether 'S' is used correctly (coherently);

(3) Therefore, it doesn't make sense for me to assert that 'S' is being used correctly (or incorrectly).

But, independently of whether or not Wittgenstein believed in verificationism at the time of the *Investigations* (step 1), he wouldn't have taken step 2: for we know all too well how to ascertain that 'S' is being used correctly. Simply, if it seems to me that it is so used, then it is, otherwise it isn't. In the case of correctness in a private language, verification, far from being precluded, is immediate. So, Wittgenstein's is not a verificationist (or, for that matter, falsificationist) argument: the point is not that 'S is being used correctly by me' cannot be verified (or falsified). It can. The point is rather that the private language thought-experiment has been so set up that it simply does not make sense to speak of my being right in the use of 'S' "whether I know it or not": for "being right in the use of 'S'" has been *defined* as 'conforming to my private convention concerning the use of "S"', and, by hypothesis, I am the only one who can "check" whether the convention has been respected or not. *There is no* being right *per se*, whether I know it or not.

2. Fodor is sensitive to this aspect of the argument. In fact, he appears to agree with Wittgenstein on two things: (a) if the notion of a language of thought is to make sense, some characterization must be given of *being right* in the LoT ("using it coherently", p. 71); (b) such a characterization must not be in terms of a private convention (pp. 73, 78). It must be said that Fodor does not set down the latter requirement in so many words: but he appears to accept it implicitly. In so doing, it seems to me, Fodor essentially accepts the gist of the PLA. What he rejects is (1) that the notion of a non-public language is essentially connected with the idea of a private convention as to its use, and (2) Wittgenstein's alleged argument from the impossibility of know-

ing whether a non-public language is being used correctly to the impossibility of its being used correctly (or of asserting that it is being used correctly). Concerning (2), I have tried to show that there is no such argument in Wittgenstein. Concerning (1), the burden of proof is entirely upon Fodor, in the sense that it is up to him to show that the notion of a non-public language can be dissociated from that of a private convention: as he himself accepts, this must be done by satisfying (a) and (b), i.e. by giving some account of being right in the use of a non-public language (such as the LoT) which is not based on a private convention. Fodor adds the further requirement that such an account must show that the sense in which a term of a PL is employed coherently is "reasonably analogous" (p. 71) to the sense in which a term of a public language is so employed (otherwise, why speaking of a *language* at all?).

According to Fodor, it is generally true, for both public and non-public languages, that terms are used coherently if there is "a stable relation" between the way they are used and the way the speaker believes the world to be (p. 71).[5] A speaker S uses a predicate 'P' coherently iff S applies 'P' to an object x just in case he believes that x is P. The stable relation that Fodor has in mind involves S's beliefs, on the one hand, and his assent-behavior on the other. Thus for instance, Fodor says, Smith uses 'Jones is sick' to represent the state of affairs in which Jones is sick iff Smith assents to assertions made by employing the *form of words* 'Jones is sick' just in case Smith believes that Jones is sick (p. 72).[6] Now, says Fodor, in the case of a *public* language, the relation between a certain form of words and a certain belief is mediated by the conventions of the language: S applies 'P' to x just in case he believes that x is P because the conventions of the language have it that such are the cases in which 'P' (rather than 'Q') is to be applied. This is to say that (in a public language) the only reason we may have to call P's 'P' is a linguistic convention. All this leads Fodor to the following general statement (relative to public languages):

(C) S uses 'a is F' to represent a's being F just in case it is conventional that (S believes that a is F iff S assents to 'a is F').

But this formulation cannot be right. For suppose S *mis*uses 'a is F' (he uses it to represent a's being G, say): then according to condition (C) it should follow that

It is not conventional that (S believes that a is F iff S assents to 'a is F').

But it clearly doesn't: whether or not anything holds as a linguistic convention does not depend on the correctness of any single person's linguistic behavior. What does follow is the falsity of the embedded biconditional: if S misuses 'a is F', he will not assent to 'a is F' just in case he believes that a is F; he will assent to it, say, just in case he believes that a is G. It thus seems that Fodor's condition is ill-phrased. What he wants is something like

(C′) (Given the conventions of the language) S uses 'a is F' to represent a's being F iff (S assents to 'a is F' just in case he believes that a is F).

or perhaps the more traditional "S uses 'a is F' (as an expression of L) . . .".

Now, of course, what Fodor is aiming at is the analogue of (C) for a non-public language such as the LoT. He believes he can get it simply by replacing linguistic convention with "nomological necessity", and assent with standing in a particular computational relation to a formula (of the inner language).

Disregarding for an instant the fact that (C) is clearly wrong, this yields

(D) S uses 'a is F' [a formula in the LoT] to represent a's being F iff it is nomologically necessary that (S R-s to 'a is F' just in case S believes that a is F).[7]

The similarity of (C) and (D) allows Fodor to claim that (1) the notion of "using a term coherently" can be made clear even in the case of a non-public language (such as the LoT) without appealing to private conventions at all, but only to nomological connections between propositional attitudes, internal representations and computational relations; (2) the notion is analogous to the corresponding notion as defined for public languages. We can have no better than an analogy, for public languages are conventional whereas the language of thought is not. If one is struck by the *dis*analogy, Fodor admits, he will be inclined to deny that the LoT is a language. Even such a denial, however, will not affect the main point: that first-language learning involves the use of an unlearned internal representational system. Which last point, as I said, I am not going to challenge.

Instead, let us consider condition (D), which is supposed to define coherent use of a language such as the LoT. First of all, it is not at all clear whether (D) can be transformed so as to make it analogous to (C′) (the right condition for public languages) rather than to (C) (the wrong condition). For (D) as it stands does convey Fodor's idea of how the expressions of LoT have meaning: 'a is F' has a certain meaning (it means that a is F) iff it embodies the content of S's belief that a is F: i.e. iff it is a fact of nature ("nomologically necessary") that S stands in relation R to 'a is F' whenever he believes that a is F. It does not seem that the same point would be made by the obvious analogue of (C′), namely

(D′) (Given the facts of nature) S uses 'a is F' to represent a's being F iff (S R-s to 'a is F' just in case S believes that a is F).

Indeed, one receives the impression that the mistaken condition (C) has been constructed after the pattern of (D), which is what Fodor wants for a language such as the LoT. (D′) just misses the point: what is a fact of nature is exactly the connection between S's R-ing to 'a is F' and his believing that a is F. So, it seems that at least part of the analogy between the LoT and plain, public languages is lost. This, however, isn't so bad in itself. The core of the analogy consists in this, that coherence of use in a public language can be expressed in terms of the coherence between a speaker's propositional attitudes and his assent-behavior; and similarly, coherence of use in the LoT can be defined as the coherence between the subject's propositional attitudes and his computational relations with respect to formulas of the "inner code". This is preserved even between (C′) and (D). Trouble comes from another source. Suppose again that 'a is F' is *mis*used. Given (D), this entails that

> It is not nomologically necessary that (S R-s to 'a is F' just in case S believes that a is F).

I.e., there are exceptions: it may happen that S believes that a is F but does not R to 'a is F', or that he does stand in such a relation to 'a is F' and yet does not believe that a is F. Well, so it may. But, would we describe such an event as a case of *misuse* of the LoT by S? Would we even describe it as a case of *incoherent use* of 'a is F' on S's part? No; we would rather say that, contrary to hypothesis, 'a is F' – a certain formula of the inner code – does not mean that a is F. Notice that this is consistent with Fodor's overt formulation of (D): for (D) says that "'a is F' is used to represent a's being F iff. . .", and in such cases we *would* say that 'a is F' is *not* used to represent a's being F. But this is just equivocating: for what we were defining was proper, or coherent, or correct use of 'a is F'. And in such cases, "S is not using 'a is F' to represent a's being F" cannot be taken to mean "S is not using 'a is F' coherently (properly, correctly)". This is the important disanalogy between (C′) and (D): for in the case of (C′), the above paraphrase is entirely in order.

3. Can there be such a thing as improper use of the LoT? Can one go wrong in the LoT? If one cannot, it may be doubted that the LoT is a language in

Wittgenstein's sense; for Wittgenstein insists that a language can be misused (by an individual speaker), that its expressions can be understood only to a limited extent, that they can be misunderstood. I don't know whether this is to be regarded as "an essential feature of language" but it does play a crucial role in the phenomena of the use of language – of the languages we know, of course. Perhaps some light upon this question can be cast by Fodor's repeated (pp. 66,67) analogy between the LoT and the computers' machine-language. Can a machine-language be misused or misunderstood? Certainly, the *programmer* can misuse a machine-language: he can make errors of syntax, or he can write a program that does not have the intended effect because the programmer has misunderstood the notation. On the other hand, what would it be *for the computer itself* to misuse the machine-language? Not operating in accordance with the intended effects of the commands (intended by the engineers who built the computer and designed its machine-language)? But in such a case we wouldn't say that the computer is misusing the machine-language: we would say that something went wrong with the design of the whole apparatus, computer-cum-machine-language.

Notice that, from the programmer's standpoint, the machine language is not "private" in any plausible sense. Indeed, it doesn't even *look* private. Many people understand the same machine language and can use it to communicate among themselves. I doubt that even from the computer's standpoint it makes sense to say that a machine-language is private, for many computers "understand" the same machine-language. This is all right with Fodor, for he claims that even the LoT is in no way "epistemically private", i.e. inaccessible in principle. Ultimately, the LoT is just neurological machinery, and "neurological events are public" (p. 70). But perhaps we have a disanalogy here: for, independently of the "ultimately it's all neurons" business, I don't know whether it makes sense to speak of *the same* language of thought being "understood" by many different minds. Anyway, to conclude this part of the discussion: from the programmer's standpoint, one can go wrong in a machine-language, but then from that viewpoint a machine language is not private, or "inner", or non-public. The computer, on the other hand, cannot be said to misuse the machine language; so that one wonders whether the machine-language *is* a language from the computer's viewpoint. For instance, does it make sense to say as Fodor says (p. 66) that computers have a machine-language "in which they talk to themselves"? Or: what do we mean exactly by saying that the computer "understands" the machine-language, as it cannot *mis*understand it? Here, Fodor does have an answer, for he distinguishes between, not two senses of 'understand', but two *ways* of understanding (p. 66). In the first (and usual) way, a device understands a predicate 'P' if it has and employs a representation of its extension, which is itself given in a language the device understands. In the other "way", the device understands 'P' in the sense that it conforms to the conditions that such a representation would specify; i.e. it behaves as *we* would behave, understanding 'P'. This is plausible: but, again, it hides the disanalogy between the two cases. In the first "way", a predicate 'P' can be misunderstood for it is possible to have the wrong representation of its extension, or a representation that is too partial for competent use, etc. Not so in the second "way": the device cannot *fail* to conform to the appropriate specifications. Perhaps it would be more appropriate to say that the computer *obeys* the machine language.

But this is not an important point. We may be willing to stretch the concept of understanding so as to cover the computer's handling of the machine language. And one may even be willing (though not so uncontroversially) to stretch the concept of *language* so as to include the machine language as seen from the computer's viewpoint (as opposed to the programmer's viewpoint, from which a machine-language is unproblematically a language, and a *public* language at that). If the analogy with the LoT holds (but we saw there are elements of disanalogy) this would entail that even the LoT may be regarded as a language, in spite of the fact that we can't go wrong in it. What matters, however, is that it cannot be so regarded for the purposes of Wittgenstein's argument against private language: for the argument crucially revolves around the fact that a language must be liable to misuse (as public languages are). So, Fodor is right in holding that the PLA does not apply to his LoT: not, however, because the LoT is not *private* in Wittgenstein's sense, as Fodor claims (p. 70), but rather, as I hope to have shown, because it is not a *language* in Wittgenstein's sense.

4. Finally, has Fodor succeeded in giving an account of "using a non-public language coherently" that would meet the two requirements of (a) not appealing to private conventions and (b) making being right in the LoT reasonably analogous to being right in a normal language? The first requirement may be said to have been met, if we take Fodor's use of 'coherent' (as opposed to 'correct') seriously. If the notion of a machine operating coherently makes sense – i.e. if we disregard Kripkensteinian problems with the notion of coherence in the use of a language – then Fodor has made sense of the notion of "using a non-public language coherently", without appealing to private conventions (or so it seems). On the other hand, it is doubtful that the other

requirement has been met. Being right in a non-public language turns out to be quite different from being right in a regular, public language: for being right in a non-public language admits of no alternative. One can of course go wrong *about* a non-public language such as the LoT, in the sense of making false hypotheses about its workings: for instance, by entertaining the hypothesis that S R-s to *p* – a formula in the LoT – iff S believes that *q*, whereas this is not the case. But one cannot go wrong *in* the LoT, except in the deviant sense of suffering brain injuries or other such inconveniences. Thus it seems that Wittgenstein was right in strictly connecting "being right *or* wrong" with the publicity of standards.

Università di Torino

NOTES

[1] Crowell, New York 1975.

[2] Here one would be tempted to object that Fodor should be speaking of 'correct' application rather than of 'coherent' application. But Fodor's choice of words is not without consequence (as we shall repeatedly see): for we speak of correct use or application of a word with respect to some standard or *convention*, whereas Fodor wants to claim that no convention is involved in the appropriate functioning of the language of thought – especially not a *private* convention, for Fodor agrees with Wittgenstein – implicitly – that there is no such thing as a private convention. I shall try to show in each case what difference is made by Fodor's choice of words.

[3] *Philosophische Untersuchungen. Philosophical Investigations* (PI), ed. by G. E. M. Anscombe & R. Rhees. Trans. by G. E. M. Anscombe, B. Blackwell, Oxford 1953, § 258.

[4] Of course, for Wittgenstein it ensues that there is really neither winning nor losing: there is no game, really – there is only the imitation of a game, or a ritual which can be mistaken for a game.

[5] The reference to beliefs (rather than to matters of fact, or shared beliefs) is puzzling, and may be one case in which speaking of "coherent" rather than "correct" use of language does make a difference. For suppose I systematically apply the predicate 'cat' to dogs: I would not be said to be using the word correctly, irrespective of whether I do so because I systematically mistake dogs for cats or because I (falsely) believe that 'cat' refers to dogs (i.e. that everybody else uses or should use the word as I do). On the other hand, perhaps I could be said to be using the word coherently.

[6] Notice that to use 'Jones is sick' to represent the state of affairs in which Jones is sick is not just using it coherently; it is using it correctly. Perhaps we should be more careful here, and make a distinction between using 'Jones is sick' *with the intention of* representing the state of affairs that Jones is sick, and using it *to* represent the aforementioned state of affairs. Fodor's definition applies to the former situation, not to the latter.

[7] According to Fodor (pp. 76–77), the biconditional "S R-s to 'a is F' just in case S believes that a is F" is not to be taken as a (mentalistic) *definition* of belief, but rather as the prospective *finding* of a cognitivistic theory of belief and other attitudes: we hope we'll be able to discover that, as a matter of fact, whenever S believes that *p* he is in a certain computational state with respect to an internal formula *q*, the translation of *p* in the internal code. Call such a relation 'R'.

Clearly, that S believes that a is F iff S R-s to ‘a is F’ (a formula in the internal code) is not a matter of convention: it is rather, according to Fodor, a matter of “nomological necessity”, a regularity of nature.

MASSIMO DELL'UTRI

MENTALESIANS AND WITTGENSTEIN'S PRIVATE LANGUAGE

In his careful analysis of Fodor's interpretation of Wittgenstein's 'Private Language Argument' (henceforth 'PLA'), Diego Marconi claims that Fodor does not entirely succeed in finding a way out of the PL marsh. What I would like to do is to present some further considerations about Wittgenstein's PLA and Fodor's Language of Thought – often referred to as 'mentalese' – in an attempt to shed some further light on the issue.

We can begin by drawing an analogy between a person who develops a PL, and the situation of a 'brain in a vat' (henceforth 'BIV'), as described by Hilary Putnam.[1] Taking into account that both situations are deemed inconceivable by their respective authors, let us first look at the man with a language exactly identical to that described in § 243 of Wittgenstein's *Philosophical Investigations,* a man who has his own terms and sentences to report his intimate feelings. Let us call him 'the PL-man', and ask whether he manages to refer to the private items of his interior monologue. Wittgenstein certainly does not deny the possibility of having an inner life, i.e. inner sensations, processes, motivations and the like, but he clearly points out that – starting from a mentalistic position – the trouble we are finally confronted with is that we lack any means for talking about what is internal and private. Because of that, it amounts to our lacking the very subject about which we are trying to talk, i.e. the private items, as § 304 reveals.

But why can't we talk about our private items? Wittgenstein's answer can be expressed by another question, namely, "How could the words in a PL acquire meaning and refer to something?" He discusses the problem of the reference of sensation terms, and finds out that neither criteria nor rules of correctness for private expressions (§§ 258, 259, 261–265) or identity of sensations or mental representations in general (§§ 253, 376 ff) are available in this internal world, stating "An 'inner process' stands in need of outward criteria" (§ 580). Rules and criteria are obtainable only on the basis of a consolidated practice or, in other words, on the basis of a social context. Wittgenstein's analysis therefore reveals the essence of language as a *social* means of communication, and stresses the importance of establishing the social context of use. It is in this context that terms acquire a meaning and their

R. Egidi (ed.), Wittgenstein: Mind and Language, 117–123.
© 1995 *Kluwer Academic Publishers. Printed in the Netherlands.*

use a reference.[2] But what our PL-man is bereft of – by definition – is just such a context, and hence cannot use his private words referentially.

In this century a more peculiar attention to the social dimension of human activity – whatever its object – has been the tenet of a number of philosophers. This dimension, in particular, is also of capital importance for Putnam's causal theory of meaning,[3] a theory which – for what concerns us here – plays a central role in his argument against the BIV, which we will now look at in order to see how far the analogy with the PL-man can be pushed.

The fantasy of the BIV is the contemporary correlate of Descartes' hypothesis of the evil genius.[4] This time however, instead of the evil genius at play, we have a mad scientist who causes a person to hallucinate the whole world. After extracting the brain from the skull of a miserable subject and placing it in a vat filled with nutrient fluids, the scientist connects the brain's nerve-endings with a highly sophisticated computer programmed to let the brain experience any situation he deemes upon it. Everything appears perfectly normal to the brain: it lives its usual life and experiences its normal bodily sensations, has a mouth for speaking, eyes for seeing and ears for hearing and, in short, seems to be the person he always was, perhaps even laughing as it listens to the amusing but horrible story of a deluded guy whose brain was kept alive in a vat!

Usually, such a story is employed in epistemology in order to raise a skeptical doubt as to the knowledge of the external world: "How do you know that you are not a BIV?" is the challenge posed by the skeptic. However, Putnam uses the BIV story for a different but connected aim: that of investigating the notion of reference. "Have your words a reference to the actual world?", or "Is the language you are using actually connected with the real world or is it a mere abstract and empty syntactic play?" is the question he engages to answer.[5] His peculiar treatment of the problem leads him to conclude that the words the BIV utter and employ in complex sentences do not have any relationship with the real objects existing outside the apparatus they live in. This conclusion partly depends on the causal theory of reference we mentioned above.

Very briefly, according to this theory the use of a word by speaker *S* refers to an object *x* if it is causally tied to *x*, either directly or indirectly, and *S*'s intentions are such that his use of that word is to refer to the object which counts as the referent of the word in question within the collective body of speakers. Hence, one of the central features of this theory is deeply Wittgensteinian. The reference of a term must be public, and whatever individual description given of the referent of a term used is no good, unless it is the description obtaining throughout the society.

But Putnam's argument against the BIV hypothesis contains something more in respect to Wittgenstein's PLA. What he tries to do is to derive the *knowability* of reference from its public and social character. He tries to show that, given Wittgenstein's argument, i.e. given that what counts as *the* reference of terms is what is socially acknowledged as such, this very reference has to be knowable at least in principle. Indeed, if reference were not knowable, it wouldn't even be public. But if reference is at least in principle knowable – Putnam argues – it follows that not every causal connection is good for reference. Only those causal connections which are part of a social practice of *knowledge* will do. Therefore, the relationship BIV could have with the world outside the vat plays no role for reference, because their knowledge is limited to what appears in the image furnished by the computer, and thus all their references are within the image. From this, then, Putnam concludes to the self-refuting character of the BIV hypothesis.

This conclusion reveals that the stress posed on the social dimension is a common feature in the arguments of both Wittgenstein and Putnam. Indeed, Putnam's philosophy has many Wittgensteinian traits, and this conclusion does not fail to show traces of Wittgenstein's criticism of mentalism, a criticism avowedly endorsed by Putnam. Explaining why mental representations do not constitute understanding, Putnam reformulates Wittgenstein's examples in §§ 159–160 of the *Philosophical Investigations*. He invites us to imagine a person who does not know Japanese, learns a Japanese text by heart, and under the influence of a hypnotist says that text in his mind with the impression of understanding it. Here it is clear that, even if this whole arrangement is so perfect as to cheat a Japanese telepath, that person does not understand what he is saying, because he is not able to give appropriate answers to any questions and use the words in the right contexts.[6] Therefore, mental representations do not bring about understanding by themselves – and moreover, as Putnam puts it:

> Mental representations no more have a necessary connection with what they represent than physical representations do. The contrary supposition is a survival of magical thinking.[7]

The step from the lack of understanding to the lack of referential power, as the quotation reveals, is at hand.

But there are obvious differences in the way Wittgenstein and Putnam arrive at realizing the importance of the social dimension. Putnam's argument is based upon his proper explanation of the mechanism of reference, while Wittgenstein's is actually based upon the notion of rule, and seems to have but a derivative impingement on the notion of reference. As Marconi says,

> The trouble with a PL is not that we have no way of assuring ourselves that a term belonging to it is being used correctly (or coherently); the trouble is that we can't be *mistaken* about whether it is being used correctly, for 'whatever is going to seem right to me is right'. (pp. 107–8)

Yet Wittgenstein's argument also has his implications as to the notion of reference – as we have been saying – and it is possible to notice this by reformulating Marconi's wording. In the reformulation we can see Wittgenstein claiming that we can't be mistaken about whether a term in a PL *has a reference*, because whatever will seem to have a reference to me has a reference. Moreover, according to § 304 which we cited above, the entire question of reference is thereby suddenly turned into non-sense for a PL theorist.[8]

Bearing this in mind, what could we say about Fodor's mentalese – the inner code of the brain – upon which Marconi's paper dwelt at length? Is it possible to involve mentalese in the same analogy we held between the PL and the language spoken by the BIV? Well, imagine a community of people which possesses a language of thought but no spoken language, perhaps because they are bereft of a phonatory apparatus. Imagine, in advance, that they have a special telepathic faculty which allows them to talk to each other. There is nothing strange in this for, according to Fodor, mentalese is what makes possible the learning of a historical language and must hence be in itself complete and fully structured.

But even if these Mentalesians – as we may call them – manage to attach a definite reference to every term of their language of thought, this would not happen by virtue of mentalese itself. Here the situation is similar to that of the people playing the language-game of the beetle in a box (§ 293). It doesn't really matter what is in the box; what matters, actually, is that all the speakers use the word 'beetle' in the same way. "The thing in the box – as Wittgenstein says – has no place in the language-game at all." Equivalently, the possible reference which Mentalesians' terms could acquire would not depend on what is in their head, but on the agreement on which their practice converges. Therefore, contrary to Marconi himself – who displays an accurate examination of some of Fodor's contentions – I think that the argument Wittgenstein develops against the PL hits the language of thought as well.

It has to be noticed that Fodor tries to safeguard himself from the start, for he claims that he is offering a *particular* reconstruction of Wittgenstein's PLA, and admits that other reconstructions could well prove Fodor's own analysis wrong.[9] So, it is Fodor himself – in a sense – that warns us against his own argument. Moreover, he claims that Wittgenstein's purpose is that of attacking the idea of a sense-datum language.[10] But the sense-datum language – as we already remarked – is taken by Wittgenstein only as a language which

– were mentalism right – would be an example of PL, where the purpose is that of showing the absence of inter-subjective criteria.

Fodor gives a two-fold reply to Wittgenstein: first, he notices that "there is no reason why a mentalist needs to assume that mental operations exhibit epistemic privacy"; and, second, he points out that the PLA isn't really any good because it follows "on the verificationist principle that an assertion can't be sensible unless there is some way of telling whether it is true, but *surely* there is nothing to be said for that principle".[11] The second point is rightly criticized by Marconi, but the first seems to miss the issue: the epistemic privacy is not something that a mentalist can assume or not, but is rather something to which a mentalist is compelled. And here resides the core of Wittgenstein's argument, the point in which his real polemical target reveals itself as *mentalism.*

If this is right, the whole project of finding a characterization of a "Language of Thought" leaves much to be desired just from the beginning – at least from Wittgenstein's point of view (and – for that matter – from Putnam's, since his criticism of mentalist theories of reference – which he labels "magical theories of reference" – is one of the core premises of the argument against the BIV hypothesis). Indeed, what Fodor tries to do, as revealed in his defence of Augustine,[12] is to revive the mentalist tradition, a tradition whose roots go back to Aristotle with many formulations in the history of philosophy. Putting these formulations aside, it could be said that the leading idea pervading mentalism is that of the existence of ready-made *representations* in the mind which give meaning to spoken words and refer to things in the world. The relation between representations and words is *conventional*, for it depends on the particular historical language which tries to convey those representations, whereas the relation between representations and things in the world is *intrinsic*, built in the very representations themselves. Moreover, the set of mental representations is structured to form a sort of language: a 'natural' language which can be seen as *private*, because in itself – without the intervention of a spoken language – cannot serve for the purpose of communication. The utility of a spoken language, in this respect, resides just in its being a conventional means by which the speaker translates his mental code in order to exchange information with a listener who, in turn, translates the spoken language into his mental one.

I mentioned before that Fodor's project is that of giving the mentalist tradition a modern style. 'Mentalese' is just the traditional inner code someway connected with the world and prior to the acquisition of any spoken language. The difference is that mentalese is explicitly located in the brain, and makes the learning of a spoken language possible, whereas this latter feature is absent from the medieval *lingua mentis*.

Marconi plausibly shows that one cannot *misuse* mentalese, or, as he puts it,

> being right in a non-public language turns out to be quite different from being right in a regular, public language: for being right in a non-public language admits of no alternative. . . One cannot go wrong *in* the LoT (p. 114).

But Marconi uses this fact in order to conclude that mentalese is not a language in Wittgenstein's sense and, therefore, that the PLA does not touch it. What Wittgenstein would have said about mentalese, on the contrary, is not that it is not a language, but that it is not a *public* language. Impossibility of "misusing", impossibility of "using a term according to a (public) rule", impossibility of "using a term (in-)correctly", are all one and the same thing from Wittgenstein's perspective. One could also grant, for the sake of discussion, that mentalese is not a language, but neither is PL. The PLA applies to mentalese because it hits its epistemological and semantical presuppositions, i.e. *mentalism*. Therefore, the very reason why – according to Marconi – the PLA does not apply to mentalese (i.e. the impossibility of misusing it), is precisely the reason showing that the PLA applies to it.

Far from constituting an autonomous and complete representational system which precedes and makes possible the learning of any historical language, far from possessing a meaning and a reference *per se*, mentalese – if there really is such a thing – inherits meaning and reference from spoken language, and therefore all the problems connected with them. Mentalese does not solve any problems: it only *duplicates* them.

Terza Università di Roma

NOTES

[1] See H. Putnam, *Reason, Truth and History*, Cambridge University Press, Cambridge 1981, ch. 1.

[2] "Our words refer to things by the way they enter in discourse; by their connexions with what people are saying and doing, for instance, and by the way they affect what is said and done". (R. Rhees, 'Can There Be a Private Language?', in *Wittgenstein. The Philosophical Investigations*, ed. by G. Pitcher, Macmillan, London 1968, p. 267).

[3] See H. Putnam, 'The Meaning of "Meaning', in *Mind, Language and Reality, Philosophical Papers*, Vol. 2, Cambridge University Press, Cambridge 1975, ch. 12, and *Representation and Reality*, The MIT Press, Cambridge 1988, ch. 2.

[4] See R. Descartes, *Meditations on First Philosophy*, Liberal Arts Press, New York 1960.

[5] See my 'Choosing Conceptions of Realism: The Case of the Brains in a Vat', *Mind*, 99, 1990, pp. 79–90.

[6] As Rhees (op. cit., p. 281) puts it: "Speaking the language. . . is not just uttering the words; any more than understanding the language is just 'recognizing' the words. It is carrying on a conversation, for instance; or it may be writing reports, or listening to a play in a theatre. It is being someone to whom the rest of us can speak and get an answer; to whom we can tell something and with whom we can make a joke and whom we can deceive".

[7] H. Putnam, *Reason, Truth and History*, cit., p. 3.

[8] The reformulation leaves it open whether terms of a PL do have a reference or not. Yet, since – according to Wittgenstein – one could be said to use a term correctly, i.e. understand it, only if one uses it following certain social standards (which state what is to count as the correct use of words and sentences), and since understanding a term involves grasping what its socially determined meaning and reference are, it follows that those standards state also what is to count as the reference of terms. Hence, failing to accord to the standards amounts to failing to have a (social) reference. Be this as it may, § 304 could be interpreted as to show that having a reference about which nothing could be said – as in the case of a PL – would serve just as well as a lack of reference. If this is true, by the way, § 304 takes on a verificationist flavour.

[9] See J. A. Fodor, *The Language of Thought*, Crowell, New York 1975, p. 68.

[10] *Ibid.*, p. 69.

[11] *Ibid.*, p. 70.

[12] *Ibid.*, p. 64.

PART III

INTENTIONALITY AND THE GRAMMAR OF MENTAL LIFE

ALDO GIORGIO GARGANI

INTENTIONALITY IN WITTGENSTEIN'S WORKS

The problem of the logical form and of its relationship with that of the understanding of propositions was worked out by Russell in his *Theory of Knowledge*, where he developed the thesis that the knowledge of or acquaintance with a pure logical form, that is with an abstract complex conceived of as a single object would be the requisite for the understanding of a sentence. In his work Russell said:

> I think it may be shown that acquaintance with logical form is involved before explicit thought about logic begins, in fact as soon as we can understand a sentence.[1]

In the *Tractatus* and in the preliminary works Wittgenstein rejected Russell's thesis and maintained that logic cannot depend on the experiencing of a fact, whether it be logical or empirical.[2]

Criticizing Russell's theory asserting the existence of an abstract formal structure the logical experience of which would precede the understanding of any proposition, Wittgenstein assigned the conditions of the meaning to the elements inherent in the very constituents of the propositions.[3] With regard to this, Wittgenstein remarked that "the precedent to which we are constantly inclined to appeal must reside in the symbol itself".[4] Formerly in a entry of his *Tagebücher* in 1914 Wittgenstein wrote that a proposition achieves its meaning "off its own bat" *(auf eigene Faust)*.[5] This immanent conception of the proposition, according to which it reaches its sense on the basis of its own resources, constitutes a thematic premise of Wittgenstein's later work, even where the main theses of the *Tractatus* have been dropped. But there was another premise connected with this conception which was bound to give rise to one of the main difficulties of the *Tractatus*, and it is the claim that every proposition is an independent picture of a state of affairs and it is not possible to infer one elementary proposition from another one. Using the metaphor of the proposition as a yardstick for reality, Wittgenstein wrote that only the end-points of the graduating lines of the scale are laid against the object to be measured.[6] This claim was to be dropped as soon as Wittgenstein realised that it prevented the proposition from being an univocal representation of reality. Logical products like "$p \cdot q$" – where p and q are sentences determining the same co-ordinate of reality in two different ways – had to be excluded.

R. Egidi (ed.), Wittgenstein: Mind and Language, 127–136.
© 1995 *Kluwer Academic Publishers. Printed in the Netherlands.*

Consequently in the course of the Twenties Wittgenstein became aware that logical constants alone cannot account for all inferences we make in language and that they have to be placed within a language now conceived of as a system, as a more comprehensive syntax of internal relations which direct our use of language, permitting some applications of it and excluding others.[7] Using the earlier metaphor again, Wittgenstein writes now that language is no longer a yardstick of which only the end-points of the graduating lines are applied to reality, but one where "all the graduating lines" are applied,[8] with a view to warranting the univocality of propositions. Wittgenstein's well-known slogan, that the meaning of a word is its use, is strictly connected with his conception of language as a system of internal relations and rules directing the proposition taken as something with which we should operate.[9] Precisely, the meaning of an expression is achieved by knowing the procedure, the method to reach it.[10] In this new conception a fundamental role is played by intentional acts (meaning, intending, expecting, wishing and the like) so that Wittgenstein goes so far as to write that "if you exclude the element of intention from language, its whole function then collapses."[11]

In his middle period, between the late Twenties and the early Thirties, Wittgenstein determined in the intention the rule governing the use of language, since it would be the intention that connects use with a system of symbols (for instance playing and score).[12] Tackling the problem of the relation of intentional acts and their fulfilments, Wittgenstein was aware that it cannot be shaped on the terms of a rough comparison between the proposition expressing an intention, a wish, an expectation and a mere fact considered from the outside, as Russell did in *The Analysis of Mind*.[13] Russell analyzed intentional acts in terms of external relations involving three factors, the thought or proposition (expressing a wish, an expectation), the fact that verifies it and a third event which would be a sort of satisfaction, "the fulfilment of the expectation." Criticizing this doctrine, Wittgenstein points out that if the fulfilment of the expectation were to consist in an external relation, that is in a third event not describable in advance, then any event or fact might be considered as the fulfilment of the expectation. In his *Philosophische Bemerkungen* Wittgenstein rejects Russell's doctrine because it does not provide any place for the inner comparison holding between an intention and its object.[14] Russell limits himself to setting up an external and rough comparison, relating a proposition to an event which is not describable in advance. For Wittgenstein, this comparison should be established on the basis of an internal relation of our language according to which expecting that p will be the case should be the same as expecting that this expectation will be fulfilled.[15]

For Wittgenstein, the fulfilment instead of consisting in a third unpredictable event, must be represented in advance as a costituent inherent in the very proposition expressing the expectation.[16] The expectation of the perception of a sense-datum, for instance, involves a sort of being prepared for the occurence of it. "If I expect to see red" – writes Wittgenstein – "then I *prepare* myself for red".[17] But, according to Wittgenstein, the connection between expectation and fulfilment doesn't fit a three-term relation (as Russell claimed), but a two-term relation.

In the period corresponding to his verificationist phase, with special reference to the *Philosophische Bemerkungen*[18] and to his conception of language as a calculus and system of internal relations, Wittgenstein rules out the notion of a judging or perceiving subject who would confront his expectation with what happens,[19] experiencing that third event, which Russell referred to, which would consist in a sort of satisfaction. For Wittgenstein, the relation of the expectation to its fulfilment no longer depends upon an act of interpretation of a judging or perceiving subject. With regard to this it is worth remarking that Wittgenstein prefers the expression "the use of signs" to "the interpretation of signs."[20] The relation between an intentional act and its fulfilment depends on the rules of a language taken as a calculus which directs our employment of expressions in their operative application to reality. With a shift which is relevant from the point of view of a new philosophical anthropology, Wittgenstein makes the connection between expectation and its fulfilment no longer depend on the so called act of interpreting or comparing peculiar to the traditional judging or perceiving subject or self, who would negotiate that connection, but directly on the system of the grammatical relations which regulate our operating with propositions; therefore the expectation must be situated in the same space as what is intended, expected and the like.[21] The fulfilment of an intentional act does not consist in a third event but directly in the immediate production of the event which taking place provides an answer or fulfilment.

> Expecting is connected with looking for: looking for something presupposes that I know what I am looking for, whithout what I am looking for having to exist . . . How do I know that I have found *that* which I was looking for? . . . I cannot confront the previous expectation with what happens. The event that replaces the expectation is a reply to it.[22]

In a way which is analogous to the case of the relation of proposition and verification, in the connection between an intentional act and its fulfilment, there isn't any go-between, any *Mittelglied*, taken as the factor which would have the task of negotiating that fulfilment or verification respectively.[23] In accordance with that, Wittgenstein in the *Philosophische Bemerkungen* puts forward a distinction between, on the one hand, the expectation of a proposition *p* considered as corresponding to the hollow shape of a body and, on the other hand, the occurence of *p* assumed as corresponding to the solid shape of a body.[24] This hollow shape translates metaphorically Wittgenstein's notion of expectation as the being grammatically prepared for something.[25] In these terms, the proposition of an expectation contains in advance the representation of what constitutes its fulfilment. But then Wittgenstein seems not to be completely satisfied with his solution. In the course of his lectures in Cambridge in the early Thirties, and then in the 'Brown Book', Wittgenstein questions his former thesis according to which the proposition expressing an expectation is applied to the point of reality which is present but at the same time it "points out beyond it . . . in the direction of the expectation".[26] Wittgenstein seems to be self-critical when he remarks during a lecture in Cambridge that "it seems to us as if the proposition was not simply itself, but pointed beyond itself and contained a kind of shadow of its fulfilment, which is neither the proposition nor the fulfilment, but something between them (e.g. my expectation that Mr. Smith will come into the room seems to foreshadow his so doing)".[27] Thus Wittgenstein is precisely criticizing his recent view according to which the proposition of an expectation or order would outline its own fulfilment. "We look for this shadow – said Wittgenstein – particularly in an expectation".[28]

In this discussion a particular role is played by the concept of similarity. In fact while analysing the embarassing consequences involved in the use of the notions of "similar", "similarity", Wittgenstein has been working out a new theoretical arrangement of his doctrine on intentional acts. "Similarity" is a treacherous notion which drives us into a blind-alley. It is strictly interlocked with the concept of proposition as an expression foreshadowing its fulfilment. If, in accordance with the *Philosophische Bemerkungen*, the expectation was "the hollow shape" of a body and the occurrence of its fulfilment was "the solid shape" of it, then there seems to be a relation of similarity between the two shapes. But it is just this relation shaped on such terms which gives rise to some difficulties. If I expect to meet Mr. Smith, then in my expectation there is something which would be similar to Mr. Smith; returning to his former example of the expectation of seeing red, which implied the being prepared for red, Wittgenstein remarks:

> If I expect a red patch, my expectation cannot contain the red patch but only something similar to it . . . But to expect something 'similar' is already to expect something different. The 'similar' element, the shadow, in the expectation, is different from the fulfilment: you are no *closer* to the fulfilment".[29]

Then, Wittgenstein underlines the paradoxical consequence that if the expectation were a shadow of the expected fact, then we couldn't expect for instance Mr. Smith to enter the room if he does not.[30] Furthermore, referring to the link between his former view on intentional acts and verification, Wittgenstein remarks that the symbol doesn't point to something outside itself, nor does it anticipate something else "in a shadowy way"; he finally remarks that a symbol doesn't imply a cognition of its being true or false, whereas

> it is supposed that the thought (belief, wish) foreshadows its verification; and this shadow we call the meaning of the proposition, and it seems to mediate between the proposition or symbol and the fact. But this assumption of an intermediate link does not help us".[31]

This shadow, which is conceived of as similar to its fulfilment, seems to correspond to what Wittgenstein in the *Philosophische Bemerkungen* had referred to as the hollow shape of the expectation contrasted with the solid shape constituted by the occurrence of what is expected. Wittgenstein's analysis of the concept of similarity was bound to bring about important consequences on the development of his latest philosophy. In fact it is connected with Wittgenstein's rejection of all those philosophical doctrines which are grounded on duplication, on the pattern of "a double", as Wittgenstein says.[32] The thesis underlying these duplications claims that an expression is in itself just a scribble and it achieves its meaning through its fitting a mould already prepared for it in the mind. Here Wittgenstein is attacking his former view on intentional acts, when he rejects the conception according to which "the words falls into a mould of my mind *long* prepared for it",[33] and when he questions the model of the word's fitting a mould which involves a comparison of the hollow shape and the solid shape before they fitted together; these are in fact the very expressions used by him formerly. Wittgenstein rules out the resorting to any duplication: in this sense recognizing a scribble as a proposition or a drawing as a face does not involve two separate steps, that is the recognizing of them as scribble or drawing and the recognizing of them as meaningful sentence or a face respectively. We get hold of the sentence or face directly in the scribble or drawing.[34] That reminds us of Brentano's remark in the *Psychologie vom empirischen Standpunkt* that "to taste a piece of white sugar does not mean to taste a piece of sugar *as white*" (*ein weißes Stück Zucker schmecken, heißt nicht, ein Stück Zucker als weißes schmecken.*)[35] What Wittgenstein is here questioning is that sort of fictitious comparison which is peculiar to philosophical attitudes. But the philosophical employment of fictitious comparisons[36] through duplications, prototype or double, is based on that notion of similarity which is mistakenly supposed to justify the use of language. Wittgenstein rejects both the false comparisons and the concept of similarity presupposed in the former by resorting to a strongly constructive conception of language. The conclusion to be drawn from his analysis is that it isn't the similarity between experiences which is responsible for our use of words but, on the contrary, it is our use of words which establishes what is to be called similar or identical. "As similarity does not explain the shadow, it does not explain the relation to fact either".[37] It is the circumstance that the same word is used to establish that an intentional act and its fulfilment have in common. The fact that I say that I expect Mr. Smith and then I say that Mr. Smith has arrived, justifies my saying that my expectation has been fulfilled.[38] "What expression and fulfilment have in common is *shown* by the use of the same expression to describe both what we expect and its fulfilment (the word 'red' occurs in both)."[39] Differently from what was claimed in the *Philosophische Bemerkungen*, now the expectation is not a hollow shape foreshadowing in advance its fulfilment; and its being fulfilled by the occurence of a certain event is testified by the use of the same expression for both. In the "Brown Book" Wittgenstein writes that "to say that we use the word 'blue' to mean 'what all these shades of colour have in common' by itself says nothing more than that we use the word 'blue' in all these cases."[40] A constructive interpretation of language practices is responsible for Wittgenstein's demolition of the traditional conception which made the use of such notions as those of similarity and identity dependent on mental or psychological representations. On the contrary, it is our constructive procedures in actual using language which define the identity or similarity of objects, or, for example, what following a mathematical rule consists in.[41] In language, said Wittgenstein, "we don't *look* for anything. We *construct* something." A constructive procedure accounts for the bridging of sign and application. In fact, no explanation ever saves the jump between them.[42] Thus intentional acts are not directed at the shadowy being of external facts and events:

> The symbol is in same sense self-contained; you grasp it as a whole. It does not point out something outside itself, it does not anticipate something else in a shadowy way. . . Thought is therefore autonomous, complete in itself; and anything not given in my thought cannot be essential to it. Thought does not point outside itself.[43]

The sense-datum, the immediately given which Wittgenstein referred to in the *Philosophische Bemerkungen* have now to be replaced by the linguistic expression of them:

> I am inclined to suggest to you to put the expression of your experience in place of the

experience. 'But these two aren't the same'. This is certainly true, at least in the sense in which it is true to say that a railway train and a railway accident aren't the same thing. And yet there is a justification for talking as though the expression "the gesture 'long, long ago'" and the expression "the feeling 'long, long ago'" had the same meaning.[44]

In conclusion, in Wittgenstein's middle period, in accordance with his conception of language as a calculus, the relation of expectation and fulfilment is identified with that of calculation and result. Wittgenstein said that "thought anticipates its fulfilment in the same sense that a calculation anticipates its fulfilment".[45] That accounts for Wittgenstein's consequential claim that meaning is set up inside language and it is given by explanation.[46] This is true of Wittgenstein's point of view to the extent that he assumes language practices to be directed and guided by rules along the rigid rails of grammar which would operate at a distance independently of our use.[47] Later, when Wittgenstein realised that no rule by itself could determine any course of action, since any course of action might be considered as conforming to the rule,[48] and worked out his doctrine of language-games, it is still exclusively within our grammar and ordinary practices that the relations of expectation and fulfilment, order and execution, wish and satisfaction, measurement and result are defined and ascertained. All these relations depend on well defined criteria of what we ordinarily call "obeying a rule" and "going against a rule" in actual cases.[49] Consequently in the *Philosophische Untersuchungen* Wittgenstein comes to the conclusion that it is in language that expectation and fulfilment make contact (*in der Sprache berühren sich Erwartung und Erfüllung*.)[50]

Now, the inclination peculiar to the traditional philosophy, to resort to what Wittgenstein calls "the double", that is duplications and processes parallel to the written signs or spoken sounds in order to explain and justify our use of words, has arisen from an ill-placed need for foundation which in fact drives us to an infinite regress. That inclination turns out to be based on the presupposition of conferring a transitive meaning on expressions actually having an intransitive use. In this sense, the philosophical metaphors concerning the hollow shape and the solid shape in the relation of expectation and fulfilment, or the image of the word's falling into our mind and fitting a mould into our mind long prepared for it,[51] are bound to represent a class of expressions as transitive, which are in fact intransitive, by means of fictitious comparisons. While trying to set up these feigned comparisons, the philosopher does nothing but compare an expression with itself, bringing out a mere reflexive comparison, without contrasting it with another. The reason for the philosopher doing that depends on his reluctance to recognize that an expression is meaningful in itself and it does not contain its meaning outside itself.

Wittgenstein conceives of language now as a sort of grammatical space which can be divided but as such is not contiguous to anything else.[52] When we say that we are getting hold of the expression of this face or the meaning of a sentence, it seems as if we were getting hold of something which is in the face or in the sentence respectively but at the same time different from them. While ruling out the primitive conception of any fictitious comparison,[53] Wittgenstein restores the meaning and understanding of expressions in terms of an intransitive use. Instead of the model of a double or prototype of expressions as mould in our mind into which the pictures fall and fit it, Wittgenstein speaks of an intransitive use through which, as he writes, "we let the picture sink into our mind and makc a mould there".[54]

> If one had asked what the colour pattern of the pansy said, the right answer would have seemed to be that it said itself. Hence we could have used an intransitive form of expression, say "Each of these colour patterns impresses one".[55] And in letting the face impress itself on me and contemplating its 'particular impression', no two things of the multiplicity of a face are compared with each other; there is only *one* which is laden with emphasis. Absorbing its expression, I don't find a prototype of this expression in my mind; rather, I, as it were, cut a seal from the impression.[56]

Thus, instead of being the requisite of the meaning of an impression, the alleged mould would be the consequence brought about by the impression. This originally intransitive form of our expressions turns out to be strictly connected with Wittgenstein's frequent comparison of the proposition with a gesture or a musical phrase which are understood by themselves.[57]

In conclusion, in spite of all the changes it has undergone, Wittgenstein has remained coherent and faithful to his original, youthful conception which made him say that the proposition represents the state of affairs, as it were, off its own bat *(so stellt der Satz den Sachverhalt gleichsam auf eigene Faust dar)*.[58] But in the new version the circumstance that a proposition represents something off its own bat is connected with Wittgenstein's conception of intention as a sort of way of using language, which in its turn doesn't find expression in language, since "the expression of the intention can't contain the intention".[59] But it is connected as well with that constructive impulse which is asserted by Wittgenstein to bridge the gap holding between the sign and its application. An unavoidable gap which cannot be saved by any explanation, nor negotiated by any reason, since, as Wittgenstein said, "no reason compels us to learn language".[60]

Università di Pisa

NOTES

[1] B. Russell, *Theory of Knowledge*, MS in the Russell Archives at MacMaster Univesity, p. 185. See R. W. Clark, *The Life of B. Russell*, Weidenfels & Nicolson, London 1975, pp. 204ff; D. Pears, 'The Relation between Wittgenstein's Picture Theory of Propositions and Russell's Theories of Judgment', in *Wittgenstein: Sources and Perspectives*, ed. by G. Luckhardt, Harvester Press Hassocks, Sussex 1979, pp. 192–212.

[2] L. Wittgenstein, *Notebooks 1914–1916* (NB), ed. by G. H. von Wright & G. E. M. Anscombe. Trans. by G. E. M. Anscombe, B. Blackwell, Oxford 1961, p. 3; *Tractatus Logico-Philosophicus* (TLP), ed. by D. F. Pears & B. F. McGuinness, Routledge & Kegan Paul London 1961, prop. 5.4731, 5.552.

[3] NB, p. 31.

[4] TLP, prop. 5.525.

[5] NB, p. 26.

[6] TLP, prop. 2.1512, 2.15121, 4.211, 5.134, 5.135.

[7] See *Wittgenstein's Lectures. Cambridge 1930–32* (WL1930–32). From the Notes of J. King & D. Lee, ed. by D. Lee, B. Blackwell, Oxford 1980, p. 84.

[8] *Wittgenstein and the Vienna Circle* (WWK), ed. by B. F. McGuinness. Trans. by B. F. McGuinness & J. Schulte, B. Blackwell, Oxford 1979, pp. 63–64.

[9] See WWK, pp. 167–168.

[10] *Ibid.*, pp. 86, 88, 167; L. Wittgenstein, *Philosophische Bemerkungen. Philosophical Remarks* (PR), ed. by R. Rhees, B. Blackwell, Oxford 1964. Trans. by R. Hargreaves & R. White, B. Blackwell, Oxford 1975, pp. 69–70.

[11] PR, p. 63.

[12] WL1930–32, p. 40.

[13] B. Russell, *The Analysis of Mind*, G. Allen & Unwin, London 1961, pp. 75–76

[14] PR, pp. 64–65.

[15] *Ibid.*, pp. 63, 65, 66.

[16] See *Philosophische Grammatik. Philosophical Grammar* (PG), ed. by Rhees, B. Blackwell, Oxford 1969. Trans. by A. J. P. Kenny, B. Blackwell, Oxford 1974, p. 157.

[17] PR, p. 71.

[18] *Ibid.*, pp. 77, 200.

[19] *Ibid.*, p. 67.

[20] *Ibid.*, p. 70.

[21] WWK, pp. 167–168.

[22] PR, p. 67.

[23] *Ibid.*, p. 87.

[24] *Ibid.*, p. 71.

[25] *Ibid.*, pp. 70–71.

[26] *Ibid.*, p. 72.

[27] WL1930–32, p. 30

[28] *Ibid.*

[29] *Ibid.*

[30] *Ibid.*, p. 44.

[31] *Ibid.*

[32] *Ibid.*, pp. 29, 31, 33, 77, 82; L. Wittgenstein, *The Blue and Brown Books* (BBB), ed. and with a

Preface by R. Rhees, B. Blackwell, Oxford 1958, pp. 163–164.
[33] BBB, p. 170.
[34] *Ibid.*, pp. 162–171.
[35] F. Brentano, *Psychologie vom empirischen Standpunkt*, F. Meiner Verlag, Hamburg 1973, Erster Band, p. 131.
[36] BBB, p. 163.
[37] WL1930–32, p. 32.
[38] *Ibid.*, p. 33.
[39] *Ibid.*, p. 35.
[40] BBB, p. 135. See L. Wittgenstein, *Zettel*, ed. by G. E. M. Anscombe & G. H. von Wright, B. Blackwell 1967, p. 53.
[41] *Philosophische Untersuchungen. Philosophical Investigations* (PI), ed. by G. E. M. Anscombe & R. Rhees. Trans. by G. E. M. Anscombe, B. Blackwell, Oxford 1953, I, §§ 208, II, p. 225; *Bemerkungen über die Grundlagen der Mathematik. Remarks on the Foundations of Mathematics* (RFM), ed. by G. H. von Wright, R. Rhees & G. E. M. Anscombe. Trans. by G. E. M. Anscombe, B. Blackwell, Oxford 1956, pp. 115, 138–139, 184.
[42] WL1930–32, p. 106; see also pp. 4, 8, 10; PG, pp. 184, 244, 256.
[43] WL1930–32, pp. 43, 45–46; see also pp. 61–63, 87; PG, p. 88.
[44] BBB, pp. 184–185.
[45] WL1930–32, p. 62; see also WWK, pp. 178ff; PR, p. 178; PG, pp. 63, 67, 70, 273, 312.
[46] WL1930–32, p. 62; PG, pp. 60, 66, 94–95.
[47] PI I, § 219; RFM, pp. 34–35, 37–38, 167–168; G. Baker, 'Following Wittgenstein: Some Signposts for *Philosophical Investigations* §§ 143–242', in *Wittgenstein: to Follow a Rule*, ed. S. H. Holtzman & C. M. Leich, Routledge & Kegan Paul, London 1981, p. 64.
[48] BBB, pp. 123ff; PI I, § 201, RFM, p. 184.
[49] *Wittgenstein Lectures. Cambridge 1932–35*. From the Notes of A. Ambrose & M. Macdonald, ed. by A. Ambrose, B. Blackwell, Oxford 1979, p. 84; PG, pp. 184, 244, 272–274; PI I, §§ 199, 202; *Bemerkungen über die Farben. Remarks on Colour*, ed. by G. E. M. Anscombe. Trans. by L. L. McAlister & M. Schättle, B. Blackwell, Oxford 1977, p. 29; see G. Baker, op. cit., p. 63.
[50] PI I, § 445; see also PG, p. 157.
[51] PG, p. 170.
[52] WL1930–32, p. 112; BBB, pp. 158ff, 178.
[53] BBB, p. 165.
[54] *Ibid.*, p. 163; see also pp. 160, 169, 170, 174–178, 180.
[55] *Ibid.*, p. 178.
[56] *Ibid.*, p. 165.
[57] *Ibid.*, pp. 184–185, 177–178; L. Wittgenstein, *Vermischte Bemerkungen. Culture and Value*, ed. by G. H. von Wright in collaboration with H. Nyman. Trans. by P. Winch, B. Blackwell 1980, p. 69; L. Wittgenstein, *Lectures and Conversations on Aesthetics, Psychology and Religious Belief*, ed. by C. Barrett, B. Blackwell, Oxford 1966, pp. 37–38.
[58] NB, p. 26.
[59] WL1930–32, p. 112.
[60] *Ibid.*, p. 67.

HERBERT HOCHBERG

INTENTIONAL REFERENCE AS A LOGICAL RELATION: A VARIATION ON A THEME IN MOORE, RUSSELL, WITTGENSTEIN, AND BERGMANN

In his 1910–11 lectures that became *Some Main Problems of Philosophy*, Moore made critical and explicit use of an intentional relation – *referring* – which he took to hold between beliefs and facts:

> Obviously this expression 'referring to' stands for some relation which each true belief has to one fact and to one only; and which each false belief has to no fact at all; and the difficulty was to define this relation. . . . and we may know both that there is such a relation, and that this relation is essential to the definition of truth. And what I want to point out is that we do in this sense know this relation. . . . Take any belief you like; it is, I think, quite plain that there is just one fact, and only one, . . . which would have no being – would simply not be, if the belief were false. And as soon as we know what the belief is, we know just as well and as certainly what the fact is which in this sense corresponds with it.[1]

The last sentence suggests that Moore held that the representational connection between "the belief" and the corresponding fact, which would "have no being" if the belief were false, is not a contingent relation that obtains or not as a matter of fact, but is a necessary connection.

Moore's use of the term 'belief', rather than 'proposition', gives his discussion a misleading air, since Moore uses 'belief' in two senses. In one sense, a belief is a particular mental state of a person, in the sense in which we speak of having a perception or a particular experience prior to another. Thus, it would make sense to speak of Othello having a belief at a particular time, just as it would make sense to speak of Othello perceiving something at a certain time.[2] In another sense, a belief is what a sentence may be said to express, as Frege speaks of a sentence expressing a thought or some speak of various sentences expressing the same proposition. While Moore explicitly denies that there are propositions, he implicitly thinks of beliefs, in this second sense, as types or characterizations of beliefs, in the first sense. In place of propositional entities, he thus implicitly appeals to universal attributes of particular mental states. Two people may be said to have the same belief at a certain time, if each of them has a particular mental state, where both mental states are instances of the same belief type, that is to say, the same universal attribute.[3]

In his way, Moore replaced propositional entities with universal attributes,

R. Egidi (ed.), Wittgenstein: Mind and Language, 137–170.
© 1995 *Kluwer Academic Publishers. Printed in the Netherlands.*

which may suitably be called "propositional attributes". He also implicitly acknowledged non-existent facts. For his referring relation connects a propositional attribute (belief) to a fact, irrespective of whether the propositional attribute is true or false.[4] Moore's view, construing propositions as attributes, implicitly acknowledging non-existent facts, and explicitly adopting a basic intentional relation, is interesting in yet another way. He may be said to take the reference relation between a propositional attribute and its "intended" fact to be logical. We noted above that he takes the connection to be necessary when he holds that as soon as we know what the belief is we know what fact it refers to. But, he characterizes the formal nature of the connection a bit more explicitly when he holds:

> So that we might say: The fact to which a belief refers is always the fact which has the same name as that which we have to use in naming the belief. . . . We can, therefore, only name beliefs by means of these expressions beginning with 'that' – 'that lions exist', 'that bears exist' and so on, or the equivalent verbal nouns. But, curiously enough, if we want to name the fact to which a belief refers – the fact which is, if the belief be true, and is not if it be false – we can only do it by means of exactly the same expressions.[5]

Thus, when we have a true belief, and hence a fact, we take a sentence to represent both the belief and the fact, and, consequently, for any true sentence 'S', the belief that-S refers to the fact that-S. But, Moore's view is not confined to true beliefs, for as he, perhaps inadvertently, says in the initial quotation: "Take any belief you like; it is . . . quite plain that there is just one fact, and only one, . . . which would have no being – would simply not be, if the belief were false."

Introducing signs formed by placing angle braces around sentences to represent propositional attributes, while taking such sentences to represent facts, Moore's analysis is then expressed by the pattern:

(1) <. . . .> refers to . . . .,

where the result of replacing both occurrences of the dots in (1) by the same sentence is taken to be true in virtue of its form. If we limit the language schema to avoid multiple references to objects and properties and, for the time, limit the discussion to atomic sentences, we can say that any instance of the pattern (1) is a formal truth when and only when the same sentence replaces the two occurrences of the dots. We may then think of (1) as something like an axiom schema, a pattern for logical or linguistic truths of the schema, and, hence, take the relation represented in (1) as a logical, linguistic, or formal relation. It may be called logical since the instances of (1) are taken to be necessary truths that are true as a matter of form. The sense in which (1), and *refers to*, may be said to be formal is clear, but the sense in which instances of (1) are necessary is not, for they involve a necessary connection between facts and propositional attributes. Since an atomic sentence occurs as a term of the relational predicate ‘refers to’, whether it is true or false, Moore is forced to recognize facts that have “no being” as terms of such a connection. Moore thus writes as if he specifies truth grounds by a pattern like:

(1*) Tr $<aR^{1}b>$ iff ($<aR^{1}b>$ refers to $aR^{1}b$ & $aR^{1}b$ exists).

(1) and (1*) embody the four aspects of Moore’s analysis; his recognition, first, of propositional attributes, second, of states of affairs which may obtain or not, third, of a basic reference relation holding between a propositional attribute and a state of affairs, and, finally, of the logical or formal nature of such a referential connection.

Russell sought to avoid reference not only to non-existent objects but to non-existent facts in his 1905 paper “On Denoting”. In that paper he suggested using a definite description of a fact and, in place of something like (1*), employed a pattern along the lines of:

(2) Tr $<aR^{1}b>$ iff the complex $aR^{1}b$ exists.

In (2), the sign $<aR^{1}b>$ can be taken to represent a proposition, a type of entity Russell still recognized in 1905. However, Russell’s analysis is clearly problematic on other grounds, since he uses an atomic sentential form in the description in (2). That is, Russell does not avoid using a variant of an atomic sentence in specifying the truth condition for the atomic sentence in question, and, hence, he does not avoid questions about another problematic entity, a non-existent or possible but not actual fact (complex), as he does avoid a corresponding question about the problematic present King of France.[6]

In the immediately succeeding years, Russell rejected propositional entities and proposed his relational analysis of intentional contexts. Construing a context like ‘Othello believes that a has F’ in terms of a three-term belief relation holding of Othello, the property F, and the object a, and expressed by a sentence like ‘B(O,F,a)’, he spoke of there being a belief-complex and took Othello’s belief to be true if there was the further complex or fact, a’s being F. He now took the correlation between the belief complex and the fact that was its truth ground to be established by the occurrence of all the constituents of the latter complex in the former. This manner of connecting complexes resembles Moore’s in that the fact represented by ‘Fa’ is the truth ground for Othello’s belief, where ‘B(O,F,a)’ represents Othello’s having such a belief. The two sentences, ‘Fa’ and ‘B(O,F,a)’, display the connection between such facts, just as a pattern like (1) displays the connection between a propositional attribute and a fact.

We can make the connection even more transparent by "abstracting" from 'B(O,F,a)' to obtain the propositional function sign 'Σ(β,F,a)' with 'Σ' and 'β' as variables ranging over attitudinal verbs and persons, respectively. In effect, in his 1913 manuscript Russell took such a function to replace the classical *proposition* expressed by 'Fa'. Given the obvious connection between the signs 'Σ(β,F,a)' and 'Fa', we can say that Russell, like Moore, takes the connection between such a *propositional* function and the complex or fact that grounds the truth of *the belief* to be logical. Since such a function was not a constituent of a belief fact, in versions of his relational analysis prior to 1913, Russell took himself to avoid acknowledging a propositional entity. It is easy to see why such a function is not a constituent of a belief fact. The function represented by 'Σ(β,F,a)' can take B and O for arguments, yielding the belief complex represented by 'B(O,F,a)'. On analysis, such a complex has the relation B, the property F, the objects a and O and the logical form Σ(β,ø,x) as constituents. The function Σ(β,F,a) is not a constituent of the belief complex. Hence, Russell will speak of propositions being "incomplete symbols", as he will speak of class abstracts (and classes) and definite descriptions (and the present King of France) in such terms.

Russell not only sought to avoid propositional entities as well as non-existent or merely possible facts. He also sought to avoid a basic relation, such as Moore's reference relation, that connects the complex that is said to be true (or false) to the complex or fact that is the ground of truth. To accomplish this he held the two complexes to be linked by their having common constituents, appropriately ordered. Such an appeal to order and to the constituents of facts raises several questions for Russell's analysis.

In the 1913 manuscript *Theory of Knowledge*, Russell fundamentally altered his relational theory by returning to a crucial theme of 1905. Prior to 1913, in *The Problems of Philosophy*, for example, he had taken the order in a belief fact to determine the order in its ground of truth:

> The relation 'loving', as it occurs in the act of believing, is one of the objects – it is a brick in the structure, not the cement. The cement is the relation 'believing'. When the belief is true, there is another complex unity, in which the relation which was one of the objects of the belief relates the other objects. Thus, e.g., if Othello believes truly that Desdemona loves Cassio, then there is a complex unity, 'Desdemona's love for Cassio', which is composed exclusively of the objects of the belief, in the same order as they had in the belief, with the relation which was one of the objects occurring now as the cement that binds together the other objects of the belief. On the other hand, when a belief is false, there is no such complex unity. . .[7]

It is clear that Russell depends on the order of the constituents in the belief or judgment fact to provide the unity for the content of the belief, and hence to determine which fact is its truth maker. Recognizing that he was appealing to the order of the constituents in facts, and not just to the constituents, to correlate judgment facts with their truth grounds, he came to believe that he must account for that order. This led to an important modification of his multiple relation theory in 1913. Convinced that he must give an account of relational order in facts, he took a non-symmetrical relation, R^1, to determine (and be determined by) two relations, R* and R**, that held between particulars and complexes (facts), and which specified the *position* of a particular in the complex. The truth condition for 'aR^1b' could then be expressed by:

(3) Tr'aR^1b' iff (℩p)(aR*p & bR**p) exists,

and Othello's belief that a stood in R^1 to b by:

(3*) B(O, (℩p)(aR*p & bR**p) exists).

But, with (3*) a familiar problem reappears. What does the belief relation hold between when the fact does not exist? While (3) offers a solution to one of the problems posed by false belief, the linking of atomic sentences to what they are about, without acknowledging non-existent entities like possible facts, situations, etc., (3*) does not provide an analysis of intentional contexts, like 'Othello believes that aR^1b', that avoids such entities.

In the *Tractatus*, Wittgenstein pronounced:

> At first sight it looks as if it were also possible for one proposition to occur in another in a different way. Particularly with certain forms of proposition in psychology, such as '*A* believes that *p* is the case' and '*A* has the thought *p*', etc.
>
> For if these are considered superficially, it looks as if the proposition *p* stood in some kind of relation to an object *A*.
>
> (And in modern theory of knowledge (Russell, Moore, etc.) these propositions have actually been construed in this way.) (5.541)
>
> It is clear, however, that 'A believes that p', 'A has the thought p', and 'A says p' are of the form '"p" says p': and this does not involve a correlation of a fact with an object, but rather the correlation of facts by means of the correlation of their objects. (5.542)

These remarks seem to be directed not so much at Russell's relational analysis or at any details of Moore's view, but at the taking of a person to be a constituent of a belief fact. Thus, Wittgenstein proceeds to remark:

> This shows too that there is no such thing as the soul – the subject, etc. – as it is conceived in the superficial psychology of the present day. Indeed a composite soul would no longer be a soul. (5.5421)

In 5.542 Wittgenstein holds that there are occurrences, thought elements or "constituents", as he writes in his notebooks, that represent the objects a and b *and* a relation between them that represents the relation R^1.[8] The occurrence of a thought-complex, an arrangement of thought elements in a relation, represents the situation that a stands in R^1 to b. (Whether a person is construed in some sort of Humean fashion, which the last sentence of 5.5421 suggests, is not relevant here.) Thus, Wittgenstein suggests something similar to Moore's pattern, as expressed in (1), in basing his discussion of intentional contexts upon his account of the representational role of atomic sentences and his correspondence account of truth, an account quite similiar to Moore's. Wittgenstein takes atomic sentences to represent situations, irrespective of whether such situations exist or obtain. Numerous passages in the *Tractatus* point to such a view, and it is such a view that Russell rejected in 1905 and 1913.

There is another aspect to Wittgenstein's view. Like Moore and Russell, Wittgenstein takes the connection between a proposition and a situation to be logical. This is one theme involved in 5.542 and in his holding that the internal or essential properties of things determine their possibilities of combination. It is also involved in his notion of shared logical form, which involves more than the idea that what represents and what is represented must share the requisite multiplicity. In this connection it is relevant to note a comment of Wittgenstein's in his *Philosophical Grammar* that virtually reproduces Moore's way of expressing matters:

> Certainly, the proposition "p" determines that p must be the case in order to make it true; and that means: (the proposition p) = (the proposition that the fact p makes true). And the statement that the wish for it to be the case that p is satisfied by the event p, merely enunciates a rule for signs. . .[9]

Wittgenstein's comment about a "rule for signs" can be seen to take a pattern like (1) to be true by linguistic rule and not as a matter of fact.

Russell was greatly influenced by 5.542, perhaps more so by this passage than by Wittgenstein's celebrated criticism of his relational theory that occurs in 5.5422, in the well-known letter to Russell, and in the often quoted passage from the notebooks.[10] For Russell continued to discuss the relational theory until he himself raised questions about it in 1918, and abandoned it in 1919 in "On Propositions", (although there are indications in his papers that he already rejected it in 1918).[11] By contrast, Russell kept returning to Wittgenstein's remark in 5.542 from the time of his discussion of it in his introduction to the *Tractatus* through his lengthy treatment of it in *An Inquiry into Meaning and Truth*; and 5.542 was the basis for the view that Russell developed in an appendix to the 2nd edition of *Principia*.[12]

In *Principia*, Russell took there to be a mental occurrence or token, a*, that "meant" or represented the object a, another, F*, that represented the property F, and a complex mental occurrence that consisted of a* and F* in a relation or arrangement. This complex belonged to both the class or series that constituted a person, on his then current view of a self, and the class that could be said to be the belief that a is F. Russell's analysis was intended to be a variant of the view in the *Tractatus*, as that view was construed above. Like Wittgenstein, Russell believed that the complex occurrence (or sentence) is correlated to a fact in virtue of its constituents being correlated with constituents of the latter, which is the point of the phrase "the correlation of facts by means of the correlation of their objects." Yet, Wittgenstein, unlike Russell, takes a propositional complex to represent, irrespective of whether it is true. Thus, such a complex is correlated to a situation or possible state of affairs, though Wittgenstein construed such possibilities in terms of constituent entities having *internal* properties. In *Principia*, Russell avoided the problem of analyzing "'Fa' is true" and 'Othello believes that a is F' without introducing possible facts.

Aside from the problems posed by explicitly taking atomic sentences to represent possibilities, it is not clear how the correlation of complexes is based on the correlation of constituents. Neither Russell nor Wittgenstein recognizes traditional propositional entities or Moore's propositional attributes. Yet, all belief complexes that have the same content – that are beliefs that a is F – have something in common. The problem is to say what that is without acknowledging propositional entities, whether in the form of Moore's attributes or Frege's thoughts. To avoid such entities Russell will eventually appeal to causal origins and behavioral dispositions. Wittgenstein, in the *Tractatus*, appealed to internal properties of the sentential or thought elements, and the correlated essential properties of *objects*, to carry the burden. The possibilities of combination of the elements in the representing combination are taken to represent the possible states of affairs, which are determined by the natures or internal properties of their constituents. Wittgenstein undoubtedly thought that one avoided shadowy entities, such as possible facts, by taking the internal or essential properties of constituents to determine the possibilities of combination. Thus, the object a could be said to have as an essential or internal property, the property of possibly being F, rather than to be a constituent of a possible fact. But, whether or not such a move avoids possible facts and provides a viable way of construing possibilities, a question remains as to just how a particular propositional fact correlates with a possibility.

The correlation of the respective constituents will not suffice to correlate the complexes of which they are constituents, unless one takes it to be internal to being a token of an element-type, an a* for example, not only that it can combine with a token of another element type, F*, in a certain arrangement, but that the resulting arrangement represents a being F. Yet, to take matters in such a way is not to correlate complexes by correlating their constituents, it is to correlate complexes directly, though the correlation of complexes is dependent on the correlation of constituents. To be dependent on, however, is not to be implicit in or to be implied by. What Wittgenstein and Russell do is to interpret an arrangement of elements, token or type, an F* combined with an a*, as stating that a is F. And, in so doing, one substitutes such a complex token or type, interpreted as it is, for Moore's propositional attribute. An occurrence of a token of such a complex type, a specific occurrence of an arrangement of an a* and an F*, represents a's being F precisely because an arrangement, whether the token or its type, is taken *as a complex* to represent a's being F. One can perhaps avoid Moore's propositional attributes by appealing to interpreted complex tokens or types, but one does not avoid linking the representing complex, as a whole, in addition to linking its parts. Hence, one does not avoid possible facts by declaring that the constituents provide the link between complexes.

In a series of articles beginning in the 1940s, Gustav Bergmann developed an analysis of intentionality similar to Moore's, while explicitly stressing the logical nature of an intentional relation and the need for propositional attributes.[13] Bergmann took mental states to exemplify such attributes, which are *logically* connected to possible facts. Expressions like '<Fa>' were construed as predicates satisfied by particulars, mental states, and his version of (1), (1b) '<. . .> M . . .', was declared to be an analytic truth, with 'M' as a relational predicate representing the intentional relation *means that*. His claiming that such propositional attributes were simple properties, while being represented by complex signs, posed an obvious problem. Yet, Bergmann was forced to do this in order to support the claim that M-sentences, instances of (1b), were true or false in virtue of their form. Replacing such complex predicates by primitive predicate symbols, as he later did, rendered specious his claim that M-sentences were formal truths. Thus, like Moore, he was forced to accept a basic, necessary connection between propositional attributes and possible facts. But, more explicitly than Moore, Russell, or Wittgenstein, he still insisted on the connection being logical.[14] One motive for his taking it to be logical was derived from Wittgenstein: logical relations were merely pseudo-relations and not *entities* that need be acknowledged in an ontological assay of intentional contexts. The sign 'M' did not represent any*thing* in Bergmann's world.

The various views we have touched on suggest, but are not solutions to the problem of specifying truth conditions for atomic sentences and belief contexts involving them. A satisfactory solution will avoid a commitment to possible facts, in any form, while specifying what mental states with the same content have in common, without appealing to unanalyzable attributes of such states, to propositions, or to mysterious internal properties. To get at such a solution, let us assume that atomic facts are the truth grounds for atomic sentences and that such facts consist, in the monadic case, of a property, a particular, and the logical form of monadic predication, a form Russell sometimes represented by the sign 'Øx'. A fact will be said to *contain* its constituents, and such a whole-part relation will be represented by the predicate 'C'. The truth condition for 'Fa' is then given by:

(4) Tr<Fa> iff (∃p)(pCa & pCF & pCØx).

In (4), the angle braces are retained, but we can now simply read them as metalinguistic quotes. As the signs 'a' and 'F' are taken to represent the object a and the property F, respectively, this can be expressed by instances of

(5) "'. . .' represents . . .",

where the dots may be replaced either by proper names or primitive predicates. When and only when tokens of the same type replace the two occurrences of the dots, we have a formally true pattern, a pattern true by the interpretation rules for the schema. Thus we have a formal expression of the referential role of names and basic predicates. Atomic sentences are another matter. Following Russell's early view, the specification of truth conditions, in patterns like (4), replaces such a role for atomic sentences. Thus, whereas basic predicates and names may be said *to represent* in a basic and fundamental way, atomic sentences represent *only* in the sense that their constituent terms refer to entities and in that truth conditions are specified for them by patterns like (4). No primitive representational relation, like Bergmann's M or Moore's *referring*, is involved in the sentential case. This eliminates Wittgenstein's situations, Moore's non-existent facts, Bergmann's potential facts, and Russell's possibilities, that may be positive or negative facts, of 1918–1919 (a view that is suggested in terms of "neutral facts" in an appendix to the 1913 manuscript).[15]

We now see in what clear and unproblematic sense representational connections are logically necessary: "'a' represents a" and "'F' represents F" are instances of a pattern reflecting interpretation rules for the schema, and (4) is an instance of the definitional pattern for a truth predicate for atomic sentences. The use of (4) might suggest modifying Bergmann's analysis by using either – (i) <Fa> M (℩p)(pCa & pCF & pCØx) – or – (ii) <Fa> M (℩p)(pCa & pCF & pCØx) exists. But, (i) will be false if 'Fa' is false, while (ii), employing a sentence as a term for the predicate 'M', raises a familiar question about its truth condition: What does M relate when 'Fa' is false? Thus, such a modification would face the same problem we noted in connection with (3) and (3*) on Russell's 1913 analysis. However, while neither (i) nor (ii) will do as unproblematic replacements for Bergmann's "'Fa' M Fa', Russell's analysis of descriptions would allow

(iii) <Fa> M (℩p)(pCa & pCF & pCØx) iff (∃p)(pCa & pCF & pCØx)

to be unproblematically taken as a formal truth, with the occurrence of the description having secondary scope. This recaptures an idea of Bergmann's regarding the logical nature of intentional reference, but not as he understood it, for in (iii) 'M' can be said to express a representational relation between an atomic sentence and an atomic fact only if there is such a fact. Bergmann, when he later acknowledged logical relations, was forced to accept possible facts as terms of his M relation–entities (iii) avoids.

Our first problem is resolved. In keeping with our earlier reading of the *Tractatus* and Russell's *Principia* adaptation of it, we can proceed to the second. Consider a mental state of belief that is a state of Othello's. We need not worry about the details of its being a state of belief, rather than of understanding or doubt, or of it being a state of Othello rather than of Iago. Our concern is with the *content* or *intention* of the state. We must state what it is for the intention of a thought being that a is F. In one sense of content or intention, we can do so quite simply. To say that a mental state or thought has a content expressed by 'Fa' is to say that it consists of constituent elements, representing F and a, in a relation such that the resulting complex or fact is true iff the fact containing a, F, and Øx exists. As Bergmann explicitly, and Moore implicitly, took propositional attributes to be true or false, and Russell and Wittgenstein, in their ways, took propositional facts to be true or false, we here speak of complexes or facts having truth values. More important, however, is that instead of recognizing possibilities, whether implicitly, as internal or essential properties of objects, or explicitly, by taking the occurrence of a propositional complex to refer to a possible fact, the present analysis takes such a complex to represent something only in the sense that the complex is taken to be true under certain conditions. While we thus recognize the need to interpret a complex, in addition to interpreting its constituents, we avoid propositions and mysterious propositional attributes, as well as situations or possible facts. To have an interpreted atomic complex, a thought content, is to have its condition of truth specified *in addition* to having its constituents *interpreted*. However, to handle the familiar puzzles posed by quantifying and substituting into intentional contexts, we shall have to acknowledge further senses of "has the content expressed by 'Fa'".[16]

In his *Philosophical Grammar*, Wittgenstein pointed to a crucial feature of the "interpretation" of thought contents:

> If a thought is observed there can be no further question of an understanding; for if the thought is seen it must be recognized as a thought with a certain content: it doesn't need to be interpreted! – That really is how it is; when we are thinking, there isn't any interpretation going on.[17]

Wittgenstein's point is correct in that there is not an interpretation in one sense, namely, an accompanying thought doing the interpreting. But this is not to say that the complex content is not interpreted. In short, the occurrence of the complex as an act of thought is an occurrence as an interpreted complex. In acts of thought one can refer to something and take a complex content to have a truth condition without also thinking that one is doing so.

Relational contents are somewhat more complex, for we need to account for the order in relational facts that involve non-symmetrical relations. There are alternative ways of doing this. One way is to introduce two relations, I and S, that, like Russell's R* and R**, obtain between elements of a fact or complex and the complex. An element stands in I to a relational complex when it is the first term of the relation, in S when it is the second. I and S are, in a sense, generalizations of Russell's R* and R**, since they indicate the place of the term in the fact without specifying the constituent relation.

To specify the truth condition for a dyadic relational sentence, where R^1 is a non-symmetrical relation, we can have:

$$(6) \qquad \mathrm{Tr} <aR^1b> \text{ iff } (\exists p)(pCa \;\&\; pCb \;\&\; pCR^1 \;\&\; pCR(x,y) \;\&\; aIp \;\&\; bSp),$$

with 'R(x,y)' representing the form of dyadic relational exemplification. A corresponding complexity enters into the case of beliefs with relational contents. (6) is based on themes from Russell's 1913 manuscript, the latter parts of which he unfortunately abandoned, supposedly in the face of Wittgenstein's criticism. Aside from Wittgenstein's objection to Russell's and Moore's earlier views regarding a self, his often cited criticism, stated in the *Tractatus*, the letter to Russell, and the *Notebooks*, focuses on the early form of the relational theory that Russell preserved for some contexts in the 1913 manuscript. The point, simply put, is that patterns like 'B(Othello,Øx,F,a)' presuppose that the subject terms 'a' and 'F' are understood to be combinable into a meaningful sentence, and the mere addition of a sign for a form does not, as a further subject term, serve to combine 'a' and 'F'. This being so, Russell's analysis is forced to implicitly take 'Fa' to be a meaningful sentential expression. Hence, Russell either implicitly employs a propositional sign or cannot viably rule out nonsensical beliefs. He is thus forced to abandon his relational theory.

Such a line of criticism is not relevant to the way the non-symmetrical relational contexts are treated on the modified theory, employing descriptions with variables ranging over complexes. But, as we noted in the discussion of (3) and (3*) above, there is another line of criticism that can be directed against the modified relational account of 1913. To consider the relevance of such problems to Wittgenstein's attacks on Russell, and Pears' recent reiteration of those attacks, we must look more closely at Russell's 1913 view. In the 1913 manuscript Russell took propositions to be "forms":

> . . . the total complex which occurs when the subject has the relation U to the objects in question may be symbolized by
>
> U(S, x, R, y, γ)
>
> If we now proceed to the "form" which results from varying U and S, i.e. to
>
> "there is a U and an S such that U(S, x, R, y, γ)"
>
> we arrive at something which is the same for all subjects and for all propositional relations which we should regard as concerned with the same proposition. Thus there is no formal obstacle to defining this as the proposition.[18]

The form, which he "defines" as the proposition, is represented by an existentially quantified sentence (the quantifiers being expressed in English) with (bound) variables replacing the attitudinal verb and the term for the person having the attitude. Taking forms, and hence propositions, to be existentially general facts posed a problem Russell recognized:

> The above definition has some merits and some demerits. Its chief merit is that it provides propositions, both true and false, as fast as we can think of them . . . Its chief demerit is that we cannot be sure that there are propositions in all cases in which logic would seem to need them. It is not necessary to our definition that there should actually be a subject which has one of the familiar mental relations to the objects, but it is necessary that there should be some term and some relation by which a complex results having the requisite form and containing the objects in question. It may be possible to prove that there always are such complexes, but I do not at present see how such a proof could be found. In its absence, we cannot know of the existence of propositions other than those that have been actually thought of. . . . we must admit, I think, that the objection in question is serious. I cannot, however,

> think of anything better calculated to fulfil the purposes for which we want propositions, and I therefore propose to continue to use the word in the above sense.[19]

It is strange that Russell did not explicitly consider propositional functions here, since they would not only fulfill his purposes but, as we noted above, they would allow him to avoid taking propositions to be basic entities. This is perhaps due to Russell's being ambivalent about propositional entities in the 1913 manuscript. Early in the manuscript, assuming that the first six chapters belong to the manuscript, he appears to reject propositions as he did in *The Problems of Philosophy* and earlier papers. And, at the beginning of Chapter I of Part II of the manuscript, he continues in the same vein:

> And this in turn, since it is repugnant to admit the reality of false propositions, forces us to seek a theory which shall regard true and false propositions as alike unreal, i.e. as incomplete symbols.[20]

This rejection was a clear motive for his relational analysis of judgment in writings prior to 1913. But, later in the manuscript, he is concerned to preserve propositional entities as "subsistents" and identifies them with existentially general forms or facts. Yet, there is something odd about his taking existential generalizations, containing constant terms, to signify *forms*, though he does not speak of *logical forms* in this connection. But, though he does not take such forms to be logical forms, his taking them to be existential facts or *truths* obviously stems from his taking logical forms to be existentially general facts – something has some property, something stands in some relation to something, and so on. Just as Russell had earlier attempted to avoid Fregean propositional entities and Meinongian contents, he also seeks to avoid a basic intentional relation connecting a belief fact, Othello's belief that a is F, with the fact that grounds the truth of the belief, a's being F. The link between the two facts is now furnished by their having F and a as common constituents and by the logical form, Øy, of the one fact, a's being F, being a constituent of the other, the belief fact.

In focusing on order in relational facts, Russell took the difference between monadic facts and relational facts, in virtue of which order was involved in the second, to be based on their different logical forms and different *types* of constituents.[21] Monadic facts were of the form Øy, while dyadic facts were of the form R(y, z). Facts of the first form did not involve order because they did not contain constituents of the same logical kind. Order *could be* involved in a fact of the second form since such a fact contained two constituents of the same logical kind. If, in addition, it contained a non-symmetrical relation, the terms of the relation must be ordered. This led to his modifying the relational theory by taking intentional (belief, judgment, etc.) facts to involve logical forms as *constituents*, and thus to be more perspicuously represented by sentences like 'B(O,Øy,F,a)' and 'B(O,R(y,z), R^1, a, b)'. The form Øy, as a constituent of an intentional fact with a monadic content, was then taken to unify the constituent object and property, a and F, without appealing to a propositional entity. But such a modification of the relational theory of judgment did not provide for an analysis of the order involved in the case of relational contents. This led to Russell's second modification of his theory in 1913, the introduction of relations like R* and R** and descriptions like '(℩p) (aR*p & bR**p)'. However, this second, major modification of his theory is not involved in Wittgenstein's critique of it. Wittgenstein's attack was directed at only the first change: the use of forms like Øy, R(y,z), etc. to purportedly provide the unity of content for a judgment fact.

Wittgenstein's criticism can be understood as two-fold. First, by claiming that a viable theory must prevent the judgment of nonsense, he is pointing out that in a pattern like 'B(O,Øy,Ø,y)' the signs that replace the variables 'Ø' and 'y' must be such that, taken in the appropriate order, they form a meaningful sentence. Thus, 'F' and 'a' are permissable replacements since 'Fa' is a sentence. But, then, Russell's analysis is purportedly misleading, for it seems that it merely employs a problematic rewriting of 'B(O,Fa)', which is the pattern of the propositional analysis. Second, to transcribe 'B(O,Øy,F, a)' as 'Othello believes that a is F' presupposes that we have interpreted the sentential pattern 'Fa', and not just the terms 'F' and 'a', along with 'Øy'. Thus, assume we have interpreted 'a' and 'F' as standing for an object and a property respectively. We must still appeal to a further rule that interprets their juxtaposition as a sentence. Thus we interpret 'Fa' as ascribing the property that 'F' represents to the object that 'a' represents. This is simply a variant of the point that a sentence is not a class of its constituent terms. By adding the form of monadic predication, Øy, to the members of the class {F, a}, to form the class {F,a,Øy}, we do not form a complex consisting of F and a standing in the form Øy, just as adding the sign 'Øy' to the class of signs 'F' and 'a' does not form a sentence. To have an atomic sentence we must take the signs to form a complex linguistic pattern that requires an interpretation rule in addition to the rules for the constituents of the complex. Russell presupposes that the coordination of the constituents serves to coordinate the complexes, but the coordination of complexes is not accomplished by the coordination of their constituents, though the former may be dependent on the latter.

We can consider the matter simply in terms of 'Fa' and 'R^1(a,b)'. Assume we have interpreted the names and predicates into objects, a monadic property, and a non-symmetrical relation. We then have, reflecting such interpretation rules: 'a' refers to a, 'F' refers to F, 'R^1' refers to R^1, and so on. Such statements can be taken to be linguistic or semantical rules of our schema. We may then add: A subject predicate sentence *states that* the object referred to by the subject term has the property referred to by the predicate term; a two-term relational sentence states that . . . ., etc. These embody the interpretation of the atomic sentences of the schema, as 'refers to' is used to interpret the subject and predicate signs of the schema. We then face two questions. Is *stating that* a relation atomic sentences stand in, as *referring* is a relation that names and predicates stand in? And, if so, what are the non-linguistic terms of the relation? The obvious candidate for the other term of such a relation is a fact, as a truth ground or truth maker for atomic sentences. But, since there can be no such ground when the sentence is false, to what is a false atomic sentence related – a possibility, a situation, a non-existent fact? The question, and its various answers, are familiar. Such a question about false atomic sentences can also be raised about purported propositional entities or mental contents that such sentences supposedly express. In early versions of his relational analysis of intentional contexts, Russell sought to avoid not only propositional entities and content properties but a basic *intentional* or *reference* relation for sentences and thoughts, along with non-existent or possible facts as purported terms of such a relation. While the 1913 analysis, at places, ambivalently acknowledges propositions as "subsistent" forms or existential facts, he still seeks to avoid a basic intentional relation, between beliefs (or propositions) and their truth grounds, as well as the possible facts that such a relation seems to involve. But his analysis faces Wittgenstein's two-fold line of criticism.

Consider the point that objects of a certain kind must be taken to be the terms of intentional relations like B. In one sense there is no difference between contexts like 'B(x,*f*, Ø, y)' (with '*f*' as a variable for logical forms) and contexts like 'Fy'. In the latter case, as in the first case, only a certain kind of sign can replace a certain kind of variable. In the first case, however, there is an additional feature. The ascription of the predicate in the "Ø-place" to the subject in the "y-place" in accordance with the form in the "*f*-place" is built into the reading of 'B'. Russell seems to have thought that this was unproblematic because of the logical differences between the form Øy, the property F, and the object a; differences that permitted only one type of combination. But the point Wittgenstein makes is that this presupposes the recognition of a *possibility* of combination that is not explicitly recognized on Russell's analysis. This raises a question about the significance of implicitly recognizing such a form of combination. Thus, in the *Tractatus* he holds that objects and signs have essential or internal or logical properties that determine their possibilities of combination.

Yet, such a move is compatible with Russell's analysis, since such logical differences are precisely what determine the role an entity can play in the fact expressed by 'B(O,Øy,F,a)'. Moreover, in addition to such specific belief facts Russell has other entities that can represent the complex composed of a, F and Øy: the propositional function Σ(x,Øy,F,a) and the form or existential fact (∃Σ)(∃x)Σ(x,Øy,F,a), identified with the proposition. On the earlier multiple relation theory Russell sought to avoid propositional entities and took belief complexes to be truths or falsehoods, and a belief *fact*, like the fact consisting of Othello, F, and a in the relation B, to represent the fact of a's being F. The analysis of 1913 focused on forms or existential facts like (∃Σ)(∃x)Σ(x,Øy,F,a), as propositions, and hence as the subjects of ascriptions of truth and falsity. Thus, like earlier versions of the theory, some facts are true, others false, on the basis of the existence or non-existence of other facts. As was mentioned earlier, it is not clear why Russell did not identify propositions with propositional functions. One can deny that such functions are basic entities in that they are not ultimate constituents of the judgment facts that are their values: they disappear on the analysis of such facts. But, while one might make such a claim for propositions construed as propositional functions, one cannot make such a claim for Russell's existentially general forms or facts.

In the case of a logical form, like monadic exemplification, whether construed as (∃Ø)(∃y)Øy or simply as Øy, we clearly have an entity that does not disappear on the analysis of the facts of which it is the form, such as the fact that a is F:

> If we take some particular dual complex xRy, this has three constituents, x, R, and y. If we now consider "something has the relation R to y", we get a fact which no longer contains x, and has not substituted any other entity for x, since "something" is nothing. Thus our new fact contains only R and y. For similar reasons, "something has the relation R to something" contains no constituent except R; and "something has some relation to something" contains no constituent at all. It is therefore suitable to serve as the "form" of dual complexes.[22]

This passage reveals that Russell does not take a logical form to be a constituent of the fact of which it is the form. If he did so take it he would face a version of Bradley's regress, for the consituents, including the form, would apparently require a further form to "unite" them. But, though the form of "dual complexes" is not a constituent of the fact that-aR^1b, it does not thereby disappear. This suggests a motive for Russell's taking logical forms to be existentially general facts. He seems to have two choices, given the symbolism of *Principia*. Either such forms are logical functions, or logical relations, represented by signs like 'R(y,z)' and 'Øy', or perhaps by 'R̂(ŷ, ẑ)' and 'Ø̂ŷ', or they are *general facts* without constituents. Russell might have thought that there is no way to decide in such a case, and, hence, there is an advantage to construing forms as general facts. For, if he did not so take them and yet required, as he does, such facts to ground the knowledge of logical truths, then he would be forced to hold that a form, like Øy, was a constituent of a fact, like (∃Ø)(∃y)Øy. He would then face the problem of specifying the other constituent or constituents of such a general fact. But, as we just noted and will discuss below, he thinks of the existential quantifier, taken by itself, as representing "nothing".

Whether it is construed as an existential fact or simply as a function, a form like Øy does not disappear from judgment facts, like B(O,Øy,F,a), of which it is a constituent. On the other hand, neither the form which is the proposition, (∃Σ)(∃x)Σ(x,Øy,F,a), nor the propositional function, Σ(x,Øy,F,a), seems to be a constituent of the belief fact. Thus, that fact does not consist of Othello standing in B to that proposition, and, while Russell recognizes propositions, they do not play a traditional role. Whether construed as existentially general facts or functions, his propositions play the role of functions mapping a subject and a propositional "verb" (attitude) onto an intentional (belief, judgment, etc.) fact. This is not to say that Russell does not recognize such an entity. The point is simply that, on Russell's multiple relation analysis of intentionality, if one expressed the fact that Othello believes that a is F by (o) 'B(O,(∃Σ)(∃x)Σ(x,Øy,F,a))', where B appears to be a two-term relation holding between Othello and the proposition (form), (o) should be elliptical for 'B(O,Øy,F,a)', where B is a four-term relation holding among Othello, Øy, F, and a. But, since Russell construes the proposition as the existential form (∃Σ)(∃x)Σ(x,Øy,F,a), which is taken to be an existential fact, it is not clear how such a fact can be held to disappear as a constituent of the intentional fact. This poses a critical problem. On the one hand, it would seem that, as in the case of the propositional function Σ(x,Øy,F,a), 'B(O,(∃Σ)(∃x)Σ(x,Øy,F,a))' should be elliptical for 'B(O,Øy,F,a)'. That is, if we take the proposition to be the propositional function, we can consider a sentence pattern like 'Σ(x,Øy,F,a)(O,B)', – where 'O' and 'B' are subject terms for the predicate expression 'Σ(x,Øy,F,a)' – instead of 'B(O,Σ(x,Øy,F,a))' to express the fact that Othello believes that a is F. And, as in Russell's early form of lambda abstraction in *Principia*, such a sentence would convert to 'B(O,Øy,F,a)'. On the other hand, 'B(O,(∃Σ)(∃x)Σ(x,Øy,F,a))' expresses the fact that a relation,

B, holds between two entities, a person and a fact, and hence should not be merely elliptical for 'B(O,Øy,F,a)'. If it is, there is little point to Russell's distinguishing propositions, taken as existentially general facts, from propositions taken as propositional functions. This poses one problem for Russell's analysis. There is another, related problem.

Since only Øy, F, and a are the constituents of the existential fact, (∃Σ)(∃x)Σ(x,Øy,F,a), one may ask what distinguishes that fact from the atomic fact that a is F? On Russell's view there are two things on which the distinction can be based. First, as the quotation just above reveals, Øy is a constituent of the existential fact, but not of the fact that a is F. It is the form of the latter fact, but not a constituent of it. Russell, to avoid problems with a Bradley-type regress, does not take the form of a fact to be a constituent of it. This leads to the second difference. The form of the one fact is then different from the form of the other. This points to a peculiarity of Russell's use of form that we noted earlier. He speaks of a form, such as (∃Σ)(∃x)Σ(x,Øy,F,a), where we not only have constituents but specific particulars and attributes or relations as constituents of such forms. This is not only a problematic use of 'form', but there is a clear sense in which such an existential fact, qua form, itself has a logical form, (∃*f*)(∃Ø)(∃y)(∃Σ)(∃x)Σ(x,*f*,Ø,y). This in itself is an oddity of the view, but the problem is compounded in that such a logical form is also the logical form of the intentional fact expressed by 'B(O,Øy,F,a)', assuming that we do not express that fact by 'B(O,(∃Σ)(∃x)Σ(x,Øy,F,a))'. That could suggest taking (∃V)(∃w)V(w,(∃*f*)(∃Ø)(∃y)(∃Σ)(∃x)Σ(x,*f*,Ø,y)) as the appropriate form, but that would be absurd.

There is another "entity" that could also be involved in Russell's analysis. Consider the class of all intentional facts containing F, a, and Øy – those obtained as values of the two-place propositional function Σ(x,Øy,F,a), i.e. the class that is the extension of that function. Such a class may also be taken as a replacement for the classical proposition and, hence, as the subject of ascriptions of truth and falsity. If a is F, we can consider such a class *to be true*, since every fact in the class *is true*, and take it, like its generating function, to represent the fact that a is F. Since classes are not entities, for Russell at this period, identifying propositions with such classes would be another way of avoiding propositions as entities.

Putting aside, for the time being, problems posed by Russell's acknowledging propositional entities, and no matter how a proposition is construed – as a class of facts, as a function like Σ(x,Øx,F,a) or as an existential fact like (∃Σ)(∃x)Σ(x,Øy,F,a) – we may note that in order to take either a proposition or a specific belief fact to have a truth value, we must acknowledge an understood connection between any of these and the fact that grounds its truth. Even if that connection is expressed simply by 'B(O,Øx,F,a) is true iff a is F', we coordinate one complex to another in addition to the coordination of their respective constituents. One of the misleading features of Russell's analysis that hides this fact is that the representing complex contains constituents that are elements of the represented complex. In the case of a sentence (or a proposition or mental content) the representing complex does not share constituents with the represented complex. If one does not see that a and F, as they occur in the representing complex, B(O,Øx,F,a), represent themselves, as they occur in the represented complex, the fact that a is F, one can easily overlook the need to connect the two complexes. Wittgenstein's criticism points to the need to connect the representing complex to the represented complex. He makes an additional critical point in that the relation B not only orders the terms, O, F, etc., but in so ordering them it takes F to function as a property, and not just as a term of a relation. Russell will explicitly recognize this in his 1918 logical atomism lectures when he will note that such facts problematically have "two verbs". However, Wittgenstein's criticism fails to recognize that Russell does have a complex representing entity. In fact, given the form (∃Σ)(∃x)Σ(x,Øy,F,a) and the function Σ(x,Øy,F,a), as well as the latter's "value range", there are diverse things that can be taken to represent the fact that a is F, in addition to the individual judgment facts. Each of these can be taken to play the role of a representing complex, though they need not be construed as complexes in the sense in which a fact or a Fregean thought is a complex. That neither such a propositional function nor its value range need be taken to be a basic entity is a feature of a Russellian type analysis that Wittgenstein and Pears ignore. There is a further ironic feature of Wittgenstein's argument.

In the *Tractatus* and the *Notebooks* Wittgenstein appealed to the internal or logical properties of the constituents of atomic sentences and of thoughts to connect atomic sentences and thoughts to their correlated situations or possibilities. Supposedly such logical properties were themselves correlated with (or identified with) internal-logical properties of constituents of possibilities. The possibilities of combination of the elements in the *representing* combinations were taken to represent the possible states of affairs determined by the natures or essential properties of their constituents. Wittgenstein seems to have thought that by having the possibilities of combination be determined by such logical properties of the combining elements, he avoided recognizing possible facts as entities, for one supposedly did not have to take one combination to represent another over and above the representative roles played by the constituents of the complexes. In a sense, the object a could be said to have the essential property of possibly being F, instead of being an element of a possible state of affairs, a being F. But, it is clear that the correlation of constituents does not suffice to correlate the complexes of which they are the constituents, unless one takes it to be a logical property of a constituent of a thought or sentence that it not only can combine with another constituent of a certain kind but that the resulting combination represents a possible state of affairs. It is then gratuitous to claim that one has correlated complexes by correlating their constituents, though the correlation of the complexes is dependent upon the correlation of constituents. Wittgenstein's alternative to Russell's analysis of intentional contexts thus faces the very problem Pears takes him to have raised against Russell's analysis.

The need to correlate complexes points to a two-fold problem for both Wittgenstein and Russell. Since the interpretation rules for terms, such as (rl) "'a' refers to a" and (r2) "'F' refers to F", must be supplemented by a rule, like Moore's (1), for atomic sentences, (r3) "'Fa' represents the fact that a is F", they cannot avoid either the connection expressed by 'represents' in (r3) or the problem about what is *represented* when 'Fa' is false. Wittgenstein believed that (r1) and (r2), along with our understanding of what sorts of things a and F are and what sorts of signs 'a' and 'F' are, given us the interpretation of 'Fa' without (r3). But this just packs (r3) into the construal of 'a' and 'F' as subject and predicate terms. For, to take them as such terms is to understand that they can combine into the sentence 'Fa', which is taken to state that a is F. Nothing is gained by his talk of internal or essential properties of signs and objects or of "shared logical form" or of the "picturing" of facts by atomic sentences. Russell's view fares no better. He must take the fact that B(O,Øy,F,a) to represent the fact that a is F. The relation B, unlike normal relations, thus plays a dual role. It combines with Othello, a, F, and Øy into a belief fact, which has a specific content since it is taken to represent a second fact composed of some of the terms that it relates. This dual feature reveals a problem with Russell's analysis.

Consider the predicate 'B(x,Øy,F,a)', formed, by abstraction from the sentence expressing Othello's judgment. To ascribe such a predicate to Othello is, then, to assert two things: that Othello has a belief and that what is believed is that a is F. Suppose we express this by the conjunction 'B#(O) & <Fa>(O)', where 'B#' is a monadic predicate transcribing 'x is in or has a belief state' and <Fa> is a monadic predicate expressing what the content of the belief is. Thus, we separate the attitudinal property – believing, judging, doubting, understanding – from the content, as Russell does. But, like

Bergmann, we now explicitly employ a predicate, <Fa>, representing a content property. What Russell does is take the occurrence of F, a and Øx in a belief fact to express the content. Thus, he does not avoid two of the things he seeks to avoid. First, he does not avoid a variant of (r3) above, since he takes the occurrence of F and a in an intentional fact (or in the form obtained by abstraction from such a fact) to represent the fact that a is F. Thus, he will still have to face the problem of non-obtaining states of affairs being represented by beliefs. Second, he makes use of a representational relation between complexes, over and above the connections expressed in (r1) and (r2). This is easily seen in the case of either the function Σ(x,Øy,F,a) or Russell's form (∃Σ)(∃x)Σ(x,Øy,F,a). Russell must correlate such a function or form to the fact that a is F. Actually, his analysis not only recognizes something that plays the role of a proposition, but several things for each classical proposition. For, aside from the class of judgment facts, containing F, a and Øx, and the relational propositional function, Σ(x,Øy,F,a), that we have introduced in addition to Russell's form, (∃Σ)(∃x)Σ(x,Øy,F,a), there will be various monadic functions like B(x,Øy,F,a) and forms like (∃x)B(x,Øy,F,a). Such entities, however, are not crucial to the present issue. What is important to note is that Wittgenstein's appeal to logical or internal properties of constituents adds nothing to Russell's analysis. In fact, it is quite natural to take Russell's view as embodying such logical properties. That is why only logically proper types of signs can be substituted for variables in contexts like 'B(O,Øy,Ø,y)'. What Wittgenstein and, many years later, David Pears fail to see is that Wittgenstein's view suffers from the same difficulty that Russell's does.[23] Wittgenstein, no more than Russell, recognized the need to explicitly invoke a pattern like (r3) in addition to patterns like (r1) and (r2) – the need to correlate complexes as well as their constituents. This is clear from 5.542 in the *Tractatus*.

Pears takes Wittgenstein to argue that the addition of a logical form like Øy to the judgment complex does not make Russell's analysis more viable.[24] This is correct, for one must still treat the relation B as being such that it can only relate certain kinds of terms, and adding a logical form only redundantly specifies that appropriate terms must be of a Ø-type and a y-type, respectively. The issue is whether the need to impose the condition that a sentence of the form 'B(x,*f*,Ø,y)' is well-formed only if the terms replacing the variables are of the requisite logical kinds reveals that Russell's analysis implicitly appeals to complex linguistic expressions as subject terms, as in 'B(x,Øy)'. Russell was obviously bothered by this question. But, it is interesting that, in spite of his dramatic reaction to Wittgenstein's criticism, he discussed the relational theory, in its original form, in the 1918 lectures on logical atomism, and he still rejected propositional entities in *The Monist* essays of 1914, which are supposedly the first six chapters of the 1913 manuscript. In 1918, as noted earlier, he expressed reservations about the theory being forced to employ "two verbs" in patterns like 'B(O,L,d,a)' and, ultimately, abandoned the theory for that reason and for the further reason, expressed in 'On Propositions: What They Are and How They Mean' of 1919 and in a 1918 entry in his papers, that he no longer believed in a *subject*, whether it be a believing person or a mental act of belief. One can take his concern over the use of two "verbs" to be a variant of Wittgenstein's criticism, as we have construed that criticism, that Russell assumes that some of the terms of an intentional fact are of appropriate logical kinds to combine into a fact grounding a belief's truth.

Pears, following Wittgenstein, focuses on the relatively insignificant and obviously problematic introduction of logical forms into the analysis of intentional facts. Russell mistakenly thought that the addition of such forms somehow connected the appropriate constituents.[25] The addition of such forms is insignificant not only in that it adds little to the earlier form of the multiple relation theory, but in that Russell introduced such forms, in part, to make a negative point: they did not suffice to specify truth grounds in the case of non-symmetrical relational facts. Moreover, in 1913, Russell no longer thought that he must account for and not simply appeal to the ordering of the constituents in a fact. For, to take facts to involve an order of the constituents requires an account or analysis of the basis for differences of order. This led to his introduction of relations like R* and R** and the analysis of order in facts that we discussed earlier.

Russell's handling of relational contents reveals a radical difference in the analysis of belief contexts involving non-symmetrical relations and those involving monadic properties or symmetrical relations. In the latter cases, the pre-1913 pattern still applies, with the relatively insignificant addition of logical forms as terms. In the former case, we now have an explicit variant of a propositional analysis, since the second subject term for the predicate 'B' in 'B(O,(℩p)(aR*p & bR**p) exists)' is a sentence. (We might also note that 'a' and 'b' occur in the embedded *sentential* forms 'aR*p' and 'bR**p'.) Russell's employment of such a radically different pattern for non-symmetrical relations is an awkward feature of his analysis, though we can understand why he makes such a move. There is no obvious way to adapt his treatment of relations for the monadic case, and if he distinguishes the positions in a symmetrical relational fact, he would recognize two such facts in any given case of the relation obtaining. The latter consequence he explicitly sought to avoid, while he saw no need to revise his earlier pattern for monadic contents. Aside from any questions about the awkwardness of employing two totally different patterns of analysis for intentional contexts, what is interesting, for our discussion, about his handling of non-symmetrical relational contexts is that Wittgenstein's criticisms are irrelevant to it. They are irrelevant since Russell employs a sentential expression as a subject term along the lines of a propositional analysis. This raises a question regarding just how much of Russell's 1913 manuscript Wittgenstein read or discussed with Russell.[26]

Russell's revised analysis, being a variant of a propositional analysis, faces an obvious problem, but one that is quite different from that raised in Wittgenstein's and Pears' criticisms. In fact, Pears, like Wittgenstein, ignores the obvious change in Russell's analysis. The problem Russell's new analysis faces is the traditional one we noted earlier: since B is now a two-term relation between a person and a complex he requires a second term for the relation when the belief is false and the complex does not exist.

As we noted earlier, Russell's new analysis thus fails, but not for the reasons Pears or Wittgenstein give. In his discussion, Pears focuses on the role of acquaintance in Russell's analysis and the question of just what logical forms are. The two questions are connected. Russell is concerned to answer the somewhat Kantian question: How is the understanding of a sentence possible? In the case of 'Fa' he takes it to be sufficient that we have acquaintance with F, with a, and with the form Øy. F and a are objects that are unproblematically terms of the relation of direct acquaintance. The form Øy poses a problem, since it appears to be necessarily connected with a complex of that form. Hence to be acquainted with it appears to require acquaintance with a complex. But, as Pears understands Russell's problem, to be acquainted with and hence understand a complex presupposes acquaintance with its constituents. Thus, Russell supposedly faces an insoluble problem, since acquaintance with the form presupposes acquaintance with a complex of which it is the form; but acquaintance with such a complex presupposes acquaintance with its component form.

Russell faces no such problem. Even if he holds that acquaintance with a complex implies acquaintance with whatever is involved in its analysis, which Pears attributes to him on the basis of a misunderstanding, there is no problem so long as one understands 'presupposes' or 'implies' in a logical and not a temporal sense. There is no more a problem with talking of acquaintance with logical forms than there is of speaking of acquaintance with universal properties by one who holds that only instantiated properties exist and can be objects of acquaintance only as instantiated. All Russell need do is hold that to be acquainted with a form is to be acquainted with a complex of which it is the form and vice versa. But the criticism itself arises, as I mentioned above, from a misunderstanding by Pears. Russell holds that to understand a proposition requires knowledge of the form and constituents. Assuming he no longer recognizes propositional entities (even though we noted an ambivalence about this in 1913), we may understand this claim to mean that an atomic sentence like 'Fa' is understood only if the object a, the property F, and the logical form Øy are known. Assume also that the form Øy is known only if an atomic fact of that form is known. To be acquainted with an atomic fact still does not require prior acquaintance with a form. All that is required, at most, is that if there is acquaintance with such a fact, and awareness of it as such, there is also acquaintance with or knowledge of a form. There is no more a problem here than there is in the case of being acquainted with several instantiations of a property. What Pears does is simply raise a variant of Plato's old objection against abstraction and, hence, of the argument for the innate knowledge of the forms. A logical form must supposedly be known in order to recognize it in a fact. But it cannot be grasped in experience, i.e. by acquaintance, for in order to experience it one must recognize it as the form it is, and, hence, it already must be known. Pears' argument is no more viable than Plato's. What gives it some credence is Russell's mingling questions about logical forms of facts with questions about logical truths.

Russell, concerned about grounding truths like '(∃Ø)(∃y)Øy', held that such a truth was not grounded by a complex. In part this was due to his concern about the constituents of such a fact. What he did, as we noted earlier, was identify a logical form like Øy with the truth ground for the existential statement '(∃Ø)(∃y)Øy', the form R(y,z) with the existential fact (but not complex) that was the truth ground for '(∃R)(∃y)(∃z)R(y,z)', and so on. The logical forms of monadic predication, dyadic predication, etc. are then existentially general facts that are *simples*. His argument for this was concise. Since neither the form nor the existentially general fact had constituents, both were simples. Hence, the existential fact could not contain the form but could be identified with it. Pears sees this as raising an inescapable regress for Russell, as he would have to be acquainted with the forms to comprehend the complexes. Yet, he could only be acquainted with the forms if he was acquainted with the relevant complexes. But, while Russell identifies the predicative forms with such existential general facts, he sees problems with the claim for two reasons. First, he is inclined to hold that for each such general fact or form to exist, there must be a basic relation and atomic fact of the appropriate logical kind, and it cannot be a matter of logic that such a relation and fact should exist. That is, the form must be the form of some fact. But it also seems evident to him that a form could be grasped even if there should be no such fact. Thus, the form of an *n*-term relational complex is grasped, or is an object of acquaintance, whether or not there are *n* terms standing in an *n*-term relation or whether or not there is such a relation. Second, identifying a logical form with a general fact reraises the problem he had earlier noted when he took propositions to be forms:

> But in saying that there are relations of which there are no instances, what we should naturally suppose that we mean would be "there are propositions of the form xRy for values of R for which such propositions are false whatever x and y may be". But with our definition of "proposition", we cannot have any reason to believe that there is a proposition which has never been understood, and we cannot know that a proposition of the form xRy has ever been understood, if the R is one of which there is no instance, and with which consequently no human being is acquainted.
>
> This instance suggests, what is also suggested by many other considerations, that our definition of a proposition is inadequate.[27]

Russell does not resolve the problem, but he does proceed to speak of "the form 'x and y in a relation'" and of 'xRy' as the "general form" of a relational proposition, and not of something having some relation to something. But this may merely be a concise form of expression.[28] Yet it suggests that he was tempted to abandon the identification of logical forms with existentially general facts and to hold that the knowing of a logical form does not involve being acquainted with a fact of which it is the form. In so far as logical forms are universals, this means that for logical forms Russell did not hold that the existence of universals requires that they be exemplified. Thus, we can know each form in the series Øy, R(y, z). . . , without it being true that there is some exemplified relation of the appropriate kind. This is in keeping with his Platonism in *The Problems of Philosophy*, where he claimed acquaintance with unexemplified universals.

It is Russell's identification of logical forms with the existentially general facts earlier in the manuscript that provides the basis for Pears posing his dilemma for Russell's analysis. But the purported dilemma, which is not really a dilemma even given that identification, completely dissolves without it. And that identification may have been abandoned by Russell in the later parts of the manuscript, or, as we shall note later, it may even have been a reason why he abandoned the manuscript. Given the problems posed by identifying the form of monadic predication with the existentially general fact that something has some property, one may wonder why Russell construes the form as a logically necessary fact. Since he takes such facts to ground logical truths, but cannot see how such general facts can have constituents, he has one motive for so construing forms. In this connection he speaks of the existential quantifier ("something") as being "nothing". Since the quantifier sign does not represent an entity there is no constituent of the existentially general fact to distinguish it from the form. This feature also provides a clear cut distinction between logical facts and non-logical facts that fits with the notion of logic being a matter of form: logical facts, being forms, are facts that have no constituents.

Russell also mentions, as another motive, wanting to take a proposition as "something which can subsist on its own account, and not only as a fictitious constituent of certain mental complexes", i.e. not as an "incomplete symbol". Both reasons fit with his taking such a form to be an object of acquaintance, rather than an object of a belief or judgment, since the latter can be true or false, but in acquaintance cases there is no correlate of falsity. There is also another, quite simple reason, which Russell hints at, but does not explicitly state, for his identifying the form with the fact. He takes the fact that a is F to be composed of a and F united by the form. We use 'Fa' to state that a is F, but we could use '(∃Ø)(∃y)(Øy & Ø = F & y = a)'. If we do so and withdraw the clause 'y = a', and hence the object a, we are left with '(∃Ø)(∃y)(Øy & Ø = F)', which we can use to say that something has the property F. It is not far fetched to think of this fact as being derived from the first by *abstracting* or *removing* the object a. We are then left with the form and the property F. If we then withdraw the clause 'Ø = F', we are left with the expression '(∃Ø)(∃y)Øy'. It is then easy to think that this existential generalization represents what remains after the other constituent of the fact, the property F, is *removed*. Hence, we can think that what remains, the form, is represented by an existentially general sentence. Thinking along such lines one might think that abstracting from 'Fa' we arrive, in successive stages, at '(∃y)Fy' and '(∃Ø)(∃y)Øy'.

Russell takes the form Øy, or (∃Ø)(∃y)Øy, to be an object of acquaintance as a consequence of his holding that understanding the proposition 'Fa' implies being acquainted with the elements of the fact that is its ground of truth. To understand a monadic atomic sentence, in short, requires that one be acquainted with a property, a subject, and the form of monadic exemplification. How one obtains acquaintance with such entities is not something Russell goes into. What Pears raises as a problem, and what he took Wittgenstein to raise, is a question about how we obtain the understanding of what it is for a subject to exemplify a property – what it is to be a monadic atomic fact – by being acquainted with the elements of such a fact, including its form. Pears argues that Russell introduced logical forms to resolve this problem but that such things do not suffice for the task. We have already noted in what sense this is correct, for it is another aspect of the criticism that the relational view is merely a verbal variant of a propositional analysis, since we must recognize that the items in a fact are of a certain kind. As put just above, the point is that a and F must be taken to be appropriate logical kinds to stand in the form Øy, and the introduction of Øy clearly does not resolve the matter. This reiterates the point that the terms in the places occupied by 'F' and 'a' in 'B(O,Øy,F,a)' must be combinable into a subject-predicate sentence, a complex sign that represents a possible situation. We noted earlier that Russell's theory of descriptions does not allow him to resolve one problem of false belief posed by 'Othello believes that a has R^1 to b', since with 'B(O,(℩p)(aR*p & bR**p) exists)' Othello cannot stand in the relation B to a non-existent complex. Thus, Russell will have to reintroduce such complexes or hold that Othello cannot believe what is false. It will not do to modify (3*) and use (3**) 'B(O, (℩p)(aR*p & bR**p))' – to employ the description as a term, in place of the sentence stating that what is described exists. For, then, on Russell's theory of descriptions, Othello can still only have such a belief if there is an appropriate fact, i.e. if the belief is true. It is interesting that this is not Wittgenstein's criticism nor is it raised by Pears. One might speculate that the failure of his theory of descriptions to provide a key for the analysis of belief contexts, not Wittgenstein's criticism, was the reason Russell abandoned the 1913 theory. For, the criticism did not apply to the contexts treated according to the real innovation of the 1913 theory, the non-symmetrical relational contexts.

Pears suggests that Wittgenstein's criticism of Russell's identification of logical forms with existentially general facts may have played a role in his abandoning the manuscript.[29] Russell, himself, raised problems about that identification in the manuscript, and it may well be that he became even more dissatisfied with it as the manuscript progressed. Pears argues that identifying forms with existentially general *propositions*, and taking these to be simple, makes "their truth unintelligible".[30] But his argument presupposes that a ground of truth for a logical truth must be a complex, as is a ground of truth for a non-logical truth. Russell is not compelled to agree. On his account there is nothing unintelligible about taking an existentially general *fact* to be a simple and a ground of truth. The problem stems, in part, from Pears' focusing on Russell's use of the term 'proposition'. This is somewhat misleading, since as Russell says, a logical form is a fact: "if a logical form, i.e. a fact containing no constants . . .".[31] The point is that if we state Russell's view more carefully than Pears does, or than Russell himself does in the passage

Pears quotes, *the proposition* would be represented by the expression 'Σ(S,(∃Ø)(∃y)Øy)', read as 'There is a Σ and an S such that Σ(S,(∃Ø)(∃y)Øy)'. *The proposition is a complex*, and *it is true*, since the object, the form, (∃Ø)(∃y)Øy, which is its ground of truth, is an object of acquaintance and, hence, *exists*. This object is not, like the true proposition, complex, but simple. The true proposition is complex since it contains the form (∃Ø)(∃y)Øy *as a constituent*. The difference between a proposition, whose ground of truth is a logical form or *logical fact*, and a proposition like 'Σ(S,(∃Ø)(∃y)Øy,F,a)' is two-fold. First, in the former case, only the form is a constituent of the proposition, and, second, its being a constituent of the proposition means that the proposition contains its own ground of truth. One may take that to "show" that it is a necessary truth, as the existentially general fact (the logical form) is a logical or necessary fact. Pears has not only not shown that Russell's view is unintelligible, he has not set it out correctly.

Pears does not discuss an obvious problem connected with Russell's identification of a form with a general fact. It is a problem that Russell undoubtedly came to notice. He cannot identify a form like (Ø)(y)Øy with a logical fact, since the corresponding statement is a logical falsehood. Yet, he has no way to analyze (Ø)(y)Øy, and he cannot deny that it is simple, by his line of reasoning in the existentially quantified case. There is no evidence in the text that Russell considered this problem, but it seems too obvious for him to have overlooked. It is useful to recall that in his book *Mental Acts*, Peter Geach criticized Russell for ignoring the problem of relational order in dealing with intentional contexts. Geach was obviously and understandably unaware of Russell's attempt to deal with such problems in detail in the 1913 manuscript. Yet, given Russell's early concern with the logic of relations, one would expect him to be aware of problems concerning relational predication and order: the problems that were taken up in detail in the sections of the 1913 manuscript that he abandoned. As in the case of relational order, it would be unreasonable to think, as Geach did, that Russell was oblivious to an obvious problem. In any case, if Russell had separated questions about logical form from questions of logical truth, he could have abandoned the view that logical forms were general facts and returned to construing them as universals or functions; a view more in line with his treatment of expressions like 'Øy', 'Øˆ yˆ ', and '(∃y)Øˆ y' in *Principia*.

Since Russell introduced a propositional complex, in the form of a complex mental occurrence, in his *Principia* adaptation of the view cryptically expressed in 5.542, he could have employed patterns like those introduced earlier in (4) and (6) to resolve both problems – the problem of specifying truth grounds for atomic sentences, without recognizing situations, as well as the problem posed by the analysis of atomic intentional contexts. For, such patterns provide a link between the complex sentential signs and their truth conditions that is applicable in both cases. Given such a link, along with the representational role of its constituent signs, an atomic sentence may be said to represent, even though there will not always be something it represents, a fact. Once complex contents are acknowledged, the same pattern can be used in explicitly intentional contexts. Though Russell took one of the two steps necessary in *Principia*, by following Wittgenstein and introducing complex mental contents, he did not then explicitly take the step of linking such states to their truth conditions via definite descriptions or existential claims. This is ironic since he had done so in 1905. What has been proposed here, as a solution to the problems posed by false belief, is thus derived from aspects of Russell's views at two distinct stages of his thought: the 1905 use of definite descriptions (or existential claims) to avoid non-existent facts and the 1927 *Principia* analysis, influenced by Wittgenstein, avoiding propositional entities.

One must still account for intentional contexts involving complex contents: truth functionally complex and quantified contexts. To do so one must first consider what entities are recognized in the specification of truth conditions for statements expressing such contents. As I have considered this in some detail elsewhere,[32] I shall simply state, without argument, what is involved. Since we acknowledge a set of atomic facts, we can specify truth grounds for atomic sentences and their negations in terms of facts with certain constituents belonging to that set. Thus, the truth ground for '¬Fa' is provided by:

$$\text{Tr '}\neg\text{Fa' iff not} - (\exists p)(pCa \;\&\; pCF \;\&\; pC\text{Ø}x).$$

Conjunctions, disjunctions, and negations of molecular statements can then be treated in the familiar reductive way without recognizing conjunctive, disjunctive, etc. facts. A conjunction, for example, is true if and only if its conjuncts are true. If the conjuncts are atomic or negations of atomic statements, the specification of the conjunction's truth condition reduces to claims about the existence and non-existence of atomic facts. If they are not, then such statements themselves involve claims about the truth (or denial) of further statements, until the claims reduce to such claims about existence and non-existence, which are claims about the members of the set of atomic facts.

One can deal with quantified statements in a similar manner, given that we acknowledge domains of facts, of particulars, and of properties. For primitive predicates, the following patterns are available:

$$\text{Tr`}(\exists x)Fx\text{'} \text{ iff } (\exists x)(\exists p)(pCx \ \& \ pCF \ \& \ pC\emptyset x)$$

$$\text{Tr`}(x)Fx\text{'} \text{ iff } (x)(\exists p)(pCx \ \& \ pCF \ \& \ pC\emptyset x)$$

with the obvious treatment of the negations. For quantification over complex contexts, the analysis can make use of the familiar construal of *satisfaction*, as in the case of molecular compounds.

We can now turn to the consideration of complex contents for intentional contexts. Consider a mental state whose content is expressed by 'Fa & aR^1b'. How are we to characterize such a state so that we may say its content is expressed in such a way? Clearly we would expect to have, at least, what would characterize states whose contents were expressed by the individual atomic sentences. The problem is to avoid taking the occurrence of '&' to represent something, as the tokens of the names and predicates are taken to represent objects and properties. But we can avoid doing that. What we have is a complex content that is taken to be true under the conditions for the conjunction being true. This, after all, is what is involved in understanding conjunction. Thus, our earlier pattern can be applied in specifying more complex truth conditions: the occurring conjunctive content will be true if and only if the constituent atomic sentence tokens are true. Similar patterns are applicable in the case of other molecular and quantified statements and, in particular, the case of negation. One need not introduce, as Russell once did, basic attitudes (characteristics of mental states, in our terms) of *disbelief* or *rejection* to link negative contexts to truth conditions. This is not to say that one cannot also recognize such states of disbelief as further psychological kinds. There would simply be a difference, corresponding to any generic difference among mental states, between states of disbelief that a is F and states of belief that a is not-F. The relation between them would be captured by the connection between the respective truth grounds for the statements expressing the contents. Thus, in one sense one may say that a person who is in a state of belief that a is F is in a state of disbelief that a is not-F. In another sense this is not so, just as one who is in a belief state expressed by 'a is F' is not, in one sense, in a state of belief expressed by 'a is F & a is F', but is, in another sense, in such a state. The present analysis can accomodate these diverse senses of belief (thought, judgment) without recognizing either possible facts or propositional entities.

The use of biconditionals specifying truth grounds might lead a reader to see a resemblance between the present analysis and attempts to offer a theory of meaning in terms of so-called *truth conditions* in the manner of Donald Davidson. Nothing could be more misleading. Davidson claims that 's means that p' can be rendered as 's is true if and only if p'.[33] Put in terms of our discussion, we would then have "'Fa' *means that* Fa", (or Moore's (1) or Bergmann's (1b)), rendered as "'Fa' is true iff Fa". Thus, Othello's understanding or knowing the meaning of 'Fa' would supposedly be analyzable in terms of his knowing or understanding that: 'Fa' is true iff Fa – where the right side of the biconditional, here 'Fa' itself, furnishes a statement of the truth condition for 'Fa'. We need only recall that the appeal to facts and situations by the logical atomists and other correspondence theorists involved the taking of truth conditions (grounds) to be facts to appreciate the utter triviality of such an approach. Davidson attempts to avoid the acknowledgment of a correlated fact as a truth ground by producing a token of a Tarskian T-sentence. But if one is to deal with the philosophical issues involved, one must probe into the role 'Fa' (or 'p') plays in Davidson's analysis and face the problem of connecting such a sentence or sentence token to a ground of truth in the sense in which a fact is a condition or ground of truth. All Davidson does is utter, as a truth condition for a sentence (sentence token), another sentence token – either of the sentence itself – 'snow is white' is true iff snow is white – or a transcription of it. On such a view one *gives* a truth condition by repeating or transcribing a sentence: one does *not specify* what it is to be a truth condition as traditional theories of truth do specify what truth conditions are – facts, coherence relations, and so forth.

If one takes up the philosophical issues involved, then the problem about the meaning of 'Fa' becomes a problem about whatever one puts for 'p' in "'Fa' is true iff p". This is one reason Davidson's analysis is trivial. It is also trivial in that it plays on the obvious fact that when one asks for the meaning of a sentence in ordinary contexts, a satisfactory answer is provided by furnishing another sentence. If we recall that Russell and Wittgenstein sought to avoid meanings, in the form of propositions, by connecting atomic sentences (or acts of thought, belief, etc.) directly to facts (states of affairs) without employing intermediary propositional entities, we see why Davidson's analysis lacks originality as well as substance. Under Tarski's influence, he alters the pattern of Russell and Wittgenstein so that in place of "Othello knows that 'Fa' *represents* Fa" explicating "Othello knows the *meaning* of 'Fa'", we have "Othello knows that: 'Fa' is true iff Fa". Davidson thus copies the basic pattern of the logical atomists who take facts as truth grounds for atomic sentences, while avoiding propositions as *meanings*. But, as he stops short by failing to finish the pattern and provide truth grounds for atomic sentences, he prematurely withdraws from the philosophical controversies involved. What he does is adopt the form of the atomists reference theory of meaning, which avoids special entities as "meanings", without its substance: the recognition of facts as truth grounds. The present analysis not only takes facts to be truth grounds but acknowledges a basic intentional or representational (meaning) relation between terms of atomic sentences (or elements of mental states that are thoughts) and the constituents of atomic facts. Yet, by not employing such a relation for atomic facts and sentences (or thoughts), it resolves the problems that Davidson dismisses.

University of Texas at Austin

NOTES

[1] G. E. Moore, *Some Main Problems of Philosophy* (SMPP), G. Allen & Unwin, London 1953, pp. 267–68.
[2] Thus Moore writes ". . . in what would this belief of his consist? What is the correct analysis of the event that would be happening in his mind?" (SMPP, p. 258).
[3] Moore does not explicitly appeal to such attributes, since, after rejecting propositional entities, he proposes "to give up the attempt to analyze beliefs" (SMPP, p. 266). But, he does implicitly appeal to such attributes in his discussion by speaking of "the belief that lions exist" and he recognizes that he must account for different particular beliefs having the same content as well for particular states of belief with different contents (SMPP, pp. 258–59). For a detailed discussion of Moore's attempt to analyse belief see H. Hochberg, *Thought, Fact, and Reference: The Origins and Ontology of Logical Atomism*, Minnesota University Press, Minneapolis 1978, pp. 3–12; 53–86.
[4] SMPP, p. 267.
[5] SMPP, pp. 256, 257.
[6] Russell obscured matters by speaking of "such an entity as the relation R" existing or not existing in such a case. What he had in mind was clearly the obtaining or existing of the relation between a and b. His speaking of "the relation" perhaps prevented him from seeing the problem. On this matter, see my 'Descriptions, Situations, and Russell's Extensional Analysis of Intentionality', *Philosophy and Phenomenological Research*, 49, 4, 1989.
[7] B. Russell, *The Problems of Philosophy*, William & Norgate, London 1956, p. 128.
[8] L. Wittgenstein, *Notebooks 1914–1916* (NB), ed. by G. H. von Wright & G. E. M. Anscombe, B. Blackwell, Oxford 1961, pp. 82, 129. There is a familiar problem concerning the interpretation of the *Tractatus* and passages of the notebooks regarding Wittgenstein's views about properties and relations, on the one hand, and predicates, on the other. For my purposes here, I will simply treat his view in the *Tractatus* as "realistic" in that he would take an ideal language to contain predicates representing properties and relations. This is not to say that the realistic interpretation is correct, as it is usually presented. Rather, as I see it, Wittgenstein thinks of predicates as both representing functions, in Frege's sense, that are "completed" by objects and as *being* functions that are completed by objects, the latter objects being names of the former. It is the ambiguous nature of Fregean functions (concepts), being incomplete entities that are contrasted with objects, that accounts for the

realistic remarks as well as for the nominalistic tone of other remarks in the *Tractatus*. See 'Frege on Concepts as Functions: A Fundamental Ambiguity', in H. Hochberg, *Logic, Ontology, and Language*, Philosophia Verlag, Munich 1984.

[9] L. Wittgenstein, *Philosophical Grammar* (PG), ed. by R. Rhees, Trans. by A. Kenny, University of California Press, Berkeley-Los Angeles 1978, p. 161.

[10] The letter is reproduced in NB, p. 121; the discussion occurs on p. 96.

[11] B. A. W. Russell, *The Collected Papers of Bertrand Russell*, vol. 8, ed. by J. G. Slater, G. Allen & Unwin, London 1986, pp. 253–284.

[12] A. N. Whitehead and B. A. W. Russell, *Principia Mathematica*, vol. 1, Cambridge University Press, Cambridge 1960, p. 662. Ironically, this analysis is essentially repeated, in a confused version, by P. Geach as *his* "theory" in *Mental Acts* (B. Blackwell, Oxford 1957). Geach, writing many years later, does not note or mention the connection, while offering criticisms of Russell's 1912 version of the multiple relation theory that, in part, repeat Wittgenstein's criticisms and, in part, complain about the supposed inability of the multiple relation theory to handle relational contents. Geach can be excused for not being aware of Russell's 1913 manuscript dealing with the latter issue, though even the first edition of *Principia* has the technical devices for handling relational contents in terms of the early form of lambda abstraction Russell employs; the failure to mention the 2nd edition *Principia* discussion is another matter.

[13] The earliest of Bergmann's papers on the subject is 'A Positivistic Metaphysics of Consciousness', *Mind*, 45, 1945. But, it is not until 'Bodies, Minds and Acts', in *The Metaphysics of Logical Positivism*, Longmans, Green & Co., New York 1954 and 'Intentionality', in *Semantica. Archivio di Filosofia*, 1955, reprinted in *Meaning and Existence*, University of Wisconsin Press, Madison 1959, that he explicitly develops the views we are considering.

[14] To preserve his claim that M-connections were analytic, Bergmann proposed a revised analysis of the concept of *analyticity*, seeking to provide an ontological ground for analytic truths. Thus he not only followed Moore's lead in his analysis of intentional contexts, but he returned to issues that Russell had tried to resolve in 1912–13, issues concerning the ontological ground of logical truth. For a discussion of the development of Bergmann's analysis of intentionality and of various problems with it, see my "Intentions, Facts, and Propositions" in *Classics of Analytical Metaphysics*, ed. by L. Blackman, Macmillan, New York 1984; 'Intentionality, Logical Structure, and Bergmann's Ontology', *Nous*, 15, 2, 1981; 'Belief and Intention', in Hochberg (1984); and 'Gustav Bergmann' in *Handbook of Metaphysics and Ontology*, ed. by H. Burkhardt & B. Smith, Philosophia Verlag, Munich 1991.

[15] B. A. W. Russell, *Theory of Knowledge*, v. 7, *The Collected Papers of Bertrand Russell*, ed. E. Eames *et al.*, G. Allen & Unwin, London 1984, p. 195.

[16] For a detailed discussion of the various senses of 'same content' involved in such matters see 'Belief and Intention', pp. 399–406 and 'Descriptions, Situations, and Russell's Extensional Analysis of Intentionality'.

[17] PG, p. 144.

[18] Russell (1984), p. 115.

[19] *Ibid.*, pp. 115–116.

[20] *Ibid.*, p. 109.

[21] Discussions of Russell's various theories of types often ignore a basic sense of 'type' in which Russell takes monadic, dyadic, triadic, etc. attributes and relations to be of logically different types. On the importance of this sense of 'type' and its relevance for Russell's paradox see my 'Russell's Paradox, Russellian Relations, and the Problems of Predication and Impredicativity', in *Rereading*

Russell: Essays on Bertrand Russell's Metaphysics and Epistemology, ed. by C. Savage & C. Anderson, in *Minnesota Studies in The Philosophy of Science*, Vol. 12, Minnesota University Press, Minneapolis 1989, pp. 63–87.

[22] Russell (1984), p. 117.

[23] D. Pears, 'Russell's 1913 Theory of Knowledge Manuscript', in *Rereading Russell: Essays on Bertrand Russell's Metaphysics and Epistemology*, cit., pp. 169–182.

[24] Pears (1989), p. 179.

[25] Russell (1984), p. 116.

[26] On a page (Russell 1984, p. 197) cited as Appendix B. 1 Folio 2, by the editors, Russell speaks of positive and negative facts and the *neutral* fact that is involved in judgment and which replaces the *form* of his old theory. The idea seems to be that the neutral fact (situation) can exist as a positive or negative fact, thus providing a truth ground for an atomic sentence or its negation. Propositions, or atomic sentences, would then represent neutral facts irrespective of their truth or falsity. This suggests Wittgenstein's influence on Russell, as culminating in the logical atomism lectures, 'On Propositions' of 1919, and the appendix to the 2nd edition of *Principia* cited above. It should be noted that Wittgenstein's 'Notes on Logic' are printed as an appendix to the NB and are dated September, 1913, when Wittgenstein dictated some of them to Russell and supplemented them with a typescript sent to Russell a "few days later". See R. Monk, *Wittgenstein: The Duty of Genius*, Jonathan Cape, New York 1990, p. 93.

[27] Russell (1984), p. 134.

[28] *Ibid.*, p. 135. In the 1918 logical atomism lectures Russell takes expressions for logical forms to employ free variables, thus returning to his *Principia* notation for such forms. Following Russell's use at different places I use 'xRy' as well as 'R(x,y)' for the dyadic relational form.

[29] Pears (1989), pp. 176–178. Russell might have been influenced by Wittgenstein's views as recorded in the latter's notebooks. Though the relevant entries date from later than Russell's manuscript, Wittgenstein and Russell had been discussing and communicating about such matters in the period prior to and during Russell's work on the manuscript. Thus, in a letter of January, 1913 Wittgenstein writes: "I now think that qualities, relations (like love) etc. are all copulae! That means I for instance analyse a subject-predicate proposition, say, "Socrates is human" into "Socrates" and "something is human", (which I think is not complex)." (Wittgenstein (1969), appendix iii, pp. 120–121.) This could have suggested the idea that forms are existentially general facts. In the NB, p. 17, Wittgenstein also writes that he "had thought that the possibility of the truth of the proposition Øa was tied up with the fact (∃x)(∃Ø).Øx", and on p. 22, "(Similarly (∃x)Øx would be the form of Øa, as I actually thought)."

[30] Pears 1989, p. 178.

[31] Russell 1984, p. 131. In the following discussion I will follow Russell's notation and use 'S' in place of 'x' for the variable that would be replaced by a constant representing a person.

[32] See 'Truth Makers, Truth Predicates, and Truth Types', in *Language, Truth and Ontology*, ed. by K. Mulligan, Kluwer Academic Publishers, Dordrecht 1991, pp. 87–117. The crucial issues concern (i) the use of a negation in specifying that no fact of a certain kind is among the atomic facts; (ii) the use of the quantifiers in specifying truth conditions for generalized sentences; (iii) the appeal to sets of atomic facts, particulars, and properties; (iv) possible regresses of the kind Frege and Bradley raised.

[33] D. Davidson, 'Truth and Meaning', *Synthese*, 17, 1967, p. 309.

ROSARIA EGIDI

WITTGENSTEIN BETWEEN PHILOSOPHICAL GRAMMAR AND PSYCHOLOGY

It is known that Wittgenstein's interest in philosophical problems of psychology goes back to *Tractatus* and his first writings. It certainly has its origins in the commitment which he shared with Frege, to challenge the long-established psychologistic solutions to the problems of logic and of the theory of knowledge. Nevertheless, the very "plan for the treatment of psychological concepts" assumed a specific form only later, when Wittgenstein lost his confidence in the anti-psychologistic strategy of Frege as well as the belief – which he had briefly held – in a phenomenological analysis of internal experience, which was in a way similar to the one already developed by Brentano and the Austrian-German schools of phenomenology. This treatment, together with reflections on the foundations of mathematics, was intended to form the II part of *Philosophical Investigations*[1] and constitute therefore a corollary or – one might say – the application of the new theory of meaning elaborated in the I part.

As we know, such a project was never carried out; only two vast fragments exist: the writings on the foundations of mathematics[2] and the later materials on the philosophy of mind collected in the two volumes of *Remarks on the Philosophy of Psychology*[3] and in the first volume of *Last Writings on the Philosophy of Psychology.*[4] Even if these ultimate texts do not contribute much to Wittgenstein's general perspective, already well known from the *Investigations* and *Zettel*[5] the range of the arguments as well as the richness and sophistication of presented details, make their study an advantageous starting point for Wittgenstein's announced conception of perceptive, emotional and cognitive language of man in the frame of non-Fregean theory of meaning.

Perhaps Wittgenstein never intended to write a systematic book that would collect his thoughts on mathematics and psychology. It is certain, however, that already in the beginning of the 1930's he had in his mind a clear picture of what such work should be like, how it should be structured and which subjects it should present. The manuscript on which he worked in these years, published more than thirty years later under the title of *Philosophische*

R. Egidi (ed.), Wittgenstein: Mind and Language, 171–184.
© 1995 *Kluwer Academic Publishers. Printed in the Netherlands.*

Grammatik can be considered an example of this ideal book, a sort of sketch book *avant la lettre* of the great work in which the two parts would be united. In an appendix to this work Wittgenstein explicitly declares what the structure and the content of a book dedicated to the philosophical grammar should be:

> If we had grammar set out in the form of a book, it wouldn't be a series of chapters side by side, it would have quite a different structure. And it is here, if I am right, that we would have to see the distinction between phenomenological and not-phenomenological. There would be, say, a chapter about colours, setting out the rules for the use of colour-words; but there would be nothing comparable in what the grammar had to say about the words "not", "or", etc. (the "logical constants").
>
> It would, for instance, be a consequence of the rules, that these latter words unlike the colour words were usable in every propositions; and the generality belonging to this "every" would not be the kind that is discovered by experience, but the generality of a supreme rule of the game admitting of no appeal.[6]

Two principal theses of the late Wittgenstein are presented in this passage; they are mutually supporting and both deserve our attention. One concerns (i) the distinction between "phenomenological" and "non-phenomenological", the other introduces (ii) the kind of higher "generality" which belongs to the rules of philosophical grammar in contrast with empirical generality characteristic of physical laws.

(i) The purpose of grammar is, then, to separate the physical, the empirical, the causally determined from the domain of the phenomenological and to attribute to non-phenomenological concepts occurring in mathematical and psychological propositions, a peculiar requisite of generality, not to be confused with the one that belongs to empirical generalisations, but one that is typical of grammar or of rules of a game. These propositions which obey the rules of grammar or of a game are in fact different from empirical generalisations; they do not have the explicative power of a deductive or inductive type and cannot be analysed by means of truth-functional logic.

It is important, however, to bear in mind that before attributing to philosophical grammar the task of analysing the constructs resistant to the treatment with truth-functional logic, Wittgenstein had thought that this could be achieved by an intermediate discipline *(ein Mittelding)* that would lie in between logic and empirical science, by a discipline Wittgenstein called in the *Philosophical Remarks* and later in the *Remarks on Colour* "phenomenology". The focal position that the distinction between phenomenological and not-phenomenological or physical occupies in the first work[7] shows that Wittgenstein's idea of grammar is still bound to the context of ideas, where the analysis of propositions concerning immediate experience was the attribute of phenomenology.

Actually, it is the function of the latter to analyse the special class of phenomena of immediate experience and constituting the language of data which Wittgenstein indicated as "primary language" *(das Primäre)*. It is distinct from the "secondary language" *(das Sekundäre)*, which consists of concepts like those of physical sciences.[8] All propositions containing psychological verbs (intend, understand, believe, expect, desire, doubt, etc.) belong to the primary language. In the *Tractatus* they had been called *Satzformen der Psychologie*, which include the propositions Russell called "propositional attitudes": "A believes/thinks/intends/hopes that p", as well as the hypotheses and modal sentences (the necessary and probabilistic propositions).[9] They form the complex of non-extensional contexts which cannot be treated by means of truth-functional logic.

The idea that phenomenology is in power to analyse primary language is discussed in the *Philosophical Remarks*[10], but the definition of phenomenology as a form of "conceptual analysis" *(Begriffsanalyse)* appears in the *Remarks on Colour*:

> Phenomenological analysis (as e.g. Goethe would have it) is analysis of concepts and can neither agree with nor contradict physics.[11]

In the same work Wittgenstein claims that phenomenology is a *Mittelding* between logic and physics and its field of application is not that of physical propositions:

> But what kind of a proposition is that, that blending in white removes the colouredness from the colour? As I mean it, it can't be a proposition of physics. Here is the temptation to believe in a phenomenology, something midway between science and logic, is very great. (p. 15)

Even though Wittgenstein admitted the existence of "phenomenological problems", those late notes of the *Remarks on Colour* denied, all the same, the existence of a "phenomenology":

> There is indeed no such thing as phenomenology, but there *are* phenomenological problems. (pp. 9, 49)

As we see, Wittgenstein qualifies as a "temptation" the belief that the phenomenology should be able to provide a language adequate to describe primary phenomena of immediate experience. Such a qualification seems still more meaningful if one realises that in the *Remarks on Colour* he is clearly alluding to a view that he himself held, when in the early phase of his thought he fell victim to the phenomenological temptation.

The broader and better articulated context for the reconstruction of Wittgenstein's conception of phenomenology can be found in the *Philosophical Remarks*. It is on the basis of the transitional works, however, that we can reconstruct the three arguments with which he renders plausible his idea of phenomenology.[12] The *first* of these arguments consists of the distinction between "phenomenological language" *(phänomenologische Sprache)* and "physical language" *(physikalische Sprache)*.[13] The *second* argument refers to the attribution to propositions of physics the power to determine "laws" *(Gesetze)* while the phenomenological propositions should have the power to determine "possibilities" *(Möglichkeiten)*.[14] The *third* argument, finally, which derives from the second, affirms that phenomenology concerns the "sense" *(Sinn)* of the propositions in contrast with the physics which deals with "truth values" *(Wahrheit* and *Falschheit)*.[15]

The three arguments interweave in the *Philosophical Remarks* and reappear in the materials concerning lectures and conversations held by Wittgenstein between 1929–1935 and collected by Moore, Waismann and others.[16] We know that most of the ideas that they were intended to convey were eventually repudiated, but this does not diminish the validity of Wittgenstein's attempt to offer an analysis of the language of internal experience free from any collision with psychologism as well as logicism. This attempt gives rise to what is often indicated as the philosophy of the intermediate Wittgenstein. One of its characteristics is the idea of a specific analysis of phenomenological sentences to be distinguished from that of physical sentences and whose reference seems to be expressed by "senses" rather than "truth values" or, better, by "sense-impressions" rather than by "physical objects". This distinction lays foundations to the conception of two languages discussed by Wittgenstein in the *Philosophical Remarks*: the phenomenological and the physical language and to the phenomenological theory of colours *(die phänomenologische Farbenlehre)* which he opposed in the same work to the physical one.[17]

It is well known that the attempt developed particularly in the II part of the *Investigations*, as well as in the writings concerning philosophical psychology, differs considerably from the context of ideas of *Philosophical Remarks* in both the elaboration and the sophistication of arguments. One of the differences lies in the fact that the idea of phenomenology disappears from this text and with it the illusion, which Wittgenstein had held for a short time, that phenomenological analysis could play a role in the anti-psychologistic strategy, or, more exactly, in the analysis of sentences expressing sense-impressions or, according to the version of *The Blue and Brown Books* "personal experiences".[18]

The problem was to determine the sense of those propositions with non-Fregean arguments and to preserve their objectivity, i.e. their not-subjective and public character with more adequate arguments than those offered in the *Philosophical Remarks*.

Even if it is an exaggeration to claim that the abandonement of the idea of phenomenology constitutes the only qualifying point of Wittgenstein's mature conception of senses and indicates something more than just a change of strategy, there is no denying the utility of its accurate examination in clarifying Wittgenstein's antipsychologistic project. Particular emphasis should be made on the reasons that made him abandon the first attempt of a phenomenological analysis and to undertake a new one, whose qualifying points would be a consequence of the rejection of the Fregean conception of *Sinn* and a treatment of intentionality and of psychological concepts in a grammatical context.

While Frege and then Carnap respectively, speaking about "senses" and "intensions",[19] were convinced to be moving within the domain of philosophical logic, Wittgenstein, talking about belief, intention, understanding, desire, doubt, hope and expectation thought to be dealing with the problems of the *Erkenntnistheorie*, which he identified with the philosophy of mind or of psychology. Wittgenstein's theses related to it can be reconstructed starting with the analysis of sense and intentionality sketched in the I part of the *Philosophical Grammar* [20] and with the theory of intentional or voluntary action, whose traces are found in the II part of *Investigations* and in the writings on philosophical psychology.[21] They can be summarised with the following theses which still today influence the contemporary theories of meaning. The one is that (a) the meaning of intentional propositions does not derive from psychological or mental phenomena or from internal experiences, but has a "grammatical", normative nature; the other thesis is that (b) the explanation of actions, the "intentional explanation" does not have the causal character typical of explanations of facts.

The thesis that the intentional sentences (or propositional attitudes) of the type "A believes/doubts/hopes/is certain that p" do not refer to internal experiences and to mental phenomena, is a doubtless testimony to the continuity of Wittgenstein's anti-psychologistic approach with Fregean aim of preserving the objectivity of senses, as well as to the affinity with Carnap's theory of intensions as expressing their senses in one of the possible worlds. It shows, in other words, that the analysis of intentionality is inspired by the common anti-subjectivistic and anti-mentalistic trend. However, Wittgenstein did not justify the objective content of intentional sentences at the cost of admitting senses or intentions. Their "objectivity" is not achieved by reference to objects, of both internal and external nature but depends on whether those sentences obey the system of rules of which they are part. According to Wittgenstein, "an intention is embedded in the situation, in human customs and institutions"[22] and the meaning of sentences expressing intentions, desires, doubts, certainties and hypotheses is based on the way in which we use them, on the manner in which "we are taught to use [them]".[23]

The leading idea of Wittgenstein's analysis is that beliefs, intentions, hypotheses, hopes, doubts and expectations are not part of a primary language, the propositions expressing them do not refer to presumed mental phenomena, both internal and private, but that they play a peculiar function of the language. Such a function takes place in linguistic games in which such sentences perform a normative, pragmatic role, different from the descriptive function characteristic of the propositions whose meaning is based on direct knowledge referred to facts of extra-linguistic nature. Their unity of sense is not a determined objective content, but a whole system of rules which they obey and, hence, not a complex of semantic but of pragmatic criteria of significance.

Within such contexts the term "pragmatic" means "referring to norms of actions", "according to rules", therefore such a requisite becomes pertinent to the sentences whose meaning is not based on reference to extra-linguistic facts or to mental, internal events, but is determined in relation to the rules or norms that govern social games, customs and institutions in which communicative behaviours or intentional or voluntary actions are embedded. They do not presuppose a direct but a background knowledge, according to which they can assume a sense, a meaningful relevance for us.

The sense is so given by the complex of activities and behaviours presupposed by "the understanding of a proposition" and not by those involved in the procedures to ascertain "the truth or the falsity" of a proposition. The experience determines the truth or falsity of a proposition but not its sense.[24] Intentional constructs do not, therefore, behave like sentences whose truth and falsity can be conclusively determined, but they reflect the institutionalised, inherited forms of behaviour that cannot be expressed within the stable and defined domain of factual knowledge:

> But I did not get my picture of the world by satisfying myself of its correctness; nor do I have it because I am satisfied of its correctness. No: it is the inherited background against which I distinguish between true and false.[25]

Exactness and completness do not constitute a requisite or an ideal for intentional sentences: rather they express a fluid and imperfect blend of criteria, at the same time ethical, aesthetic, economic, social, inspired by consolidated habits of a society or of a group and by considerations of an essentially pragmatic nature:

> We judge an action according to its background within human life, and this background is not monochrome, but we might picture it as a very complicated filigree pattern, which, to be sure, we can't copy, but which we can recognize from the general impression it makes. (RPP II, § 624)

This is the reason why it is necessary to refute the deep-rooted conviction that the analysis of psychical or mental processes which should constitute the reference of intentional propositions, can be realised by a phenomenological language in the same way as the analysis of facts performed by means of physical language. It is, in particular, the idea of a phenomenology as the analysis of internal experiences, referring to the domain of *sui generis* propositions that must be given up. It is necessary to resist to the temptation to mythologize the sense of expression deprived of a direct denotation, to admit the existence of mental processes as their objects and thus to avoid not only the mythology of logicism but also the subtler mythology of psychologism. In this way Wittgenstein's flight from the Fregean senses serves the aim to avoid the double threat: the ontologization of mathematical concepts and numbers and the parallel reification of psychological concepts and mental states.

As far as the other thesis of Wittgenstein is concerned, the one about the not-causal character of the explanations of propositional attitudes and of sentences expressing intentional actions, one can say that it is a kind of corollary of the preceding thesis, i.e. of the idea to deny the identification of intentions, desires, beliefs, doubts, etc. with mental phenomena and to claim the grammatical and pragmatic nature of the class of propositions which express them. Such a thesis is an alternative to the arguments which distinguish the causal theory of intentionality maintained, above all, by neoempiristic epistemology. As it is known, according to such a theory, the explicative relation between the intentions of an agent and his actions could be substantially analysed as a causal link.

In the *Investigations* Wittgenstein develops his thesis of the causal inefficiency of intentions in respect to the actions[26] and the concept of "intention" is contrasted with that of "prediction" *(Voraussage, Prognose)* as an inference of the future actions of a person on the basis of a presumed psychological or mental process. In empirical sciences prediction is in fact a proposition based on inductive procedures and formulated in terms of extensional logical models. In intentional contexts, on the contrary, what determines the possibility of a prediction of future actions – or of an explanation ("justification") of our past actions – is separate from all experiences and depends on the system of our linguistic conventions, institutions and rules. Wittgenstein emphasizes in *On certainty*:

> The squirrel does not infer by induction that it is going to need stores next winter as well. And no more do we need a law of induction to justify our actions or our predictions. (§ 287)[27]

This is the reason why Wittgenstein in *On Certainty* calls them, in contrast to empirical propositions, "systematic propositions" or propositions which have "the character of a rule".[28] A systematic as well as an intentional construct can have the linguistic form of an empirical proposition, "but of course it is not one that is based on experience".[29]

Wittgenstein's insistence on the *sui generis* character of such a relation, seems to create the need to distinguish between the methods of explanation of intentional and empirical propositions. It is impossible to give a causal explanation of intentional propositions, since intentions, desires, hopes and beliefs, i.e. the "motives" of an action are of a different category than the so called "causes" of a physical event are.[30] In the former case the link between cause and effect is not of logical or rational nature; *vice versa* the relation between intention and action – the intentional explanation *(die intentionale Erklärung)* – is systematically or logically linked to the intended action.[31] However, for Wittgenstein this does not mean to propose a new dualism between facts and actions, between physical and mental events, in a word, between causality and intentionality. The difference is not ontological but a grammatical one; it simply indicates that the sentences constituting contexts in which intentional, rather than descriptive functions obtain, are not true or false statements but normative sentences or value judgments. The old purpose attributed to the phenomenology to distinguish between the phenomenological and the non-phenomenological is now absorbed by the philosophical grammar, not in the sense that a peculiar domain of propositions is attributed to grammar but in the new sense of distinguishing or clarifying different functions of language, "stable" and "fluid" forms of our modes of expression.

According to the line of thought that continues up to *On Certainty* the complex of not-empirical, intentional contexts does not constitute a peculiar and independent ontological sfere. To Wittgenstein it is always possible that expressions of rules or norms, whose meanings do not depend on experience but on the system in which they occur, in determined contexts and for determined communicative purposes, achieve a stability that makes them similar in use to empirical propositions. Sometimes, the latter ones are used for communication purposes and are given the status of rules or prescriptions for actions, that is of propositions of grammatical rather than empirical nature. The fact that "the same proposition may get treated at one time as something to test by experience, at another as a rule of testing"[32] is underlined in another famous passage of *On Certainty*:

> It might be imagined that some propositions, of the form of empirical propositions, were hardened and functioned as channels for such empirical propositions as were not hardened but fluid; and that this relation altered with time, in that fluid propositions hardened, and hard ones became fluid. (OC, § 96)

Analogically, the distinction between intentional and causal constructs is viewed not in dualistic terms since their distinction is not ontological but functional.

(ii) According to Wittgenstein the entire conceptual tool set to work by mental sciences of his time to describe and explain the "real machinery of mind" and so to give a meaning to intentional contexts ceases to belong to the domain of experimental methods and is attributed to the system of rules governing the use of propositions containing mental concepts. The treatment of psychological concepts shifted to the level of grammar, whose rules have a higher kind of generality than that attributed to empirical methods. Any attempt to describe, understand and explain a psychological phenomenon, an individual behaviour, a mental process by means of empirical methods fails because when we believe to describe or explain a determined behaviour or process, what we actually have described or explained is the general context against which it emerges in its interplay with a number of other behaviours and processes. In a passage of the *Remarks on the Philosophy of Psychology* and *Zettel* Wittgenstein says:

> How could human behaviour be described? Surely only by sketching the actions of a variety of humans, as they are all mixed up together. What determines our judgment, our concepts and reactions, is not what *one* man is doing *now*, an individual action, but the whole hurly-burly of human actions, the background against which we see any action. (RPP II, § 629; Z, § 567)

As we know, the rules of grammar do not apply to single words and sentences but to every word or sentence where "every" is not empirically determined. Analogically, mathematical and psychological concepts do not function as descriptions of singular states and processes but can be applied in a multiplicity of cases. Our concepts and conceptual schemes do not exercise absolute power to represent a single case but a whole network of cases, do not describe the behaviour of an individual but serve to insert it in a behavioral scheme shared by many individuals. Wittgenstein says that by means of one of his fascinating similitudes:

> Seeing life as a weave, this (pretence, say) is not always complete and is varied in a multiplicity of ways. But we, in our conceptual world, keep on seeing the same, recurring with variations. That is how our concepts take it. For concepts are not for use on a single occasion.
> And one pattern in the weave is interwoven with many others. (Z, §§ 568–569)[33]

In this sense the normative (rule-guided and aim-directed) character of our conceptual world immunizes Wittgenstein's idea of interdependence and variety of our concepts from subjectivism and relativism.

What is the aim, one might ask, of Wittgenstein's provocative thesis on the grammatical character of concepts and propositions of mathematics and psychology and what is the meaning of his idea of generality which appeals to the higher rules of the game? At the first sight, his aim seems polemical, as it results from the severe evaluation in the well known passage of the *Investigations*:

> The confusion and barrenness of psychology is not to be explained by calling it a "young science"; its state is not comparable with that of physics, for instance, in its beginnings. (Rather with that of certain branches of mathematics. Set theory). For in psychology there are experimental methods and *conceptual confusion*. (As in the other case conceptual confusion and methods of proof). The existence of the experimental method makes us think we have the means of solving the problems which trouble us; though problem and method pass one another by. (PI II, p. 232)

The fact that the idea of the grammatical nature of psychological concepts was presented without a precise argumentation, but emerges from dialectical confrontation with the science of his time is not of secondary importance in the frame of Wittgenstein's thought. Polemic is rather a constitutive character of the Wittgensteinian philosophizing. It is linked with the therapeutical and not doctrinal aim of philosophy and indirectly also of the philosophical grammar. Philosophy and grammar are not systems of doctrines created in splendid isolation but activities obeying a system of rules and developing in a communicative context whenever there are problems to solve, mistakes to correct and misunderstandings to clarify.

When Wittgenstein says that psychological propositions – and we know that it applies to mathematical ones as well – have grammatical nature underdetermined by experience and not governed by explanation models of causal kind, he clearly alludes to the peculiarity of a language, that of immediate experience, whose aims and intentions are profoundly diverse from physical language and experimental sciences. But in what way could grammar appear adequate, in Wittgenstein's gaze, to exemplify these aims? His appeal to grammar is more a programmatic manifest than the mention of a determined discipline. Moreover, Wittgenstein is not the first to present the relationship between grammar and psychology in a polemic function. This relationship was the focus of Austrian-German schools of linguistics and philosophy of language between the eighteen and nineteen hundreds and was recalled in the *Sprachphilosophien* inspired by Brentano to juxtapose intentional and pragmatic use of language to traditional expressive, semantical functions.[34]

In contrast to Fregean logic and Carnap's syntax, Wittgenstein's grammar is not really a formal discipline; its propositions are not logically determined but obey a received system of rules; it does not follow inflexible criteria of truth or falsity of a particular proposition but is ready to submit to the need of multiple use and adjust to a variety of situations. The so called psychological or intentional propositions express basically the rules of behaviour, codices, habits and institutions which as such refer to general, not singular cases. So they have a higher character than the empirical and inductive generality, as they depend on the standards established as rules or norms by the entire community.

As a result of rule-guided and intentional (aim-directed) character as well as of peculiar generality of grammar the psychological concepts do not aspire to be correct and precise as it is required for physical concepts:

> The pedigree of psychological concepts: I strive *not* after *exactness*, but after a synoptic view. The treatment of all these phenomena of mental life is not of importance to me because I am keen on completeness. Rather because each one casts light on the correct treatment of *all*. (Z, §§ 464–465)

The fact that philosophical psychology is part of grammar sheds light on the peculiarity of the notion of meaning in the application to psychological concepts and propositions and show that it is controversial to connect to them the double aspect of meaning Frege called *Sinn* and *Bedeutung*. The meaning of propositions in which psychological concepts occur not only cannot be determined by the means of truth functional logic but is also independent of any reference to mental objects, to internal psychical processes. Wittgenstein refuted the idea of objective contents of thinking as reference of psychological or intentional propositions, of *Gedankeninhalte* of the type Frege called "senses" *(Sinne)*. For Wittgenstein *Sinn* is not in the mind, but it is an eminently intentional, rule-guided, aim-directed notion. Probably he saw in the Fregean objective notion of *Sinn* a hidden tendency to psychologism. For this reason he did not accept it and rejected, at the same time, the idea of a phenomenology as a temptation to acknowledge an extra-linguistic meaning to senses. From the *Philosophical Grammar* to the last writings on philosophical psychology Wittgenstein's aim is to reformulate the problem of the sense of a proposition and, in general, the problem of understanding and of intentionality within the context of linguistic uses and conventions.

The Wittgensteinian appeal to grammar is first of all polemical to Frege's strategy to give to the senses an ontological status and to Russell's thesis that propositional attitudes have the structure and the meaning of "external relations".[35] The task of grammar is not to express thoughts and to determine the meaning they carry. The not-extensional propositions do not express internal experiences, mental activities or processes. So the *Bedeutung* is lost. Grammar aims at the reconstruction of rules which allow for accomodating of our linguistic means to the aim of communication and understanding. And so the Fregean *Sinn* is lost as well. The meaning is not a ready made notion but it is the result of deliberate activity interconnected with aims, thoughts, feelings and desires which constitute our form of life. The meaning is not "found" but "given", because it belongs to the intentional, pragmatic element without which language cannot work. This is the way, according to Wittgenstein, through which we can in some way maintain the objective character of our concepts, the inner rationality of our conceptual world. Wittgenstein espresses this in the following way:

> Our investigation does not try to *find* the real, exact meaning of words; though we do often *give* words exact meanings in the course of our investigation. (Z, § 467)

Terza Università di Roma

NOTES

[1] *Philosophische Untersuchungen. Philosophical Investigations* (PI), ed. by G. E. M. Anscombe & R. Rhees. Trans. by G. E. M. Anscombe, B. Blackwell, Oxford 1953.

[2] *Bemerkungen über die Grundlagen der Mathematik. Remarks on the Foundations of Mathematics*, ed. by G. H. von Wright, R. Rhees & G. E. M. Anscombe. Trans. by G. E. M. Anscombe, B. Blackwell, Oxford 1956.

[3] (RPP I), ed. by G. E. M. Anscombe & G. H. von Wright. Trans. by G. E. M. Anscombe, B. Blackwell, Oxford 1980; (RPP II), ed. by G. H. von Wright & H. Nyman. Trans. by C. G. Luckhardt & M. A. E. Aue, B. Blackwell, Oxford 1980.

[4] (LWPP I), ed. by G. H. von Wright & H. Nyman. Trans. by C. G. Luckhardt & M. A. E. Aue, B. Blackwell, Oxford 1982.

[5] (Z), ed. by G. E. M. Anscombe & G. H. von Wright. Trans. by G. E. M. Anscombe, B. Blackwell, Oxford 1967.

[6] *Philosophische Grammatik. Philosophical Grammar* (PG), ed. by R. Rhees, B. Blackwell, Oxford 1969, p. 214. Trans. by A. J. P. Kenny, B. Blackwell, Oxford 1974.

[7] *Philosophische Bemerkungen. Philosophical Remarks* (PR), ed. by R. Rhees, B. Blackwell, Oxford 1964, §§ 1–9, 57–66, 218–224. Trans. by R. Hargreaves & R. White, B. Blackwell, Oxford 1975.

[8] PR, §§ 11, 19, 53, 216.

[9] *Tractatus Logico-Philosophicus* (TLP), ed. by D. F. Pears & B. F. McGuinness, Routledge & Kegan Paul, London 1961, 5.541–5.542.

[10] §§ 1, 53, 57, 60. In the *Philosophische Bemerkungen* Wittgenstein assimilates phenomenology sometimes with "grammar" (§ 4), sometimes with "theory of knowledge" or *Erkenntnistheorie* (§§ 57, 60). po

[11] *Bemerkungen über die Farben. Remarks on Colour*, ed. by G. E. M. Anscombe. Trans. by L. L. McAlister & M. Schättle, B. Blackwell, Oxford 1977, p. 16.

[12] Wittgenstein's transitional works include a considerable quantity of writings composed betweeen 1929–1935. The first of them is the conference 'Some remarks on logical form', *Proceedings of the Aristotelian Society*, Suppl. Vol. 9, 1929, pp. 167–71; the manuscripts later published as PR and PG; the texts of his Cambridge lectures: *Wittgenstein's lectures. Cambridge 1930–32*. From the Notes of J. King & D. Lee, ed. by D. Lee, B. Blackwell, Oxford 1980; *Wittgenstein's lectures. Cambridge 1932–35*. From the Notes of A. Ambrose & M. Macdonald, ed. by A. Ambrose, B. Blackwell, Oxford 1979; *The Blue and the Brown Books* (BBB), ed. and with a Preface by R. Rhees, B. Blackwell, Oxford 1958.

[13] PR, § 1; 51–53; 57–61; 68–69; 75.

[14] PR, § 1. See also PG, §§ 45; 82–84 and PI I, § 90.

[15] PR, 10–19; 28; 39–46 and PG, §§ 45; 69–84.

[16] See in particular G. E. Moore, *Wittgenstein's Lectures in 1930–1933*, in *Philosophical Papers*, G. Allen & Unwin, London 1959, pp. 252–324 and *Wittgenstein und der Wiener Kreis* (WWK). Aus dem Nachlaß hrsg. von B. F. McGuinness, B. Blackwell, Oxford 1967.

[17] PR, §§ 218–24. See also WWK, pp. 63–67.

[18] See in particular the arguments presented in the 'Blue Book', pp. 44–74.

[19] See R. Carnap's *Meaning and Necessity*, Harvard University Press, Chicago 1947, chap. I -II.

[20] See in particular §§ 1–44, 85–108.

[21] See PI I, §§ 611–629; II, xi (see also parallel arguments in LWPP I, §§ 183, 848, 893, 905–913); RPP I, §§ 705–715, 830–851; RPP II, §§ 116, 222, 258, 267– 275, 624–631.

[22] PI I, § 337.

[23] *Ibid.*, § 190.

[24] PR, § 23.

[25] See *Über Gewißheit. On Certainty* (OC), ed. by G. E. M. Anscombe & G. H. von Wright. Trans. by D. Paul & G. E. M. Anscombe, B. Blackwell, Oxford 1969, § 94.

[26] PI I, §§ 629–647. See also 188–208

[27] See also PI I, §§ 624–28; 650; II, p. 224.

[28] OC, § 494.

[29] *Ibid.*, § 385.

[30] See PI II, pp. 224–225 (LWPP I, § 907–9087; RPP I, §§ 631, 633).

[31] See in particular PI I, § 220, where Wittgenstein underlines "a difference between being causally determined and being logically determined".

[32] OC, § 98.

[33] See also PI II, p. 174 and RPP II, § 624.

[34] See for example Anton Martys theory of the intentional origin of the language *(der absichtlichen Sprachbildung)*, first outlined in his 1875 *Über den Ursprung der Sprache* (Minerva GMBH, Frankfurt) and later developed in die *Untersuchungen zur Grundlegung der allgemeinen Grammatik und Sprachphilosophie*, M. Niemeyer, Halle 1908.

[35] See the alternative proposal of "internal relations" to Russell's conception in PR, §§ 20–30.

ROBERTO CASATI

NOTES ON PHENOMENOLOGY AND VISUAL SPACE

The *Philosophical Remarks*[1] belong to those intermediary works of a philosopher, which seem to be fatally destined to remain unknown due to their experimental character and a sketchiness which lacks a precise aim. On reading Wittgenstein, one often retains the impression of an immense effort oriented towards details and nuances, as well as of possible hidden consequences of an argument or a hypothesis. Nonetheless, critical literature directed towards the most important works, such as the *Tractatus* and the *Investigations*, has aptly shown the structural force of the former and the coherence and thematic unity of the latter. Even a series of short amorphous notes such as the *Bemerkungen über die Farben*, has shown proof of an impressive originality; the edition of this work was without doubt an important cause for a renewed interest in the classical theme of colour in contemporary philosophy.

None of this can be said of the *Remarks* and with good reason, or so I believe, for this is a work we may term experimental, in which it is difficult to find a thematic line pursued till the end.

I do not believe I am proposing anything new by substituting the chronological classification of Wittgenstein's works with a thematic division. Roughly speaking, this is a division between two fundamental foci: language and psychology of mental states respectively. The *Tractatus* and a good part of the *Investigations* center around the first focus, whereas the *Philosophical Remarks* and the *Remarks on Colour* center on the second one. Having said that, I shall analyse in some detail one theme amongst those of the *Philosophical Remarks* which belong to the second focus, which has not been extensively treated, though its originality deserves some interest.[2]

1. My topic is the problem of the visual field, treated in §§ 205–217. This group of remarks consists of an argument designed to show that visual space is provided with absolute coordinates and a similar argument in favour of the demonstration of the existence of an absolute distance in the visual field.

The second argument is directly linked to the discussion of the problem of quantities and measures, which is subject to a long series of remarks in the book and is mostly treated in the *Remarks on the Foundations of Mathematics*. Given the fact that there exists a literature specifically dedicated to the

R. Egidi (ed.), Wittgenstein: Mind and Language, 185–192.
© 1995 *Kluwer Academic Publishers. Printed in the Netherlands.*

Wittgensteinian treatment of measures, my central question is therefore the first one, i.e. whether the visual field can be considered an absolute space. This question is introduced during the development of another problem, namely the complexity or the simplicity of the patch in the visual field.

> Can we say that a patch is simpler than a larger one? (§ 205)

Indeed, the problem is not so much to know whether one can have simplicity in the visual field, but rather whether one can have composition: whether the smaller of the two patches can be considered a part of the bigger one, in other words, whether there is composition in the visual field. Wittgenstein has an argument for giving a negative answer to this question, even though the answer which first comes to mind seems to be exactly the opposite. His argument is quite direct:

> It seems as if it is impossible to see a uniformly coloured patch as composite, unless you imagine it as *not* uniformly coloured. The image of a dissecting line gives the patch more than one colour, since the dissecting line must have a different colour from the rest of the patch. (§ 205)

At this point some of us will probably be tempted to object that Wittgenstein mixes considerations about what there is in the world with considerations about what we can represent of the world to ourselves. Even if we cannot represent to ourselves the parts of a patch, it would be presumptuous to conclude that the patch is not composed of parts. If it is not, we need a better argument.

This objection allows us a better evaluation of what is at stake for Wittgenstein. What interests Wittgenstein is the patch as seen, the visual patch. He suggests that the logic of the operations which one can perform on this object differs radically from that which governs the operations which can be performed on material objects. Thus, according to Wittgenstein, this object – the patch – belongs to a completely different category.

Wittgenstein opposes the case of the non-composition of a visual patch to the case of the composition of complex configurations, such as a chessboard. Contrary to the previous case, here we have to account for composition. The figural unity is not a primitive element, it is inseparable from the composition of the figure.

> An intellect which takes in the component parts and their relations, but not the whole, is a nonsense notion [*ein Unding*]. (§ 205)

Once more I use a proposition which at first sight is quite surprising, in order to better isolate the central thesis. Every philosopher knows about the existence of the phenomenon of composed colours, a phenomenon which is striking when we look at pointillist paintings: that which from afar appears like a uniform orange patch turns out to be a chaotic heap of small red and yellow points. It is therefore possible to perceive only the component parts of a composition. Hence the intellect which apprehends the whole without apprehending the parts is not an absurdity, it is actual.

Nonetheless, we must maintain once more that the simplest and most coherent interpretation of this remark is the one which considers visual objects as the only referents of the Wittgensteinian discourse. A visual orange patch cannot be composed of small red and yellow visual patches, as the latter are not visible: they are not visual patches themselves.

2. These preliminary remarks introduce our subject: the visual field as absolute space. Let us see the three arguments Wittgenstein gives for the absoluteness of position in the visual space. If the visual patch has no visual parts, *a fortiori* one cannot predicate a colour of a part of the patch. Nevertheless, we would say that the corner at the bottom left of the patch is the same colour as the corner at the top right. What is it that permits us to make the predication implicit in this identity statement? A simple solution would be to accept the existence of absolute places in visual space, that is to say, to accept that we can individuate places like the top right of the visual field without directly or indirectly referring to visual objects occupying that place. Previously, we were talking about the possibility of distinguishing one part of the patch in relation to other parts, this being a necessary condition for a predication concerning the patch itself. We were therefore obliged to use a relational style of individuation, without which no predication would have been possible. It is only now that we avoid relational individuation and the problems which are linked to it by maintaining that the individuation of a place in visual space is absolute, that is to say, that it does not depend on the individuation of other places. I shall call this argument (a) for the absolute properties of visual space.

A second argument (b) is an argument from movement. Wittgenstein says:

> It's obviously possible to establish the identity of a position in the visual field, since we would otherwise be unable to distinguish whether a patch always stays in the same place or whether it changes its place. (§ 206)

This argument is insufficient, unless we make a restriction, for we can grasp the movement as long as it concerns a movement in relation to another object. The intervening restriction thus concerns the presence of a single patch in the field. If one patch only[3] changes in place, we can see it, even if there is nothing else to see. Therefore we retain the possibility of individuating places independently of the presence of objects which either occupy them or occupy other places. We could have a better grasp of the difficulty if we were to imagine the visual field as equipped with an asymmetric form. In this case the determination of a place could be obtained by naming certain sides of the field with proper names, and by indicating towards which side the patch moves. Now, the visual field is *de facto* symmetric in relation to two axes and this renders such a method inapplicable. Thus we must provisionally accept the possibility of absolute position in the field.

(c) We cannot imagine

> a visual space in which we would only perceive certain spatial positions but no absolute position. (§ 206)

Here a good example is that of the watch the dial of which shows only dots without numbers. If it were impossible for us to distinguish a dot in relation to its successor, it would be equally impossible to notice the movement of the pointer.

I think the three arguments are valid. But the fuller consequences drawn by Wittgenstein – which I believe to be false – are worthy of discussion. The central remark is the following: since visual space is an absolute space,

> [w]e can also say visual space is an oriented space, a space in which there is an above and below and a right and left.
>
> And *this* above and below, right and left have nothing to do with gravity or right and left hands. It would, e.g., still retain its sense even if we spent our whole lives gazing at the stars through a telescope. (§ 206)

In this dense passage there are three points to discuss. Wittgenstein suggests (1) that to say of visual space that it is absolute is the same as to say that it possesses a system of absolute coordinates. (2) That these coordinates are associated with the directions of the field of bodily orientation: left, right, top, and bottom. (3) This association is however said to be independent of bodily orientation directions.

I believe that none of these three points is correct in itself, that each requires specification, as I shall now try to show. First, the present state of the discussion: owing to (a), (b) and (c) we have been forced to admit the existence of absolute position in visual space. Is it possible to say that this is equivalent to admitting the existence of a system of coordinates? My answer is no. There is no doubt an equivalence between the capacities or the behaviour of two beings, one of which disposes of a system of reference in which places are individuated separately, whilst the other disposes of a system in which there are basic individuals, such as the point on the two axes of references, as well as individuals which are individuated by means of a reference to basic individuals. It matters little whether I give you a map with numbers spread in all places (used as proper names of the places) or a map with an abscissa and an ordinate. Nevertheless an operational equivalence does not imply anything interesting on the ontological level. In particular, it does not imply anything concerning the existence of a system of individuation of places by means of coordinates in the visual field.

I must here emphasize that I am not talking about the factual presence of a visible abscissa and ordinate in the visual field. This is not a fact and, moreover, it concerns a possibility which is envisaged and criticised by Wittgenstein himself at § 206:

> But even that isn't an accurate representation, since if we really saw such a set of axes of co-ordinates [. . .], we would in fact be in a position, not only to establish the orientations of objects relative to these axes, but also the position of the cross itself in the space, as though in relation to an unseen co-ordinate system contained in the essence of this space.

What I am talking about is thus rather that orientation which is contained 'in an invisible way' in the very essence of space. And what I object to is that the existence of *such* an orientation should be implied by the existence of absolute places in the visual field. My argument is very simple: As far as individuation is concerned, the existence of any system of coordinates is operatively equivalent to the existence of absolute places in visual space. Thus even if it were possible to positively infer the existence of a system of coordinates by basing ourselves upon the existence of absolute places, it would not yet be possible to infer the existence of *this* system of coordinates to which Wittgenstein refers. But this goes for *any* system of coordinates. Therefore we cannot infer the existence of *any* particular system of coordinates.

Curiously, Wittgenstein appears to grasp this difficulty when he says that "this above and below, right and left have nothing to do with gravity or right and left hands" (§ 206). This is quite puzzling, but either Wittgenstein wants to say that this system is arbitrary, and therefore there is no reason to use specific proper names for these directions – and names such that they resemble the names for left and right; or he intends to keep the analogy with the directions of the orientation field; and in both cases the previous argument applies.

Indeed, Wittgenstein seems to choose the second interpretation, which is to consider visual coordinates as being somehow analogous to orientation coordinates. His example is:

> If I hold a book the wrong way, I can't read the print at all, or only with difficulty. (206)

This takes us to the discussion of point (2): I should like to say that the visual field is oriented according to our *Bewegungsgefühlsraum*. If this is true, the association of such coordinates with the structure of the visual field (although extrinsic, as the previous argument shows) is strongly conditioned by the fact that the visual field belongs to a being which has a body and a capacity of movement similar to ours. Let us think about what happens when we look at a wall in front of us as we slightly balance our head from left to right. If there were a line on the wall and if this line could be considered a visual line, we would say that it has kept its orientation, even if the visual field has changed position in relation to the orientation of the line. By this I mean that the geometrical axes of the field turn with thc balancing of our head, but the axes of orientation are not modified.

Hence, factual orientation of the visual field is (at least in one case) linked to that of the orientation field. In other words, a change in position of the visual field does not always entail a change of orientation of the field in relation to the environment, but it *does* entails a change of the positions of the different parts of the field relative to the orientation of the field itself. Therefore, it entails the impossibility of linking the absolute position of the parts to orientation, because orientation remains, even if the absolute position changes.

Nevertheless, we have the possibility of making sense of the idea that the visual field has coordinates which move in concert with the movements of the field. This possibility is rendered effective by associating the system of coordinates of the visual field with the system of *head-centered* coordinates.[4] If we accept a pictorial version of this association, we could consider "visual right" as synonymous with "the parts of the visual field closer to the right ear", "visual top" as synonymous with "the parts of the visual field closer to the forehead", etc.

In my opinion, the visual field *can be given* this other oriented structure, which is mainly head-oriented, but this depends on the fact that this visual field belongs to an animal which has a head with such and such capacities of movement. The orientation does not result from the fact that there are absolute positions (as I tried to show against (1)): and therefore it has to base itself (so to speak) on another system of orientation. Hence orientation is not independent of this other system, contrary to the point Wittgenstein tries to make in stating (3).

Consequently, the last paragraph of the remark 206 presents an unsound argument:

> I can obviously see the sign V at one time as v, another as an A, as a 'greater than' or 'less than' sign, even if I were to see it through a telescope and cannot compare its position with the position of my body.
>
> Perhaps someone might reply that I feel the position of my body without seeing it. But position in feeling space[5] (as I'd like to put it for once) has nothing to do with position in visual space, the two are independent of each other, and unless there were absolute direction in visual space, you couldn't correlate direction in feeling space with it at all.

The false premise of this argument is the last one, which pretends to establish that the fact that the directions in visual space are not absolute would hinder us in correlating them with the directions of the orientation space. We have indeed no need of an absolute direction in visual space: it is sufficient that there are absolute *positions* in order to *construct* an orientation of the field.[6]

3. This concludes my examination of Wittgenstein's argument for the absoluteness of some properties of visual space. Let me briefly comment upon Wittgenstein's attitude towards this kind of problems.

We know that one of the reasons which motivated Wittgenstein to go to Cambridge to continue his philosophical work in 1929, was precisely his interest for the visual field. As he writes in a letter to Schlick:[7]

> Ich habe mich entschlossen ein Paar term hier in Cambridge zu bleiben und den Gesichtsraum und andere Dinge zu bearbeiten.

Wittgenstein's interest concerns, I believe, the discussions on the phenomenal languages envisaged in several texts by Russell and the logical positivists. Wittgenstein's phenomenological approach thus belongs to a reaction to this type of approach to the problem of perception and perceptual description; and not exclusively to a critical return to the metaphysical problems left open in the *Tractatus*. I think that the meticulousness of Wittgenstein's argument on this point indicates a profound effort of reflection which seems to me exceedingly difficult to square with the received historiographical schema.

In the 1930's Wittgenstein is not only the philosopher who extends the horizons of his philosophy of language by refusing the monolithic model of the *Tractatus* in favour of the multiplicity of the language games; he is moreover interested in phenomenological problems in themselves.

CNRS
Aix-en-Provence

NOTES

[1] *Philosophische Bemerkungen. Philosophical Remarks* (PR), ed. R. Rhees, B. Blackwell, Oxford 1964. Trans. by R. Hargreaves & R. White, B. Blackwell, Oxford 1975.
[2] The two short commentaries dedicated to the *Philosophische Bemerkungen* do not treat the topic I am concerned with here. N. Malcolm gave a survey of the fundamental themes of the *Bemerkungen* in 'Wittgenstein's Philosophische Bemerkungen', *Philosophical Review*, 76, 1976, pp. 220–229 by underlining above all the connections with other works by Wittgenstein. He does not mention Wittgenstein's interest in visual space. H. Spiegelberg also commented on the *Bemerkungen*: 'The Puzzle of Ludwig Wittgenstein's *Phänomenologie*', *American Philosophical Quarlerly*, 5, 1968, pp. 224–256, but from a metaphilosophical point of view. These two articles have been re-edited in *Ludvig Wittgenstein: Critical Assessments*, ed. by S. Shanker, Croom Helm, London 1986, Vol. I, pp. 213–221 and 222–240.
[3] For this reason the following paragraph, concerning the two stars which pivot around each other is irrelevant, since it concerns relative movements.
[4] A possibility envisaged by Husserl. See *Ding und Raum. Vorlesungen 1907*, ed. by U. Claesges, M. Nijhoff, The Hague, 1973, Abh.B. 'Stufen der Raumkonstitution', pp. 309–321.
[5] *Gefühlsraum*. It is precisely that which in the context of the present essay I called orientation space.
[6] Perhaps one of the reasons which led to Wittgenstein's error was his use of the example concerning the letters of the alphabet. In my opinion it is difficult to do without a component of the *Gefühlsraum* in the perception of the letters.
[7] 18.2.1929, quoted in the introduction by Brian McGuinness to vol. III of *Schriften*, Suhrkamp, Frankfurt 1967, p. 17.

PART IV

BELIEF, DOUBT, AND RELATIVISM

NORMAN MALCOLM

DISENTANGLING MOORE'S PARADOX

In the two volumes of *Remarks on the Philosophy of Psychology*[1] and in the first volume of *Last Writings on the Philosophy of Psychology*[2] Wittgenstein wrote approximately one hundred and thirty remarks on the verb 'to believe'. This concentration on a single verb was chiefly provoked by what Wittgenstein called 'Moore's Paradox'. What is Moore's Paradox?

G. E. Moore noticed that such a sentence as, 'I went to the pictures last Tuesday, but I don't believe that I did,'[3] would be an absurd thing for anyone, yourself for instance, to *say*. Moore remarked that this

> is a perfectly absurd thing to say, although *what* is asserted is something which is perfectly possible logically: it is perfectly possible that you did go to the pictures and yet you do not believe that you did. . .[4]

In a later essay Moore referred again to this 'absurdity'. Taking the example of someone's saying 'I believe he has gone out, but he has not', Moore's view continued to be that this is an absurd thing to *say*, but that *what* is said is 'logically possible'. As he put it, 'This, though absurd, is not self-contradictory; for it may quite well be true.'[5]

The central idea in Moore's view of the Paradox is that such a sentence as 'It is raining and I don't believe it is raining', is an *intelligible* sentence. A number of philosophers have taken the same position. Like Moore, they have thought that the sentence itself is intelligible; it makes sense; we understand it; it expresses a logical possibility; it could even be true. What is 'absurd', or 'strange', or 'logically odd', is to *say* it.

I think this position is plainly wrong. If I asked someone what the weather is outside, he might reply 'It's raining', or 'I believe it's raining', or 'I believe it's not raining'. I would understand these replies. But if he said, 'It's raining and I don't believe it is', then I would not understand the sentence. I would think that I had misheard him. If he insisted on repeating the same words, and if he seemed to be serious, I would be stunned. Certainly I could not take his sentence as a report about the weather; nor as information about anything. I might react to it as to a piece of deliberate nonsense, a grammatical joke.

Why should a philosopher think that this combination of words is an intelligible sentence? There are a number of considerations that I shall take up one at a time.

R. Egidi (ed.), Wittgenstein: Mind and Language, 195–205.
© 1995 *Kluwer Academic Publishers. Printed in the Netherlands.*

'It Isn't Self-Contradictory'. Moore was obviously influenced by the idea that a sentence of the form 'p, and I don't believe p', is not self-contradictory. It is true that it doesn't *look* self-contradictory. But perhaps this appearance is deceptive. Before we can be sure that such a sentence is not self-contradictory, we must reflect on the ordinary meaning of the expressions 'I believe p' and 'I don't believe p'. I will return to this point.

The Past Tense. The sentences that figure in the Paradox are sentences in which 'believe' occurs in the first person present tense. Examples are, 'It's raining and I don't believe it is', and 'I believe it's raining and it isn't raining'. The corresponding *past* tense does make sense: e.g., 'I believed it was raining and it wasn't raining'. The fact that the past tense makes sense, inclines one to think that the present tense *must* make sense.

One could reply that there is no justification for this 'must'. But far more illuminating is Wittgenstein's observation about the way in which the present tense, 'I believe p', is actually used. What he says is:

> 'I believe p' means roughly the same as 'p'. (RPP I, § 472)

He goes on to say:

> 'I believe it's going to rain' means something like: 'It's going to rain', but 'I believed then it was going to rain' doesn't mean anything like 'It rained then'. (RPP I, § 473)

Wittgenstein asks the following question:

> What does it mean to say that 'I believe p' says roughly the same as 'p'? We react in roughly the same way when someone says the first and when he says the second; if I said the first, and someone didn't understand the words 'I believe', I would repeat the sentence in the second form, and so on. As I would also explain the words 'I wish you would go away' by the words 'Go away'.[6]

Wittgenstein adds the following important explanation of why 'I believe p' can have approximately the same meaning as 'p'. He points out a significant logical difference between the first person and the third person of 'believe':

> That he believes such-and-such, we gather from observation of his person, but he does not make the statement 'I believe . . .' on the basis of self-observation. And *that* is why 'I believe p' can be equivalent to the assertion of 'p'. (RPP I, § 504)

I will return to this remark later. Right now I will say that Wittgenstein's statement that 'I believe p' means roughly the same as 'p', is an accurate observation of the use of 'I believe'. Most of the confusion about Moore's Paradox is due to the failure to notice this fact. On the surface it seems that 'p' and 'I believe p' have entirely different 'content'. But if we go beneath the surface and remind ourselves of the way we actually use the expression 'I believe p' in daily speech, we see that what it commonly amounts to is a *tentative* or *hesitant* assertion of 'p'.

Suppose I am indoors preparing to go out. I wonder whether I should take an umbrella; so I ask a person near the window to look out to see if it's raining. It is dark outside and hard to see through the window-glass. The person says, 'I believe it's raining'. I would take this as a cautious assertion that it's raining, and would fetch my umbrella. My reaction would be roughly the same as it would be if the person had simply said, 'It's raining'.

Wittgenstein later puts this point in a slightly different way:

> Consider: the words 'I believe it is raining' and 'It may be raining' (*Es dürfte regnen*) can say *the same: inasmuch* as in certain contexts it makes no difference which of the two sentences we use. . . That is: in order to say that perhaps it is raining we do *not* need the concept 'believe', although we can use it for that purpose.[7]

Suppose there were a characteristic tone of voice in which we said things of which we were not entirely certain. 'I believe . . .' is used to play the role of such a tone of voice. We have other linguistic devices that do the same. For example, 'It may be raining', or 'Perhaps it is raining'. Of course these sentences are also used in other ways. But they can be used to make the cautious assertion that it's raining. 'I believe it's raining' is typically used to do the same thing. As Wittgenstein says 'I believe' does not *have* to be brought in to do that job. But it *can* be, and often is, used for that purpose.

A statement of the form 'I believe p', is not always a hesitant assertion of 'p'. There are contexts in which it is an *emphatic* assertion of 'p'. Suppose that in a certain town there have been rumours that the Mayor is engaged in fraudulent practices. A supporter of the Mayor, speaking before an audience, declares 'No one here believes that the Mayor of our city is dishonest'. Someone shouts from the audience: '*I* believe he is dishonest'. In that context this would be a forceful assertion that the Mayor is dishonest, not a hesitant one.

The fact that 'I believe p' is commonly an assertion of 'p', destroys what might be called 'the argument from the past tense'. This argument assumes that 'I believe p' and 'I believed p', differ *only in tense*. This is an incorrect assumption. 'I believe p' is normally used to assert 'p', sometimes hesitantly, sometimes emphatically. But the past tense, 'I believed p' is never an assertion of 'p'. Moore's Paradox exists only in the present tense.

The Supposition. When Moore declared that a sentence of the form 'p, and

I don't believe p' or 'I believe p, and p is not true', expresses 'a logical possibility' and 'may quite well be true', could be have confused an *assertion* with a *supposition* ? For the corresponding supposition *does* make sense. There is no absurdity involved in saying, 'Let us suppose that I believe it's raining and in fact it isn't raining'. What this supposition proposes *is* 'logically possible'. In this connection Wittgenstein makes the following remark:

> Moore's Paradox may be expressed like *this*: 'I believe p' says roughly the same as 'p'; but 'Suppose I believe p. . .' does not say roughly the same as 'Suppose p. . .'.[8]

In saying that Moore's Paradox "may be expressed" in this way, Wittgenstein does not, of course, mean that Moore expressed it in this way. What Wittgenstein might mean by this curious remark is that if it were not for this logical difference between 'I believe p' and 'Suppose I believe p', then the sentences that Moore spotted would *not* be nonsensical.

Another thing that Wittgenstein's remark may suggest is that this difference between the assertion and the corresponding supposition could be called 'paradoxical', not because there is anything logically dubious about it, but because it is *surprising*: it goes against our expectations. For in other cases the relation between an assertion and the corresponding supposition is that what is asserted and what is supposed is *the same*. The assertion 'The ship sails at dawn', is the assertion of what is supposed by 'Suppose the ship sails at dawn'. The assertion says neither more nor less than what is supposed in the supposition. But this is not so in the case of 'I believe'. This is why Wittgenstein says:

> So it looks as if the assertion 'I believe. . .' were not the assertion of what the supposition 'I believe' supposes. (RPP I, § 493)

To suppose 'I believe it's raining' is *not* to suppose 'It's raining'. But to assert 'I believe it's raining' *is* to assert 'It's raining'. The assertion asserts something that is not supposed by the supposition. The assertion does not exactly *match* the supposition; it brings in another element.

But would it be right to say that 'I believe it's raining' is *not* the assertion that corresponds to the supposition 'Suppose I believe it's raining'? Surely not. What *other* assertion would be the corresponding one? There is no answer. We must simply *accept* this peculiar feature of the logical grammar of 'I believe'.

In this connection Wittgenstein makes a penetrating remark:

> With the words 'Suppose I believe that' you already presuppose the whole grammar of the word 'believe'. We are not supposing something that, so to speak, is given to you unambiguously through a picture, so that you can tack on to this supposition something other than the ordinary

> assertion. *You would not know at all* what you were supposing here, if you were not already familiar with the use of 'believe'.[9]

If you substituted some assertion other than 'I believe p', as the assertion that 'correspond' to 'Suppose I believe p', then you would not understand what you were supposing.

We have seen already that the *past tense* of 'believe' presents the same 'paradox' as does the *supposition*. The present tense, 'I believe p', *is* an assertion of 'p'; the past tense, 'I believed p', is *not* an assertion of 'p'. So it *looks* as if 'I believed it was raining' is not the 'correct' past tense of 'I believe it is raining'. But is there a 'more correct' match? Certainly not. We can apply here the same point that Wittgenstein made about the *supposition*. The assertion 'I believed' presupposes the whole grammar of 'believe', including that of the first person present. Otherwise you would not know what you were asserting in saying 'I believed'.

Is 'I Believe P' a Description of a Mental State? We come now to the main source of perplexity about the absurd formulation, 'p, and I don't believe it', or 'I believe p, and p is not true'. On the one hand these conjunctions strike us as nonsensical. On the other hand, it seems that they should *not* be nonsensical: for the two parts of such a conjunction appear to be dealing with separate subject matters; they appear to be logically independent of one another. Their conjunction in a single sentence should not produce any logical conflict. As regards the conjunction 'I believe it's raining, and it isn't raining', one wants to say that the first part is about a mental state, while the second part is about the weather. So why shouldn't I be able to put together, without absurdity or inconsistency, a report of a mental state of mine, and a report of the weather?

Let us agree that when I say 'It isn't raining', I am speaking about the weather. The problem lies in the assumption that when I say 'I believe it's raining', I am reporting or describing a mental state of mine.

We have already seen that 'I believe it's raining' is a cautious assertion of 'It's raining'. Now someone might agree with *that*, but hold that 'I believe it's raining' *also* reports or describes a mental state or attitude of the speaker. Wittgenstein imagines someone putting this view as follows:

> At bottom, with these words I describe my own mental state – but here this description is indirectly an assertion of the state of affairs that is believed.[10]

At first sight this appears to be a very plausible view. It amounts to saying that the statement 'I believe it's raining', does two jobs at once. It is *directly* a report or description of a mental state, and *indirectly* an assertion about the weather. This view would provide an explanation of why the whole conjunction is absurd: for the 'indirect' assertion of 'It's raining', carried by the first member of the conjunction, 'I believe it's raining', contradicts the explicit assertion by the second member of the conjunction, 'It isn't raining'. The interpretation is attractive: for it attributes the absurdity to a logical contradiction; yet it also preserves the inclination to think that 'I believe p' describes or reports a mental state.

But a pressing question arises. *Why* should what is primarily, or 'directly', a description or report of a mental state, be at the same time an 'indirect' assertion about the weather? What connection is there between a description of a mental state and a description of the weather? Why should 'I believe it's raining', construed as a description of a mental state, be an indirect assertion of 'It's raining' – rather than an indirect assertion of 'Today is Tuesday', or of 'Turkey wants to join the Common Market', or of anything whatever?

I think it is this problem that leads Wittgenstein to suggest that the interpretation in question is treating the sentence 'I believe it's raining' as if it were a *photograph*. Commenting on the idea that 'I believe it's raining' is an 'indirect' assertion of 'It's raining', he says:

> As in certain circumstances I describe a photograph, in order to describe what it is a photograph of.[11]

But if one does this one is assuming that the photograph is *reliable*. Wittgenstein turns this point into a profound criticism of the view at issue. He says:

> But if this analogy holds good, then I should have to be able to say further, that this photograph (the impression on my mind) is trustworthy. So I should have to be able to say: 'I believe that it's raining, and my belief is trustworthy, so I trust it'. As if my belief were a kind of sense-impression.[12]

We are sometimes aware of having illusory sense-impressions. It looks as if the railroad platform is moving, but we know it is only our train that is moving. So often we trust our sense-impressions, but sometimes not. Now I can trust or distrust another person's belief. But can I do this with *my own* belief? Can I firmly hold a belief, but at the same time distrust it? Wittgenstein delivers this smashing remark:

> One can mistrust one's own senses, but not one's own belief. (LWPP I, § 419)

I think that Wittgenstein has completely destroyed the conception that 'I believe p' is, first of all, a description or report of a mental state, and secondly, an assertion of 'p'. For why does this mental state refer to 'p' in particular? How does it pick out 'p' from all of the things that might be asserted? Presumably, because this mental state is or contains a 'picture' or a 'representation' of 'p'. If I asserted 'p' I would be assuming that my mental state (my belief) is a *reliable* picture, a trustworthy representation. This in turn implies that I can think of a *present* belief of mine as being reliable or unreliable – which is nonsensical. Wittgenstein has undermined the notion that the sentence 'I believe p' is 'primarily' or 'directly' a description of a mental state.

The chief source of our confusion about a sentence such as 'I believe it's raining and it isn't raining', is our inclination to think that the first part is an assertion about what is *within* me, while the second part is an assertion about what is *outside* me. It *seems* that there can be no conflict here; yet we feel that the whole sentence is absurd. Wittgenstein fixes his attention on this image of the contrast between something *within* me and something *outside* me. He says:

> 'It is raining and I believe it is raining'. Turning to the weather I say that it is raining: then, turning to myself, I say that I believe it. – But what do I do when I turn to myself, what do I observe? Suppose I say: 'It's raining, and I believe it will soon stop' – do I turn to myself at the second part of the statement? – Of course, if I want to find out whether *he* believes that, then I must turn to him, observe him. And if I wanted to find out by observation what I believe, I should have to observe my *actions*, just as in the other case I have to observe his.
>
> Now why don't I observe them? Don't they interest me? Apparently *not*. I hardly ever ask another person, who has been observing me, whether he has the impression that I believe such and such: that is, in order in this way to make inferences to my future actions. . .[13]

From the actions and utterances of other persons I continually make inferences about their beliefs and likely future actions. But I don't do this in my own case. Why not? Wittgenstein says:

> How is it that I cannot gather that I believe it's going to rain from my own statement 'It's going to rain'? Can I then draw no interesting conclusions from the fact that I said this? If someone else says it, I conclude perhaps that he will take an umbrella with him. Why not in my own case?
>
> Of course there is here the temptation to say: In my own case I don't *need* to draw this conclusion from my words, because I can draw it from my mental state, from my belief itself. (RPP I, § 704)

The temptation to give this 'explanation' is nearly irresistible. One wants to say: 'I *know* what I believe; I don't have to infer it from my actions or my words'. This attempt at explanation can be criticized in various ways. One thing Wittgenstein does here is to scrutinize the meaning of the words, 'I know what I believe'. He begins with a sample statement of the 'explanation':

> 'Why should I draw conclusions about my behaviour from my own words, when in any case I know what I believe'? And what is the expression of my knowing what I believe? Isn't it

expressed in this: that I do not infer my behaviour from my own words? That is the fact.[14]

This brief criticism exposes the emptiness of the proposed explanation. And the criticism applies equally to the idea that I don't have to infer *what I believe* from my actions or words, since I *know* what I believe without inferring it from anything. The point can be put like this: the only *meaning* that the words 'I *know* what I believe' will bear in this context, is just that I do *not* infer what I believe from my words and actions. So the proffered explanation is circular.

That a person can determine by observation what others believe, but rarely, if ever, by self-observation what he himself believes, is something that applies not only to belief but to many psychological concepts. Wittgenstein's insight into this feature of the grammar of those concepts, is one of his great contributions to the understanding of what it is to be a human being.

I want to try now to give an account of the part that this difference between the first person and the third person of 'believe', plays in the disentangling of Moore's Paradox. I will return to a remark of Wittgenstein's that I quoted previously:

> That he believes such-and-such, we gather from observation of his person, but he does not make the statement 'I believe . . .' on the basis of self-observation. And *that* is why 'I believe p' can be equivalent to the assertion of 'p'.[15] (RPP I, § 504)

This compressed reasoning can be expanded as follows: Since the statement 'I believe it's raining' is not based on self-observation, the only observation it is likely to be based on is an observation of the weather. So it can be roughly equivalent to the statement 'It's raining', when this is based on an observation of the weather.

Let us return to the problematic sentence, 'I believe it's raining, and isn't raining'. Moore thought that *'what* is said*'* makes sense, expresses a logical possibility, could be true. The 'absurdity' lies only in *saying* it. It is as if one could *think* it, but not *say* it.

Wittgenstein's reflection on the way in which the expression 'I believe p' is commonly used, leads to an entirely different view of the Paradox. 'I believe p' is a cautious or hesitant assertion of 'p'. The statement 'I believe it's raining, and isn't raining', would be both an assertion and a denial of 'It's raining'. It would have the normal incoherence of a self-contradictory sentence. Each part of the sentence cancels the other part. The outcome is nonsense. Moore was wrong in claiming that '*what* is asserted is something which is perfectly possible logically'. For nothing is being 'asserted': in a clear sense, nothing is being *said*.

Moore insisted that the sentence is not self-contradictory. But this was because Moore did not perceive those surprising features of the logical grammar of 'believe' which Wittgenstein brought to light.

I want to consider further the problem of whether a belief is a 'mental state'. Some philosophers have assumed that this is so. David Armstrong, for example, says that 'Belief is a dispositional state of mind which endures for a greater or lesser length of time. . .', and that 'there must be some difference in A's state of mind if he believes p from his state of mind if he does not believe p'.[16] I find this a puzzling assertion. Armstrong is speaking in the most general terms. He is not saying just that *some* beliefs are mental states (or 'states of mind'), but that *all* of them are. And when Armstrong says that there must be some difference in a person's mental state if he believes p from his mental state if he doesn't, Armstrong is speaking with complete generality. Apparently he is holding that to believe p (regardless of what p is) *is itself* a mental state. And this would appear to imply the view that every statement of the form 'I believe p', describes or expresses a mental state.

I do agree that sometimes when people say things of that form they are describing or disclosing mental states. If a friend said to me, 'I believe I am going mad', that would tell me something about his mental state. Or if he said, 'I believe that all the people I see in the streets are spying on me'.

There can be a different kind of case in which a statement of the form 'I believe p' expresses, not a 'mental state', but an *attitude* toward a person. Suppose a man is accused of intending to avoid payment of a debt. A friend of his says in his defence, 'I believe he will pay it'. I am supposing that he does not say this on the basis of any special information, but instead that he is declaring his faith in the integrity of his friend. As I am imagining the situation, this would not be a cautious or hesitant assertion, but a firm statement of trust. His statement, 'I believe he will pay the debt', does two jobs here: it asserts 'He will pay the debt', and it expresses an attitude towards the accused person.

Whether a statement of the form 'I believe p' describes or expresses a mental state, or mental disposition, or an attitude, depends on the subject matter and the particular circumstances. *Sometimes* sentences of that form are used in that way. What is wrong is the assumption that this is *always* so: that the words 'I believe p', *by their very meaning*, describe or disclose or express, some mental state or disposition or attitude. This is an example of *thinking without looking*. Wittgenstein's treatment of Moore's Paradox should teach us better.

When a farmer studies the sky and then says, 'I believe it will snow this afternoon', his remark is tantamount to the cautious assertion, 'It will snow this afternoon'. In saying, 'I believe it will snow this afternoon', he need not be ascribing to himself, nor expressing, any mental state, disposition, or attitude. He need not even be saying that he is in a hesitant frame of mind about whether it will snow. As Wittgenstein points out,

> A hesitant assertion is not an assertion of hesitancy. (LWPP I, § 522)

Although I have seen emphasizing the fact that there is a common use of sentences of the form 'I believe such & such', in which the speaker is not talking about himself, is not describing some attitude of his, but is talking about the such & such – yet it is true that in certain situations someone who says 'I believe such & such', *is* talking about himself. I do not wish to give the impression that Moore's Paradox arises *only* when the speaker is talking about the such & such, and not when he is talking about himself.

Let me give an example. Suppose that a committee of a trade union has been debating the question of whether the members of the union should go on strike. The advantages and disadvantages of a strike have been thoroughly discussed. There seems to be no point in continuing the debate, even though there still is disagreement. The chairman then says: 'It would be fruitless to hear again the reasons for or against a strike. What I want to hear from each of you is whether you personally believe there should be a strike'. He puts to each member the question, 'Do you believe there should be a strike?' When a member replies, 'I do believe there should be a strike', it seems right to say that here he is talking about, or for, *himself* – that the primary thing he is doing is expressing his own attitude on the issue. Yet it would be a logical absurdity for him to say, 'I do believe there should be a strike, but there shouldn't be one'.

The important question, therefore, is not whether when one says 'I believe p' he is talking about himself, but rather *how* he is talking about himself, when that is what he is doing. Wittgenstein has reminded us that usually I say of someone, 'He believes p', on the basis of observing *him*, but that usually I do not say 'I believe p' on the basis of observing myself. 'He believes p, but not-p' is not an absurd form of words. Moore's Paradox exists only in the *first person* of the present tense. But it does not exist even there, when (as sometimes happens) the statement 'I believe p' does arise from self-observation and recollection. If I remember certain things I have said and done in the recent past, or certain ways I have reacted, and if solely on the basis of these recollections I exclaim, 'It appears that I believe p', or 'I seem to believe p', or 'So I believe p' – then there would be no absurdity in adding, 'but not-p'. It would not be absurd, for there I would be speaking of myself *as if* I were another person.

Thus, at the most fundamental level, Moore's Paradox arises from the fact that there is an essential difference between the normal use of the first person singular present tense of 'believe', and the use of the third person form. The grammar of the word 'believe' reflects a striking feature of human life, namely, the fact that there is an attentive observation which we direct on other people, and not on ourselves.

King's College
London

NOTES

[1] Vol. I (RPP I), ed. by G. E. M. Anscombe & G. H. von Wright. Trans. by G. E. M. Anscombe, B. Blackwell, Oxford 1980; Vol. II (RPP II), ed. by G. H. von Wright & H. Nyman. Trans. by C. G. Luckhardt & M. A. E. Aue, B. Blackwell, Oxford 1980.
[2] Vol. I (LWPP I), ed. by G. H. von Wright & H. Nyman. Trans. by C. G. Luckhardt & M. A. E. Aue, B. Blackwell, Oxford 1982.
[3] See G. E. Moore, 'A Reply to My Critics', in *The Philosophy of G. E. Moore*, ed. by P. A. Schilpp, Tudor Publishing Co., New York 1952, p. 543.
[4] *Ibid.*
[5] *Philosophical Papers*, G. Allen & Unwin, London, 1959, p. 175.
[6] RPP I, § 477 [The translation from German by Norman Malcolm]
[7] RPP I, § 821 [The translation from German by Norman Malcolm]
[8] RPP I, § 478 [The translation from German by Norman Malcolm]
[9] RPP II, § 416 [The translation from German by Norman Malcolm]
[10] RPP I, § 481 [The translation from German by Norman Malcolm]
[11] *Ibid.*
[12] RPP I, § 482 [The translation from German by Norman Malcolm]
[13] RPP I, § 715 [The translation from German by Norman Malcolm]
[14] RPP I, § 744 [The translation from German by Norman Malcolm]
[15] RPP I, § 504 [The translation from German by Norman Malcolm]
[16] See D. M. Armstrong, *A Materialist Theory of the Mind*, Routledge & Kegan Paul, London 1968, p. 214.

EVA PICARDI

MALCOLM ON MOORE'S PARADOX

As I find myself in agreement with most of what Professor Malcolm has said in his paper, my task as a discussant is comparatively easy. Accordingly, I shall confine myself to a few brief remarks and pause on a statement of Professor Malcolm's which in my opinion stands in need of qualification.

Let me first mention what I particularly liked about Professor Malcolm's interpretation, namely the emphasis he has laid on the decisive role which the asymmetry between the first and the third person present tense plays in Wittgenstein's argument to the effect that the words "I believe" in their natural setting are not *descriptive* of the state of mind of the speaker. This however does not rule out that they can be used to say anything about the speaker: indeed, as Professor Malcolm points out in the last section of his talk, they are sometimes used for *disclosing an attitude* towards people or events. Disclosing an attitude, however, is very different from *describing a mental state*: we can account for this use of the words "I believe" without endorsing the naive picture of mental states as something which is there to be inspected, recorded or photographed by the mental eye of the speaker with his privileged access. When I say "I believe it's raining" I am talking about the weather and not about me. If one wants to say (and Moore surely wanted to claim at least this much) that "I believe" illuminates *(beleuchtet)* the state of the speaker, then we have to say the same thing of the assertion "p".[1] This is why "I believe that p" as an *isolated assertion* can mean the same as "p". The office of these words is to hint (*andeuten*, instead of straightforwardly saying) that an assertion is being made. As Frege stressed, in language we do not have any special mark to signal assertoric force, and Frege's assertion sign is no better off in this respect. It is only seen but not heard and this is why you could not detect its presence in Professor Malcolm's quotation. Here I am just drawing your attention to the fact that the notions and the terminology which Wittgenstein uses for formulating Moore's paradox derive from Frege. In the *Remarks on the Philosophy of Psychology*, but not in the *Philosophical Investigations*, both Frege's name and his assertion sign are explicitly mentioned. This sudden resurrection of Fregean notions should be found perplexing in the extreme and alert the reader to the very complex gambit Wittgenstein is trying to play here. For as we know from the *Tractatus* and the *Philosophical Investigations*,

R. Egidi (ed.), Wittgenstein: Mind and Language, 207–210.
© 1995 *Kluwer Academic Publishers. Printed in the Netherlands.*

Frege's idea of splitting an assertion into a thought (functioning as *Annahme)* and a force indicator has never satisfied Wittgenstein because of the accompanying naive picture of an inner mental theatre in which a thought is first grasped, then judged and finally asserted. Luckily, many of these Fregean connections have been pointed out by J. Schulte in his paper 'Es regnet, aber ich glaube es nicht'.[2] But something still remains to be said. Wittgenstein's chief move is that first he uses Frege's embedding test to detect the source of Moore's puzzle and then uses this result to controvert both Moore's and Frege's naive belief that words function in one way only. And Wittgenstein also argues for the converse of his claim in *Philosophical Investigations*, § 28 according to which it is not the case that every assertion contains an assumption. This rather mysterious statement is illuminated by some of his later remarks. Here he wants to show that not every assumption contains a potential assertion, an assertion which matches the content put forward for consideration. To appreciate the first part of Wittgenstein's claim it is sufficient to see that which is asserted in "I believe that the train is late" is not the same as that which is assumed when these words are used in the antecedent of a conditional. As isolated assertions "The train is late" and "I believe that the train is late" can say the same thing but in the sentences "If I believe that the train is late, then we'll miss the plane" and "If the train is late, we'll miss the plane" obviously do not.

As regards the second claim – that not every assumption contains a potential assertion – Professor Malcolm says:

> The assertion asserts something that is not supposed by the supposition. The assertion does not exactly *match* the supposition; it brings in another element.
>
> But would it be right to say that 'I believe it's raining' is *not* the assertion that corresponds to the supposition 'Suppose I believe it is raining'? Surely not. What *other* assertion would be the corresponding one? There is no answer. We must simply *accept* this peculiar feature of the logical grammar of 'I believe'.[3]

I am not entirely happy with this remark. The answer "Surely not" sounds a bit too sweeping, for it suggests that we have already got hold of an explanation of the remarkable grammatical fact, whereas at this stage of the discussion that is not yet the case. We still need an argument which shows that we do not stand to our words in the same way in which we stand to other people's words. This argument is in fact that which Professor Malcom provides in the second part of his talk.

My next point is a little bit more substantial and concerns the observation that the "assertion does not quite *match* the supposition; it brings in another element". Now it seems to me that this is not a correct description of the results achieved so far. For it seems to me that our understanding of these conditionals presupposes that we are conversant not only with the use of the verb "to believe" in the first person present tense but with a good deal more, indeed with the whole grammar of the word "believe", as Wittgenstein says. The argument given so far has established that "I believe p" can say the same as an assertion of "p"; moreover, in order to understand what it is to make an assertion we need the concept of belief. Thus for an understanding of the assumption embedded in such a conditional we need a fuller grasp of what is it to ascribe beliefs to our fellow creatures, a grasp which is not exhausted by the observation that to assert p and to assert "I believe that p" can amount to the same thing. We have not yet found the appropriate contexts whence our grasp of the words "I believe" as they occur in the assumption might have been derived. I think that we derive this understanding from our practice of ascribing beliefs to other people, an understanding which we then derivately apply to our own statements. Here we must not suppose that it is by introspection that we come to grasp the notion of belief needed to formulate the thought expressed by the assumption. We should ask what it means to ascribe beliefs to others and on which basis such ascriptions count as correct. And in this context we must advert to the outer criteria which every ascription of mental states stands in need of.[4]

Thus the first lesson to be learned is that in grasping the words "I believe" occurring in the assumption we also bring to bear our understanding of "he believes . . ." used as an ordinary assertion, that is, our understanding of what it is to attribute beliefs to others on the basis of what they say. But now you may ask: What is the difference? Do not we proceed in the same way when ascribing beliefs to other people and when ascribing them to ourselves? The answer is: No, we do not. I do not normally infer my beliefs from my utterances, which is what we tend to do concerning the beliefs of others. We do not guess our future actions from listening to what we are saying, nor do we treat beliefs as sense impression. This answer in turn leads to the second important lesson implicit in Wittgenstein's argument, namely that we should look in vain for an assertion in the first person present tense which exactly corresponds to that which the assumption assumes. In this sense also the converse claim seems vindicated, viz. that not for every assumption there is a corresponding assertion. For if this were the case it should be possible to make an assertion and in the same breath to take a distance from it in the way we do with respect to the utterances of other people. This mistaken image is the source of Moore's paradox which I find successfully dispelled by Wittgenstein's analysis.

Università di Bologna

NOTES

[1] *Philosophische Untersuchungen. Philosophical Investigations* (PI), ed. by G. E. M. Anscombe & R. Rhees. Trans. by G. E. M. Anscombe, B. Blackwell, Oxford 1953, II, p. 515.
[2] *Teoria* , 2, 1985, pp. 187–204.
[3]See p. 198.
[4] PI I, § 580.

WOLFGANG CARL

WITTGENSTEIN'S REFUTATION OF SCEPTICISM IN 'ON CERTAINTY'

Since the days of antiquity the refutation of scepticism is a highly estimated occupation of philosophers, but it was not till Descartes that it was regarded as a condition of any justified claim to knowlege. Whoever did not attempt to refute the sceptic and succeed in doing so could not be certain to be in a position to reject his doubts and was, therefore, for Descartes a concealed sceptic.[1] The refutation of scepticism became the best way ("optima via") of reaching knowledge which meets philosophical standards[2] and, in this way, scepticism itself changed radically: it didn't stand any more for a way of life free of all dogmatic claims to knowledge, but it was a philosophical standpoint taken for tactical reasons. It was not the end, but a means of philosophical reasoning. The doubt of Descartes is a "staged", methodological doubt. There are other modifications of the sceptical doubt and the strategies to reject it connected with this change in the role of the sceptic.[3] I don't pay attention to them because one can make sense of the idea 'no knowledge without refutation of scepticism' without Descartes' distinction between the mental and the physical and without claiming any epistemological priority of the mental. It is this idea which makes his 'First Meditation' so attractive and which seems reasonably plausible, as Stroud has shown rather recently.[4]

Since Descartes it seems to be a truism that the sceptic doubts the reality of an external world; for Moore and Wittgenstein this kind of doubt is the characteristic mark of the sceptic. But how is it to be understood? Wittgenstein observes: "'Doubting the existence of the external world' does not mean for example doubting the existence of a planet, which later observations proved to exist".[5] Such a doubt does not concern the existence of a certain object or a certain kind of objects. If Descartes attempts to justify a doubt whether he is sitting by the fire or is touching a piece of paper, he doesn't want to doubt some fact or other. Rather, he is trying to doubt what is presupposed by every critical examination of our beliefs and thus cannot be itself under consideration; he is trying to doubt what "stands fast" in all our revising and examining of beliefs. The sceptical doubt doesn't concern any "white spots" of the increasingly comprehensive and detailed map of our knowledge; it aims at those beliefs the truth of which is obvious to everybody and presupposed by all critical examination of our beliefs. So, the doubt of the reality of an

Rosaria Egidi (ed.), Wittgenstein: Mind and Language, 211–222.
© 1995 *Kluwer Academic Publishers. Printed in the Netherlands.*

external world is not remarkable with regard to its subject; only by taking account of the sceptic's intentions can one understand his doubt. He wants to keep a distance between himself and what everybody who entertains belief takes for granted.[6] But why does he seek this distance?

An answer may be found by considering the reasons given by Descartes for doubting his beliefs. The so-called "argument from dreaming" is an attempt to justify a doubt concerning beliefs which are acquired under ideal conditions of observation. I will not discuss whether he really succeeds in doing so, but confine myself to pointing out that this argument doesn't directly attack my beliefs about myself and my situation, but only indirectly by way of a critical consideration of my cognitive capacities. The question 'How do I know that I am not dreaming?' amounts, from Descartes' point of view, to the question, rather offensive for the reader, 'How do I know that I am not mad?', – i.e. to a question which is intended to undermine my confidence in beliefs acquired under ideal epistemic conditions. The faith in the assumptions is shaken that these conditions are satisfied and that I am able to know anything at all. This doubt doesn't concern my beliefs about what I see or come to know otherwise, but my beliefs about my epistemic capacities. What is at stake is my conviction that I am able to know anything. This is the subject of the doubt called by Descartes "metaphysical" or "hyperbolic doubt". You can't dissolve this doubt by referring to your established knowledge of what is going on in the world; you have to challenge this doubt by raising the fundamental question whether I am in a position to know anything at all.[7] The sceptic makes an attempt to raise this question. Because he wants to raise it, he keeps his distance from the established standards of making and justifying claims of knowledge. But why does he want to raise this question?

The 'first meditation' gives an answer to this question too. The well-known *genius malignus* is said to have arranged the world in such a way that whatever I take to be true is false. Nowadays this thought-experiment, lacking its theological content, is replaced by the assumption that I am a "brain in the vat" manipulated by a malicious scientist and his computer. Descartes considers the possibility of man being so imperfect that he is always wrong.[8] This imperfection does not only concern the truth of my beliefs but also their material adequacy. Descartes considers the possibility that his representation of the world doesn't fit it at all. However, to be able to consider such a possibility he has to assume that the world is totally independent from our representations of it and from the kind of representations we have. This assumption states a condition which seems to be essential and indispensable for our idea of what knowledge really is. Following Stroud, it shall here be called the "very idea of objectivity".[9] This seems to be a trivial claim: what is known is conceived as being independent of our knowledge. What we know is not produced by our knowledge and doesn't depend upon it. It is possible that what is known is itself some kind of knowledge or an effect of it; but in this case too, knowledge doesn't produce what is known. To understand the sceptical doubt we don't have to consider this complex case; we may confine ourselves to the simple case of the knowledge of the world, i.e. the knowledge that something which is what it is, independently of our beliefs and knowledge. It is the subject of our knowledge which is independent in this way. The fact *that* we know that there is no life on the moon, of course depends on our other beliefs and knowledge. It requires a system of beliefs and assumptions giving that opinion its proper place. But this doesn't apply to *what* we know. This independence inspires the sceptic to take seriously the possibility considered by Descartes, and to wonder how we do know that the world isn't totally different from the best of our knowledge. The trivial claim for the objectivity of the subject of our knowledge admits this possibility. Since this claim is essential to a realist view of truth and knowledge, Stroud's and Nagel's thesis that scepticism and realism are two sides of the same coin, suggests itself.

So far, I have tried to give a rough sketch of the motives and intentions which stimulate the "modern", "Cartesian" sceptic. I haven't given an argumentative exposition of scepticism, and I have refrained from a critical examination. My intention was to make clear the meaning of the sceptical question and thus to indicate a material condition of adequacy for any refutation of scepticism. This condition may be expressed as follows: You can't challenge the sceptical doubt concerning what is obvious to everybody by just putting the obvious in its proper position, as Moore, for example, did. You have to account for the intention of the sceptic to call in question the very possibility of knowledge; and you can't do that without considering the idea of objective knowledge underlying that question. Without such a discussion the refutation of scepticism can't succeed. Whether Wittgenstein, in *On Certainty*, managed to refute scepticism, must be discussed with reference to this idea. I will sum up his remarks with regard to their subjects and concentrate on Wittgenstein's considerations on the concept of doubt and his analysis of knowledge.

"The child learns by believing the adult. Doubt comes *after* belief" (§ 160). The connection between believing and learning makes it quite clear that Wittgenstein's remark does not concern only the *genesis* of doubt. The child learns "to react" (§ 538); it acquires a piece of "knowledge of natural history" (§ 534), consisting in certainties which are not only earlier than justified knowledge and critical doubt but establish the very foundations of them. Doubt, seen from a genetic point of view as well as from a conceptual point of view, is an epiphenomenon which appears only after having "got hold" of a "world-picture" by one's cultural and intellectual socialization. This world-picture is, as Wittgenstein emphasizes, not deliberately chosen (§ 167). A genetic dependence of doubt upon belief won't puzzle the sceptic very much. Descartes himself repeatedly underlined this dependency. Not wanting to confine himself to believe what he believed he claimed that one should, once in a life, try to doubt every belief hitherto held for more or less obscure reasons. This attempt – *eversio opinionum* – presupposes that one already has opinions, that one believes something or other. But Wittgenstein does not only want to state the genetic posteriority of doubt; he rather tries to draw the boundaries of doubt and to refute Descartes' project in this way. For, the "game of doubting itself presupposes certainty" (§ 115). Or, "doubt itself rests only on what is beyond doubt" (§ 519). Thus the thesis 'no doubt without belief' must be understood as stating that we only can doubt something if we don't doubt something else but believe it.

This thesis confines the sceptical doubt twofold: it is not possible only to doubt, and it is not possible to doubt everything. Let's consider first the assertion that doubt without belief is not possible. Wittgenstein argues as follows:

> Admittedly, if you are obeying the order "Bring me a book", you may have to check whether the thing you see over there really is a book, but then you do at least know what people mean by "book"; and if you don't you can look it up, – but then you must know what some other word means. And the fact that a word means such-and-such, is used in such-and-such a way, is in turn an empirical fact, like the fact that what you see over there is a book. (OC, § 519; see § 114–115; § 370)

The belief which is in question here is a knowledge consisting in the understanding of a language; and the argument says that doubting presupposes the understanding of a language and thus the knowledge of the meaning of the word.[10] The sceptic who, by relying on the method of doubt, poses the question of the possibility of knowledge thus has to presuppose some kind of knowledge. Doesn't this project refute itself? I don't think so. Wittgenstein claims that understanding a language is some kind of empirical knowledge, – i.e. no knowledge of analytic truths. The sceptic will agree but call attention to the difference between "empirical facts" depending upon our linguistic usage, and empirical facts being part of a world which, independently of our talking and thinking, is what it is.[11] We saw that the sceptic poses the question of the possibility of knowing something about the world; and this question can't be answered by appeal to linguistic usage and linguistic intuitions.

In other words: the sceptic won't deny that the English word 'moon' is intended to signify a celestial body circling 'round the earth; but it doesn't follow from this that the moon exists. The kind of knowledge which is mentioned by Wittgenstein and has to be considered as a condition of doubt, doesn't belong to that kind of knowledge the possibility of which is called into question; and the sceptical doubt isn't yet removed by the belief which Wittgenstein takes to be its condition.

Wittgenstein's second objection to the sceptic says that doubt essentially belongs to a language-game and can only be understood within the context of a language-game (see § 24). Two anti-sceptical consequences are supposed to follow from that: scepticism doesn't correspond to any conceivable form of life; and every doubt is an "internal" doubt. Let's consider the first consequence. Doubting requires a certain "behaviour" (§ 354). The usage of linguistic expressions articulating a doubt or indicating that it has been removed has to be connected with an identifiable, specific "*practice* of the language" (§ 524). Doubting something or being certain of something requires specific epistemic circumstances. If these circumstances are not given and if the doubter's behaviour can also be understood without these sceptical "modifiers", his way of talking is "merely a bit perverse, and does not affect the issue" (§ 524). If the declaration of a doubt or the indication that it has been removed doesn't touch upon my acting and behaving, no form of life corresponds to the sceptical attitude and it is nothing but the "fifth wheel of a coach": "it does not infect practice" (§ 524; see § 428). Such a doubt is "hollow" (§ 312), it is no doubt.

Doubting is part of a language-game; hence there must be a form of life corresponding to it. This may be understood as an explication of a connection of knowledge and action which Wittgenstein critically uses against Descartes' project of universal doubt. "Would my knowledge still be worth anything if it let me down as a clue in action? And *can't* it let me down?" (§ 409). These questions are rhetorical ones and, moreover, not well-expressed ones. For, the knowledge referred to is a kind of knowledge which Moore considers in "A Defense of Common Sense"; it consists in certainties which cannot be separated from my acting and which thus can't justify those activities. They constitute the "basis of action, and therefore, naturally, of thought" (§ 411; § 414; see § 87). My actions, however, reveal – "with a certainty that knows no doubt" (§ 360) –, that I have those certainties. This acting doesn't refute the sceptic's doubt, it doesn't even let it arise: "My life shows that I know or am certain that there is a chair over there, or a door, and so on. – I tell a friend e.g. "Take that chair over there", "Shut the door", etc. etc." (§ 7). As there is no reason or justification for this life, there are none too for these certainties: they are "there" (§ 559). It isn't difficult to answer the sceptical question; being a senseless question it can't even arise.

Wittgenstein regards the language-game as a symbiosis of certainties and actions; the actions of men, therefore, show "that they believe certain things definitely" (§ 284). The sceptical question concerns the possibility of knowledge and it is asked from a standpoint which doesn't lie – as Stroud, following Clarke, says – "within the practical contexts which give our words their social point".[12] Descartes always stressed the specifically theoretical nature of the sceptical question.[13] But does such a question make sense? Wittgenstein denies this: acting is not possible without knowledge and this knowledge must be usable as "a clue in action" (§ 409). It seems to me that Wittgenstein himself rather critically comments on this connection of knowledge and action; he says: "So I am trying to say something that sounds like pragmatism. Here I am being thwarted by a kind of Weltanschauung" (§ 422). But irrespective of this the sceptic may reply that it is far from clear that the certainties necessary for action are more than fast-standing opinions and unshakeable beliefs. Why must it be a knowledge of that kind which the sceptic is looking for? Why isn't it enough here "to know 'for all practical purposes'"?[14] However, the sceptic isn't interested in that kind of knowledge. In other words: Wittgenstein's antisceptical objections concerning the connections of actions and certainties is convincing only if the debate is transformed into a fundamental discussion of the concept of knowledge. I'll come back to this issue soon.

Let's now consider the second objection to the sceptic which rests on the concept of a language-game. It starts with the claim that a doubt must be justified (see § 122; § 458). Reasons can justify knowledge in showing that one "is in a position" to know such a thing (§ 555); and respectively the reasons justifying a doubt must show that error is possible. A doubt makes sense if the correction of an error is conceivable; thus doubt presupposes the possibility of error. Wittgenstein maintains against Descartes first that reasons are part of a system of beliefs in which "some things stand unshakeably fast and some are more or less liable to shift" (§ 144). Thus a reasonable doubt must regard many things as "absolutely solid" (§ 151). The thesis discussed above – 'no doubt without belief' – had after all to put limits on what can be doubted. The claim 'no doubt without the possibility of justification' implies that this is what must stand fast relative to all questions and justifications – "as it were like hinges on which those turn" (§ 341). Thus, every doubt is an "internal" doubt, is relative to a frame of fast-standing beliefs which cannot be doubted. Second, Wittgenstein says that not all of our beliefs can in the same way be exposed to the risk of error: "For it is not true that a mistake gets more and more improvable as we pass from the planet to my own hand. No: at some point it has ceased to be conceivable" (§ 54). To assume the possibility of error in this case is a symptom of "a mental disturbance" (§ 71); to have this doubt is to be a "half-wit" (§ 257).

The sceptic will reply to this reasoning by asking a further question. What stands fast? What is the status of that which must stand fast in order to render justification and doubt possible? Is this, again, a matter of truth and knowledge concerning our linguistic usage and our habits of thought? Wittgenstein asks: "Who decides *what* stands fast?" (§ 125). This question is obviously treated as senseless as the possibility of "learned judges being asked whether a doubt was reasonable or unreasonable" (§ 453). Rather, "regarding it as absolutely solid is part of our *method* of doubt and enquiry" (§ 151). But does this imply its truth? Let's once more go back to the "argument from dreaming". The conviction that I am standing here is doubted by asking whether I know, after all, that I am not dreaming. Saying that we only know something about the external world if we know that we are not dreaming implies, as Stroud has tried to show, that every belief acquired under normal conditions of sense-perception may be doubted.[15] I'm not going to discuss Stroud's argument except to point out a weakness of Wittgenstein's thesis. The assumption that every doubt is "internal" doesn't meet the point of the sceptical strategy; for this strategy doesn't proceed by balancing the pros and cons within a homogeneous class of sentences which have the same error-proof conditions. The sceptic picks up one of these conditions which are supposed to be fulfilled if one knows the truth of any sentence of this class. How can Wittgenstein criticize this challenge? How can he justify the thesis that it must "stand fast" that this condition is fulfilled? He says that assuming the contrary would be a symptom of a "mental disturbance" (§ 71) and would seem to him "madness – of course, . . . also in agreement with other people" (§ 281). But are these little favourable estimations sufficient to reject the sceptical question whether I could know that I am not dreaming? Wittgenstein remarks on error: "In order to make a mistake, a man must already judge in conformity with mankind" (§ 156). This may also be said about doubt. But then the truth of what is taken "to stand fast" rests upon its being held true; and the sceptic has good reasons to reject this. So the thesis that every doubt is "internal" seems to be confronted with the difficulty that it either leaves open the truth of what "stands fast" – thereby not at all answering the sceptical question – or that it justifies its truth by a collective holding to be true – thereby giving an unsatisfactory answer to the question. There remains the possibility to deny the thesis by adopting the sceptical strategy of the "argument from dreaming". So did Descartes and Stroud.

Let's now consider the "anti-sceptical potential" of Wittgenstein's conception of knowledge; let's concentrate on his thesis that "the concept of knowing is coupled with that of the language-game" (§ 560). I can't explain this thesis here at length and with the necessary attention; so I shall confine myself to selecting two characteristics of the concept "language-game" which are of particular importance to the understanding of the thesis. According to it, knowledge must be regarded as holistic: it is not possible that one should only have one belief but there always is a manifold of inter-connected beliefs forming a "system" (§ 102). This system is the "element" in which they "have their life" (§ 105), – i.e. it determines the roles of the single beliefs in the totality of knowledge. Second, our knowledge is connected with a form of life which lies "beyond justified or unjustified" (§ 318-319) and thus "has to be accepted", "is given", "like our life" (§ 559).[16] In the following, these two theses about knowledge will be explained in more detail.

Since our knowledge forms a system, not all of our beliefs have the same "status" (§ 167): not all of them are "equally subject to testing" (§ 162). Some of them are the "foundation of my research" (§ 167) and serve as the "background against which I distinguish between true and false" (§ 94); they have a "peculiar logical role" (§ 136) and can't be identified by semantic criteria (see § 308) or by rules in general. Their role consists in determining a "world-picture" which serves as a condition and a framework of my knowledge. Wittgenstein compares such sentences with the rules of a game (see § 95). His analysis of how we know or are familiar with such sentences is of crucial importance to his criticism of scepticism. Here he is using the term 'certainty', having in mind a peculiar, non-epistemic attitude or mode of presentation. With reference to Kant's terminology one could speak of transcendental conditions of knowledge; for Wittgenstein, however, as opposed to Kant, there is no knowledge of such conditions which is open to justification and thus there is no transcendental philosophy.

The certainties which reveal that I have a world-picture, and what kind of world-picture I have, are certainties which I have "in agreement with other people" (§ 281) and can only have insofar as I "belong to a community which is bound together by science and education" (§ 298). The certainty I have is a certainty *we* have: it is a certainty I can't choose but which we receive through our cultural tradition and our socialization in which we "grow up": those certainties are the historically conditioned truisms of a community which in our case includes some knowledge of the various sciences.

The critique of scepticism based upon Wittgenstein's concept of knowledge says that those certainties constitute a form of life and can't be conceived as a kind of knowledge. In the words of Marie McGinn, they constitute "a non-epistemic attitude to judgments".[17] The non-epistemic status of the certainties follows from the fact that the world-picture which I share with others is the "inherited background against which I distinguish between true and false." (§ 94) This background is, therefore, neither true nor false (see § 205) and lies beyond what is accessible by a knowledge which rests on reasons (§ 243). So, "Moore's mistake" in his critique of sceptical doubt consists in 'countering the assertion that one cannot know that, by saying "I do know it"' (§ 521). The sceptic, as Wittgenstein says, is quite right in pointing out that I can't know whether that is my hand; but he gives the wrong interpretation of it. He regards it as a deficiency of justification whereas it is important to conceive those certainties as something beyond knowledge which rests on reasons. Our having such certainties doesn't mean believing certain propositions which "strike us immediately as true, i.e. it is not a kind of *seeing* on our part"; rather it reveals itself in our "*acting*, which lies at the bottom of the language-game" (§ 204). With regard to the language-game of knowledge this acting consists in our method of calculating (§ 212) or in our practice of judging: "From a child up I learnt to judge like this. *This* is judging. This is how I learned to judge; *this* I got to know *as* judgment" (§ 128-129). The quest for reasons for my judgments doesn't arise any more, because our practice of judging is exactly what we understand as judging. Our concept of judging consists in our making such judgments; we would not have this concept of judgment if we didn't judge this way. And we can't choose this concept or any language-game; we have no alternative to it. Wittgenstein's remark "The *truth* of my statements is the test of my *understanding* of these statements" (§ 80) must be formulated precisely: My taking something to be true reveals my having understood or not having understood *what* I take to be true. The sceptic, in this case refraining from judgment, doesn't understand the language he uses; and Wittgenstein won't try to refute him but will treat him like a child having to learn "to react in such-and-such a way" (§ 538) –, like a child having to learn the use of the expressions "before" it is able to pose the problem of the justification of its use (see § 548). As doubt is later than belief, knowledge only starts on a later moment (see § 538) – after we have already got certainties.

How will the sceptic react to this pedagogical therapy? Supposing that teachers are open at all to arguments (see § 314-315), he will try to evade it with the following argument. Marie McGinn said about Wittgenstein's critique of scepticism "that it has provided a model for understanding certainty as a non-epistemic attitude to judgments that have a non-empirical role".[18] This "model" implies that our knowledge fits into a world-picture we have since we have certainties which we aren't able to justify or give reasons for. The "groundlessness of our believing" (§ 166) should not be regarded as a deficiency of knowledge[19], but as being prior to all our knowledge and at the same time as being its "foundation" (§ 167; op. cit., p. 135). How is this possible? If certainties really shall serve as the foundation of our knowledge they can't be conceived as quite independent from it. Wittgenstein himself concedes the possibility that we compare "our system of knowledge" with that of other people and that theirs "is evidently the poorer one by far" (§ 286). It is far from clear what the criteria are according to which we can compare and evaluate different systems of knowledge; but Wittgenstein himself mentions the fact that our system is acquired "by means of observation and instruction" (§ 279), and that we "belong to a community which is bound together by science and education" (§ 298). It is evident that the richness or poverty of a world-picture is not, or is not only a matter of faith. The superiority of our knowledge has many sources; and it's not necessary to be "converted" or "persuaded" (§ 612, 262) to realize this. Therefore, it must be possible not only to accept the world-picture which our knowledge fits on faith but also to justify it with regard to its adequacy and explanatory power. What serves as the foundations of knowledge can't be totally independent from what is rendered possible by these foundations. What only can be accepted by faith, cannot serve as the foundations of our knowledge. Thus, insofar as certainties are considered as the foundations of our knowledge, they can't lie beyond what is rendered possible by them. Therefore, we can't consider certainty, as Marie McGinn proposed, as a "non-epistemic attitude to judgements" and justify in this way the exclusion of the possibility of doubt.

However, if one sticks to the idea that certainties are prior to all knowledge and, following Marie McGinn, interprets Wittgenstein as advocating a "conception of certainty that is connected with, and prior to, the idea of knowing", one gets into other kinds of trouble.[20] Concerning those who don't ask the physicist but the oracle whether water boils at 100°C, he remarks: "Where two principles really do meet which cannot be reconciled with one another, then each man declares the other a fool and heretic" (§ 611). "Conversion" seems to be left open: it is possible to be "persuaded" to accept certain beliefs (§ 262, 612), to adopt another view of the world (see § 92). But how shall we understand the possibility of a plurality of world-pictures having the same status and leaving only the possibility of conversion open? I'm not going to treat the objection raised by Davidson and others against the assumption of such a possibility: how are we, within and with the help of our world-picture, able to *recognize* that other people have another world-picture? For the sceptic, the certainty of the world-pictures and the possibility of a plurality of certainties corresponding to the possible plurality of world-pictures is more interesting. Wittgenstein's distinction between knowledge which is open to justification and may be doubted, and certainty where this is not the case, may be regarded as an epistemic version of Carnap's semantic distinction between internal and external existential statements.[21] As the existence of a piece of paper on my desk depends for Carnap on the acceptance of a thing language,[22] for Wittgenstein the certainty that the earth existed before I was born depends on the belief in a world-picture in which our knowledge fits. Other world-pictures correspond to other certainties. This, however, means that the epistemic status of statements varies or can vary from one world-picture to another. In other words: The historical background and the cultural socialization of the knowing subject determine what is knowledge, open to doubt and justification, and what are certainties which have to be accepted on faith. The idea that our knowledge gives a more or less adequate and unfeigned representation of the world loses its sense; and this too parallels Carnap's view that being true or wrong depends on a chosen language-system. Following Carnap, we have to give up the idea of a world which is what it is, no matter how and whether we talk about it; following Wittgenstein, we have to renounce the idea that our knowledge of the world is determined by something which exists independently of the social and historical standpoint of the knowing subject. As Carnap misses what is "factual" about the world, Wittgenstein totally loses sight of the objectivity-claim of knowledge. The sceptic will neither accept Carnap's ontological idealism nor Wittgenstein's epistemological relativism as a refutation; on the contrary, he may justify the sceptical question by appealing to the implications of these attempts to reject the question as senseless.

Let's conclude! I have been concerned with Wittgenstein's discussion of scepticism in *On Certainty*, raising two issues: his critique of doubt and his conception of certainty which he developed in connection with his analysis of knowledge. Whether his remarks are to be considered as a refutation of scepticism depends on how one understands the sceptic. I have described him as raising the fundamental question of the possibility of knowledge, assuming that what is known is independent of our knowledge and the conditions under which, for us, knowledge is possible. Wittgenstein's attempt to draw the boundaries of doubt either doesn't meet this question or doesn't meet the objectivity-claim of what knowledge is about. His conception of certainty as a "foundation for my research" (§ 167) doesn't offer good reasons for rejecting the possibility of doubt; if we take certainty, however, as a "background" or a framework of our knowledge (§ 94), the problems of an epistemological relativism arise and he will not meet the objectivity-condition of knowledge without having discredited it. Contrary to the claim that "doubt gradually loses its sense" (§ 56), the sceptic's doubt still hasn't lost its sense.

Universität Göttingen

NOTES

[1] *Meditationes*, Object. Sept. (AT VII.550).
[2] *Principia Philosophiae* I. 8; *Meditationes*, Synopsis (AT VII.12).
[3] See M. Burnyeat, 'Idealism and Greek Philosophy', *Philosophical Review*, 91, 1982, pp. 3 ff.
[4] B. Stroud, *The Significance of Philosophical Scepticism*, Clarendon Press, Oxford 1984.
[5] *Über Gewißheit. On Certainty* (OC), ed. by G. E. M. Anscombe & G. H. von Wright. Trans. by D. Paul & G. E. M. Anscombe, B. Blackwell, Oxford 1969 § 20.
[6] See M. McGinn, *Sense and Certainty: A Dissolution of Scepticism*, B. Blackwell, Oxford 1988, p. 3.
[7] B. Williams, *Descartes: The Project of Pure Enquiry*, Humanities Press, Atlantic Highlands, N.Y. 1978, pp. 61ff.
[8] *Meditationes* (AA VII. 21).
[9] See Stroud, op.cit., pp. 81–82.
[10] See A. Kenny, *Descartes. A Study of his Philosophy*, Random House, New York 1968, p. 21.
[11] See Descartes, *Principia Philosophiae*, I. 10.
[12] Op. cit., p. 81.
[13] *Principia Philosophiae* I. 3; *Meditationes*, Respon. Quintae (AT VII. 350/1).
[14] See Stroud, op. cit., p. 68.
[15] Op. cit., pp. 19–23.
[16] See also *Philosophische Untersuchungen. Philosophical Investigations*, ed. by G. E. M. Anscombe & R. Rhees. Trans. by G. E. M. Anscombe, Oxford, B. Blackwell 1973^3, II Part, p. 226.
[17] Op. cit., p. 138.
[18] *Ibid.*
[19] See C. McGinn, op. cit., p. 146.
[20] Op. cit., p. 135.
[21] See *Empiricism, Semantics, and Ontology*, in R. Carnap, *Meaning and Necessity: A Study in Semantics and Modal Logic*, Chicago University Press, Chicago 1964, pp. 205 ff.
[22] Op. cit., pp. 206–207; see Stroud, op. cit., pp. 192 ff.

RUDOLF HALLER

WAS WITTGENSTEIN A RELATIVIST?

Let me first confess a certain uncertainty about the title of my paper: it is a *simple question*, and a simple question may be bewitching, in that one might seek an answer as simple as the question appears to be. So it is necessary first to get clear about the question – and *then* to try and find an answer, which I hope will also help us to understand the position of Wittgenstein better.

Relativism, as we know, has many faces. To start with I shall mention only some of them. Relativism is also not highly esteemed among philosophers, since it is very often presented only in a negative form. Mostly it has been presented in a rather distorted form, for example, by Richard Rorty, who states that 'relativism' is the view that "every belief on a certain topic, or perhaps about any topic, is as good as any other." And he adds, "no one holds this view".[1] Clearly, if you present a view held by 'many relativists' in such a way that its characterization does not fit any actually existing relativist you are indeed entitled to say that no one holds this view. But this could not be said of the view propounded by Protagoras. Perhaps it *is* this view which many take for granted, and which we find expressed in the famous words of the old Sophist. According to this view, what is true is not what is true *per se*, but what seems to be true to anyone to whom it seems to be true. This outstanding example of the attempt to invest the individual with the power to decide what is true or false has been attacked – as have been almost all the other forms of relativism – by way of arguments which attempt to exhibit the essential inconsistency and even absurdity of such a view.

The main argument against Protagoras' thesis was that if no belief or no judgment is objectively true (or true *per se)*, since truth (and falsehood) are only functions of what is believed or judged to be so, then the thesis itself is either self-refuting or absurd. If the thesis itself is taken as non-relativistic, that is, as *absolute*, then it is self-referentially inconsistent, because there would be an exception or a counterexample to it before the thesis itself can be considered.

If, however, the thesis is taken to be *relativistic*, then it does not falsify the claim of the objectivity of truth *per se*, but presupposes it.

However, relativism has – as I said – many faces. Indeed, it seems that we are nowadays confronted with several new forms of relativism which are

R. Egidi (ed.), Wittgenstein: Mind and Language, 223–231.
© 1995 *Kluwer Academic Publishers. Printed in the Netherlands.*

perhaps less objectionable than those held hitherto. Perhaps best known in present-day philosophy are two other kinds of relativism: we might call them ontological and paradigm-relativism.[2] The first is concerned with the question how we individuate the constituents of our ontological framework if we are hindered in principle by "the indeterminacy of translation of identity". The way out of the prison of one language is the translation into another one. If we stick to the question what objects (of a theory) – absolutely speaking – are, we are lost, because there is no fact of the matter and no method which enables us to decide in favour of one interpretation of the reference of a term or utterance or against another. There may exist different, mutually incompatible translations, better systems of translations which accord with the behavioural data.

This point of view is akin to the (successful) attempt to convince the scientific community that scientific theories are very often not only mutually inconsistent, but not even commensurable. And the main reason is that the conceptual means, the conceptual apparatus, is essentially different, and not translatable from one into the other. If we project paradigms to the systems of languages then the words of *one* language may not even be translatable into another; and if neither facts or minds can decide between them, they remain incommensurable. The upshot of the incommensurability thesis is – among other things – that he or she who "moves" within one conceptual system will not be able to understand what is meant in the other. In order to find a way to bridge the gulf between two incommensurable systems, be they theories or languages, one would require the complete switch of the systems one has learned by entering, that is, learning the other from scratch.

To the extent that truth is taken as a topic in these views, it certainly remains a predicate relative to the system, and has no meaning outside a theory or paradigm. But this does not mean that the distinction and application of truth and falsehood turns out to be entirely subjective, like personal taste. Learning one from the other *ab ovo* remains the *only* way to bridge the gulf. Even Paul Feyerabend,[3] the jolly anarchist and relativist, does not accept the idea that the simple individualistic belief "existent or true for me" could be substituted for objective existence or truth *per se*. This is and remains a *non sequitur*. But when I mentioned other faces of relativism, I tacitly referred to some well-known and widely accepted views, respected and used even by those who strongly oppose relativism in any of its subjective versions. Actually the rise of *conventionalism* in mathematics and physics at the beginning of this century was a first big step into the direction in which I am alluding. That the deepest principles of science – axioms – are neither synthetic apriori, as Kant thought, nor empirically true, but are the result of human decision, and therefore subject to change, was a step towards relativism. The idea that the different geometrical systems which may be used to describe physical space do not contradict one another, but are to be viewed as more or less convenient means of our mathematical or physical language, the idea that free choice does not mean arbitrariness or anarchy, marked a turn in our understanding of the conventional character of the language of theories. Moreover, conventionalism had a much wider impact than we usually assume. I shall not embark on the journey needed to show or remind you of the places where one can hear the echo of its main theses: in Neurath, Hahn and Frank; in Schlick, Reichenbach and Carnap, not to mention Einstein, it is easy to identify this train of thought. All of them were fascinated by the new idea that decisions and conventions can do the work which, according to the traditional view, only a priori intuition, synthetic judgments a priori or, at best, tautologies could do.

An important consequence of the same line of thinking was Wittgenstein's understanding of what grammar and grammatical rules are. Like the conventionalists, Wittgenstein stated that, for instance, an "axiom about number is a postulate or rule according to which we proceed".[4] Here he takes rules on a par with postulates. Rules of grammar, like postulates, do *not describe* reality: they neither conform to nor contradict facts.[5]

It is not my intention here to impose on anyone the belief that Wittgenstein should have been a conventionalist, in spite of the fact that I cannot see how one can put some of the main ideas of his philosophy of mathematics on the opposite (non conventionalist) side. I only want to stress the fact that from the time of his return to philosophy in 1929, Wittgenstein defended the possibility of turning any empirical proposition into a rule, and insisted that it is our decision whether we do the one or the other. It is "the *roles* in the language-game, the kinds of use" to which necessary statements and non-necessary statements are respectively put.[6] When Wittgenstein points to decisions in explaining the acceptance of a proof as a proof, he merely extends the idea that any move in a language-game might be taken as the subject of a decision.

Since postulates and rules do not come into existence out of the blue, it is not at all out of place to ask where they do come from, and to whom they are addressed. A quick answer would be that it cannot be individuals, since individuals proper, that is, individuals taken in isolation, could not be the subjects of rules of meaning invented by themselves, since – as Wittgenstein has said and Kripke has shown – for the individual proper, any actual decision to mean or to use a word in a certain way cannot determine its meaning (in the future) except in so far there are outer criteria or methods applied by other members of the language community to ensure the correctness of the use.

Kripke's so-called "sceptical solution" points to one side of the answer to the question of where rules and postulates – to mention only two of the main pillars of our language-games – get their regulative power from – the answer is: the community. The main feature of the connection between the individual and the language community to which the individual is linked is an agreement in regard to the meaning of words and signs.

It is not the purpose of this talk to enter into a discussion of the so-called sceptical paradox and its solution. But there is one particular idea to be found in Kripke's approach which is of the utmost importance also for our question here, namely, the idea that the individual taken as a private human being in its privacy could not follow a rule on the basis of his or her own beliefs, since he or she is in principle unable to distinguish by any facts between meaning something and meaning nothing at all.

Wittgenstein says that "following a rule is *analogous* to obeying an order" (PI I, § 206); but from the examples given, it is clear that the analogy works only when there exists a certain regularity in the *reactions* of those *following* a rule (or obeying an order). Without regularity we could not 'see' the pattern of behaviour which connects the rule-following actions with the rule or the obeying with an order. Regularity is thus presupposed when we attribute to individuals that they are following rules as well as when we say of them that they are giving themselves orders, ask themselves questions or are answering them.

But regularity is not enough. Wittgenstein's answer to the question where the rules themselves come from not only *points* to the practice but states that it *is* the practice together with the drill and training belonging to it which establish the link between meaning and use. A practice consists of established rules for doing something in a certain way by doing it in a certain way, and of provisions for the continuation of this way of doing that thing by teaching others how to do it. At a deeper level the practice comprises within itself, therefore, the harmony between words and deeds. Thus, when Wittgenstein says "The practice gives words their meaning",[7] he points to the difference words make at various situations in life. If saying something does not have a relation to our actions which are – as Wittgenstein says – "interwoven" with the language, then the words used do not mean what they should mean according to the rules for their use. Thus the meaning of the term 'practice' can be restricted neither to a technique of bringing something about according to a model, nor to actual behaviour, without taking into account the historical point of view. Because to state that some kind of behaviour is customary, or rather, is a custom in a society, necessarily brings a historical dimension into the analysis. Calling a form of behaviour a *custom* presupposes that in a society or community it has been, from some time onwards, usual to act according to this form or model of behaviour, either always or on certain appropriate occasions. Thus, for example, if the custom concerns only some sub-group of a society it usually remains unnoticed or even unknown by the majority of members of that society. But if the custom is one like the use of the common language, shared by all the members who learnt to speak the language, then the custom is known to all of them, even though the knowledge is in most cases only implicit or practical. To say that there is *agreement* concerning the application of the language-rules, however, does not mean that such an agreement has been achieved or has to be achieved. Agreement, as Wittgenstein uses the notion in the relevant text, is not something the language-user has to attain, rather, it has to exist in order that one may play a language game. Thus – except in the period of learning a language–agreement is a condition for the application of the rules. This is important and often misunderstood. 'Agreement' *(Übereinstimmung)* and 'rule', Wittgenstein says, are related to one another. And the main point, as I read the passage, is: that it is same means which is concerned by communicating in the institution named "the language-game": namely that our agreement and the rules we are obeying are working when we act. The "queer" statement of Wittgenstein – he himself called it such since it may sound so – is then that in order for human discourse to function properly there should exist agreement not only in definitions but also in judgments.[8] This is puzzling, because it is not true that people do agree with one another in their judgments, and since judgments are either true or false like the sentences expressing them, it is not true that they agree about the same when one person states the opposite of another.

Thus, one has to find out what is really at stake, when it is stressed that agreement in judgments should be part of the communicational undertaking. The methods we apply to measure something may certainly be subject to descriptions which may be shared by those who enter a discourse. So, what has to be described may be judged in the same way. But not necessarily so. Rather, two people communicating about a state of affairs may (and very often will) judge it differently and still there will not be a dispute over the question whether both have obeyed a rule or not. Therefore, agreement in judgment can be asked for if agreement in concepts is at stake, because the grammar does not deal with ready-made judgments, but with the composition of them in the language, and with their application. So, the definition of words is one thing, and if there is not agreement about it, we can not meaningfully proceed in communication without making clear in what sense a word or term is used. And if we accept this, we may think that the agreement about the rules of grammar together with the definitions suffices to decide what is true and what is false. But Wittgenstein strongly denies that any agreement does decide what is true or false (PI I, § 241). It is on the basis of agreement in definition, judgments and rules, that people say things which are true or false, but they do not agree in what to say. 'Agreement' is used in a much wider sense, which leads to the very problem I have used as a title of this consideration. "Agreement", Wittgenstein tells us, does not refer to opinions but to a form of life. Thus, truth pertains only to sayings, agreement in this context to much more, namely, the sharing of language and the sharing of a form of life.

Now, I cannot go into an extended and deeper analysis of this concept. But this much seems clear to me, that it does not refer to "the presumption of a common nature", as C. McGinn has suggested.[9] 'Lebensformen' or 'forms of life' is a plural, therefore there are more than one. On the other hand, there is only one common way of human acting; therefore the expression: 'form of life' should not be taken as meaning 'the common way of human acting'. Neither should one fall into what Hintikka has dubbed the "fallacy of verbal language-games".[10] Discourse and speaking a language is only "Part of an activity or a form of life" (PI I, § 23), neither itself the form nor identical with the language-game.

According to Wittgenstein's descriptions, forms of life are at the very bottom of any analysis of language, and that *without an understanding* of the ways people act and live, language cannot be understood, or better, fully understood. When Wittgenstein uses the example of an explorer coming to a land in which the language spoken is totally unknown to him, Wittgenstein states that "he [the explorer] could learn to understand it only through the connection with the rest of the life of the inhabitants".[11] Thus, even if you have only mastered the language but have not reached the connection with the actions and the life of the people, you may still remain unable to understand the people in a country with totally different traditions.[12] Here we touch on the extreme possibility which Wittgenstein has put into the simile: "If a lion could talk, we could not understand him".[13]

In his last writings, namely those which are published under the title *On Certainty* and *Remarks on Colour*, Wittgenstein seems to be replying to the question, "On what basis or ground do we decide between true and false propositions?" But I do not see here a big change of attitude, amounting to a new phase of Wittgenstein's philosophy. It is true that some of the already familiar notions, like the central concepts of a "picture" or of a "form of life", have been given a wider application. For example, when we are told that our picture of the world "is the inherited background against which I distinguish between true and false".[14] So the following question suggests itself: What if the background – e.g. your picture of the world – is different? This I think is one door where relativism concerning truth and falsity may enter. If distinguishing between true and false presupposes the acceptance or assumption of certain beliefs, then the acceptance or assumption of beliefs, which – as the case may be – are contrary to those, cannot serve as background for the same distinctions between true and false.

Since in learning a language we are not only taught the use of words, rules of their use and examples, but since we are also taught judgments, *other* judgments may make up another system of beliefs when other judgments are taught. Since Wittgenstein says that neither experience nor success is the ground for the game of judging when we cannot rely on the former experiences and not wait for success we remain without any ground for the decision between conflicting judgments based on different world pictures. Of course we may try to invent some instruments for comparing them, like simplicity, but there is no guarantee that the invention will work and, if it works, that it will be accepted as a criterion for deciding between conflicting judgments. If we want to convince someone who does not share our picture of the world, we may try to instruct him, teach him all that we believe, but since no ground for a decision is there, our last resort will be *persuasion.*

And, whether we like it or not, this *is* the end of the line of arguments. If we do not believe in an eternal discourse, which will make clear which part of our prejudice has been destroyed by the grace of nature, we shall have to rest our system of beliefs on a presupposition, which normally will not be called onto question, but which is of the utmost importance. It is the fact that *all* human beings speak a language, even if the language spoken may not be understood by others: "We talk and act. That is already presupposed in every thing that I am saying".[15] There is after all a fact which is common to all human beings and which in our language may be expressed in the statement "We talk and act". To fill the gap between the unbridgeable systems of beliefs and their different backgrounds or forms of life, there is the common way of human acting, which comprises assent to and dissent from the action and talk of others. Even when we imagine a Wittgensteinian language-game of the most primitive form, like the one with the four words for building-stones ('block', 'pillar', 'slab', 'beam'), we have to presuppose that it is used among human beings and therefore that they may fail. So dissent is needed and assent is tacitly built into the actions. How else could they have learned the language?

Now, if the main point of relativism is that judgments presuppose some test procedure or criterion for deciding whether they are true or false, and that since there are different procedures, relying on different basic beliefs or systems of beliefs, no one of which may be called the fundamental one, *the acceptance* of one cannot be grounded on any other belief than those, which are presupposed by the *acceptants*: There are different language-games, different forms of life.

So it seems we are bound to the relativity of our belief-systems, which rest on or may be embedded in different forms of life. If someone does hold such a view, a view we may also ascribe to Wittgenstein's conceptions of it, it would be hard to deny that he or she is a relativist.

Stephen Hilmy in his rich book on the later Wittgenstein,[16] a book which uses so much material from the manuscripts and typescripts, has named a chapter 'Wittgensteinian Relativism and the Dynamic View of Language'. He describes the fundamental shift in Wittgenstein's philosophy of language as a shift away from an absolutistic and static to a relativistic and dynamic conception of language. More important is even that Hilmy reminds the reader that Wittgenstein himself on different occasions uses the label "relativity theory" in regard to his own conception of language: "A step is necessary which is analogous to that of relativity theory" (MS 108, p. 270).

So, when Wittgenstein does associate his own findings to that of a relativity theory, surely he has in mind the Einstein paradigm which changed the meaning of the key concepts and therefore of the systems by reshaping the question. The question now concerned also the relativity of the observer to the system itself. Abandoning the old idea of *absolute concepts*, like space, time and motion, was only a first important step. At least as important is the change in the general outlook covering concept-formation and theory-building. And, indeed, if the reference to language-games – the only way we act and use language – has a deeper role than just serving as a substitute for *calculus*, the role will be reflected in its use. Here I think the analogy with relativity theory comes in: If it is true what Wittgenstein says that everything is to be done in language, we do *not* have a resort to our inner life, be it a *lingua mentalis*, a private language, or the inner world of impressive emotions. The 'Nothing is hidden', indeed, is not a battle-cry but the signpost which points to the new relativistic point of view. Whether from this a theory emerges is another question or a question for another occasion.

Universität Graz

NOTES

[1] R. Rorty, 'Pragmatism, Relativism and Irrationalism', in *Proceedings and Addresses of the American Philosophical Association*, 53, 1980, p. 727.

[2] See W. V. O. Quine, *Ontological Relativity and Other Essays*, Columbia University Press, New York 1969, 1971[2]; N. R. Hanson, *Patterns of Discovery: An Inquiry into the Conceptual Foundations of Science*, Cambridge University Press, Cambridge 1958; T. S. Kuhn, *The Structure of Scientific Revolutions*, Chicago University Press, Chicago 1962; S. Toulmin, *Human Understanding*, I: *The Evolution of Collective Understanding*, Princeton University Press, Princeton 1971; P. K. Feyerabend, *Against Method*, New Left Books, London 1975; N. Goodman, *Ways of Worldmaking*, Hackett Publ. Co., Indianapolis-Cambridge 1978; H. Putnam, *Reason, Truth and History*, Cambridge University Press, Cambridge 1981; M. Hollis & S. Lukes (eds.), *Rationality and Relativism*, B. Blackwell, Oxford 1982; J. Meiland & M. Krausz (eds.), *Relativism: Cognitive and Moral*, Notre Dame University Press, Notre Dame 1982; P. K. Feyerabend, *Farewell to Reason*, Verso, London 1987; R. Egidi (ed.), *La svolta relativistica nell'epistemologia contemporanea* (with bibliography), F. Angeli, Milano 1988; W. Sauer, *Aspekte und Probleme des Begriffsrahmen-Relativismus*, Habilschrift, Graz (unpublished) 1989.

[3] See *Farewell to Reason*, cit.

[4] *Wittgenstein's Lectures. Cambridge, 1932–1935*, From the Notes of A. Ambrose & M. Macdonald, ed. by A. Ambrose, B. Blackwell, Oxford 1979, p. 224.

[5] *Ibid.*, p. 65

[6] C. Wright, *Wittgenstein on the Foundations of Mathematics*, Duckworth, London 1980, p. 365.

[7] L. Wittgenstein, *Bemerkungen über die Farben. Remarks on Colour*, ed. by G. E. M. Anscombe. Trans by L. L. McAlister & M. Schättle, B. Blackwell, Oxford 1977, III, § 317, p. 59. (I diverge from the English translation, which translates the singular "*Die Praxis* gibt den Worten ihren Sinn" with "*Practices* give words their meaning".)

[8] *Philosophische Untersuchungen. Philosophical Investigations* (PI), ed. by G. E. M. Anscombe & R. Rhees. Trans. by G. E. M. Anscombe, B. Blackwell, Oxford 1953, I, § 42.

[9] C. McGinn, *Wittgenstein on Meaning: An Interpretation and Evaluation*, B. Blackwell, Oxford 1984, p. 55.

[10] M. B. Hintikka & J. Hintikka, *Investigating Wittgenstein*, B. Blackwell, Oxford 1986, p. 218.

[11] L. Wittgenstein, MS 165, p. 97f, cit. by S. St. Hilmy, *The Later Wittgenstein*, B. Blackwell, Oxford 1987, p. 291f.

[12] PI, II, xi, p. 223.

[13] *Ibid.*

[14] *Über Gewißheit. On Certainty*, ed. by G. E. M. Anscombe & G. H. von Wright. Trans. by D. Paul & G. E. M. Anscombe, B. Blackwell, Oxford 1969, § 94.

[15] *Wittgenstein's Lectures on the Foundations of Mathematics*, ed. by C. Diamond, Harvester Press, Hassocks-Cornell University Press, Ithaca, N.Y. 1976, VI, § 17, p. 321.

[16] Op. cit.

PART V

TOPICS IN THE PHILOSOPHY OF PSYCHOLOGY

S. STEPHEN HILMY

WITTGENSTEIN ON LANGUAGE, MIND AND MYTHOLOGY

The notions of language, mind and mythology are intertwined in certain passages in Wittgenstein's later writings. Even when these intertwined themes are not expressly identified, their intertwinement often seems to serve as a general thematic backdrop for many of Wittgenstein's later philosophical deliberations. However, there is a danger of misinterpreting the intertwinement of these themes in Wittgenstein's later philosophy as an indication that he shared the outlook of others who have tended to intertwine these themes in their writings. This danger is especially acute, because the intertwinement of these themes in the writings of others has tended to be intimately linked to an outlook which runs counter to some of the central critical thrusts of Wittgenstein's later philosophy. This paper offers some reflections on the nature of the intertwinement of the themes of language, mind and mythology in Wittgenstein's later writings, and examines the import of that intertwinement for an appreciation of his later philosophy, and in particular his fairly recently published volumes of remarks in the philosophy of psychology.

Among those in whose writings one finds an interweaving of the themes of language, mind and mythology is Wittgenstein's contemporary Bertrand Russell. So Russell can conveniently serve both as a springboard and as a foil. In the context of a discussion of the concept of 'meaning', Russell once wrote that he believed "there is one method of acquiring knowledge, the method of science; and that all specially 'philosophical' methods serve only the purpose of concealing ignorance." His goal, he indicated, was to "endeavor to construct a theory of meaning after the model of scientific theories".[1] Likewise, he regarded "thought as merely one among natural processes, and hope[d] that it may be explained some day in terms of physics".[2] His ambition, as he explained in the context of a discussion of his view of 'belief', was to "reject everything mythological" and instead present a "theory" of belief which is empirically verifiable.[3]

This is certainly an alluring outlook. It was sufficiently alluring and adaptable that it remained relatively undisturbed as Russell periodically shifted in and out of materialist, phenomenalist and behaviourist phases during his protean career. A recent form of this outlook, although rejecting the specific theories of the mind that Russell propounded, has gone so far as to suggest

R. Egidi (ed.), Wittgenstein: Mind and Language, 235–248.
© 1995 *Kluwer Academic Publishers. Printed in the Netherlands.*

that the "primitive psychological taxonomy in ordinary language" ought to be eliminated and in its stead we ought to replace our ordinary psychological concepts with a "neuro-psychological taxonomy" which would "allow us to better describe, conceive and introspectively apprehend one's inner life".[4] The psychological concepts of ordinary language themselves constitute, as it were, bad scientific theories, myths if you will, about the nature of our inner life, and we ought to replace this nexus of pseudo-scientific theories about 'mental states and processes' with modern neuro-psychological theories and vocabulary about the states of our brain and nervous system.

For reasons which I will explain presently, I will call this overall outlook, shared by Russell and many others, the 'taxonomic view' of psychological concepts. One cannot deny the hold that this outlook has over us. It would seem that surely science will accomplish with respect to man's psychological constitution what it has accomplished with respect to the forces of the physical world around us – as it were, debunking mythologies about the 'mind' just as it has debunked mythologies about the forces of nature.

The question arises as to whether Wittgenstein subscribed to a form of the taxonomic view of psychological concepts, and what implications that would have for an appreciation of the status of his last writings and remarks on the philosophy of psychology. If he accepted the taxonomic view, it would seem that the topics which are the subject matter of his last writings in the philosophy of psychology really ought to be the province of empirical psychology, not philosophy. And yet, if he rejected the taxonomic view of psychological concepts, that would seem to entail that his remarks in the philosophy of psychology serve, as Russell might put it, "only the purpose of concealing ignorance", because Wittgenstein, it would seem, employs a philosophical method which is futily in competition with the theories and discoveries of the empirical sciences.

It is important for appreciating Wittgenstein's last writings in the philosophy of psychology that one both recognize that he rejected what I call the 'taxonomic view' of psychological concepts, and, even more importantly, understand *why* he rejected that view. An understanding of why he rejected the taxonomic view would pave the way for an appreciation of his last writings in the philosophy of psychology as categorically different from but yet at least potentially of constructive value for empirical psychology.

There are, though, some striking, but misleading, parallels between what I have called the taxonomic view of psychological concepts and general remarks Wittgenstein himself has made about the nature of the philosophical problems that concerned him. For instance, he often wrote of there being a 'mythology' in language, and of language presenting us with false pictures which hold us captive. In his chapter on "Philosophy" in the 'Big Typescript' he suggests that "In our language lies a whole mythology".[5] In that same typescript one also finds a version of the similar familiar remark, which has since found its way into *Zettel*, namely, the suggestion that "In philosophy one is in constant danger of producing a mythology of symbolism, or of psychology [mental processes]. Instead of simply saying what anyone knows and must admit".[6] More often than not, such remarks are made in the context of his attacks on mentalism, and it is very tempting to read such remarks as involving a form of the taxonomic view.

It is quite possible that very early in his career Wittgenstein may have entertained an outlook not very far removed from a form of the taxonomic view, one of the symptoms of which is the attitude that ordinary language is defective and in need of reform. In one of his later notebooks he seems to admit that as he first expressed himself, "it seemed as though ordinary forms of expression wouldn't really accomplish what we wanted of them. As if they were an (old) form of historical rubbish. . .that would be replaced by something direct, modern".[7] However, it is clear from his last writings in the philosophy of psychology, as it is clear in his later writings in general, that the goal of refining or reforming language was abandoned by him. This is obvious from *Philosophical Investigations*, § 98, where he expressly indicates that

> every sentence in our language is 'in order as it is'. That is to say, we are not *striving after* an ideal, as if our ordinary vague sentences had not yet got a quite unexceptionable sense, and a perfect language awaited construction by us.[8]

This theme is echoed in the notebooks from which *Remarks on the Philosophy of Psychology* was drawn, where in the context of a discussion of the concept of 'seeing' Wittgenstein advises that we "must just take the concept of 'seeing' as one finds it; not want to refine it. . . Because it is not our problem to change it, to introduce (as science does) [a sense] adapted for some purpose or other, but rather to understand it; in order not to produce a false conception of it".[9] He goes on to explain that we need to "*accept* the familiar language-game, and to note *false* explanations of the matter as false. The primitive language-game we originally learned needs no justification, and false attempts at justification, which force themselves on us, need to be rejected".[10] Those to whom Wittgenstein wishes to direct his advice clearly are *not* empirical psychologists per se, since there is implicit recognition here of the legitimacy of the psychologist's task in its own right, as distinguished from the concerns of philosophy. Rather, Wittgenstein is directing his advice to *philosophers*, by suggesting that the problems that ought to be the concern of philosophers, as he envisions their proper task, are confusions *about* our ordinary concepts, not confusions of language itself, which he claims is "in order as it is". At the root of this, lies both the basis for Wittgenstein's repudiation of the 'taxonomic view' of psychological concepts, and the possibility of constructively construing the potential relationship between his last writings and the tasks of empirical psychology.

It was not Wittgenstein's view that ordinary language is in some sense an inherently inadequate and flawed repository of false theories, as it were, a repository of myths or false pictures about psychological processes. Quite to the contrary, the mythology and the 'false pictures' of which he speaks are misconceptions or confusions conjured through our *misunderstandings* of the actual function of our expressions. Thus the mythology would consist essentially of *our* confusions *about* language, not confusions *of* language – and this would be the case even if our confusions *about* language were spawned by misleading but inessential analogies within language. Misleading parallels in language may give rise to our confusions, but it is the latter confusions which are the problem, not language itself. This is why Wittgenstein can consistently claim that philosophy is a struggle against the bewitchment of our intelligence by means of language,[11] and yet without contradiction also claim that language is in order as it is. Both of these claims are made in the *Philosophical Investigations* (§ 109 and § 98 respectively), and indeed they were originally drafted on the same page in his notebooks.[12] The myths and false pictures are thus constructs of our misunderstanding of our language, not defects in language per se. As Wittgenstein quite explicitly put it in one of his manuscripts: "philosophy struggles against myths [*Mythen*], pictorial tendencies in our *understanding* " [my emphasis].[13] This is the real import of his frequent references to the "false pictures" and "myths" conjured by language – it is an indictment of our understanding, not an indictment of language itself.

However, simply recognizing certain myths as by-products of our misunderstandings or confusions about our language is not enough to avoid the risk of falling prey to what I have called the taxonomic view. Even if one acknowledges these myths as misunderstandings of our language, one might nevertheless make the further mistake of simplistically dismissing those myths as the false theories of mentalism, and arguing that our psychological concepts do not function as reports of mythical mental events at all, but rather merely as reports of our behaviour or our tendencies to behave in certain ways.

It is tempting to interpret Wittgenstein as having held such a view, particularly if one is inclined, as many have been,[14] to interpret him as advocating a form of behaviourism – in which case he would be advocating a behaviouristic understanding of our psychological concepts in place of a mentalistic one. The exegetical ease with which this interpretive move can be made is perhaps exactly why so many have tended to interpret Wittgenstein as a behaviourist, and why, in spite of his apparent protestations to the contrary in the *Investigations*, it is so easy for it to *seem* that he is advocating a behaviourist position. I am alluding here to the *Investigations*, § 307, where Wittgenstein has an imaginary interlocutor ask him: "Are you not really a behaviourist in disguise? Aren't you at bottom really saying that everything except human behaviour is a fiction?" And Wittgenstein replies: "If I do speak of a fiction, then it is of a *grammatical* fiction." The import of *Investigations* § 307 is more explicit in the broader context of Wittgenstein's initial draft of the remark:

> But aren't you really only a behaviourist in disguise? For you say that nothing stands behind the expression [*Äußerung*] of experience. Aren't you at bottom really saying that everything except behaviour is a fiction? So do I therefore believe that we don't really feel pain, but rather only make faces?! But the fiction *is* the object behind the expression [*Äußerung*]. It is a fiction that our words in order to be meaningful must refer to a something that I nevertheless can exhibit to myself even if not to someone else. (Grammatical fiction.)
>
> The statement "Nothing stands behind the expression [*Äußerung*] of experience" is a grammatical one – it doesn't therefore claim that we experience nothing. . . .
>
> My critique consists in that I ~~show~~ bring out that the whole (common, primitive) view of the ~~functioning~~ function of words (in the language game //in the use of language//) is too narrow.[15]

By calling a statement "grammatical", Wittgenstein means that it is a statement about the use or function of a linguistic term. He is not denying the existence of mental entities *per se*. Thus when he makes the *grammatical* claim that 'nothing stands behind a given *expression* of experience', he is not denying that there is anything going on inside us as we utter the expression, rather he is making a comment on the function or use of the 'expression of experience', namely, that it does not *function* to report a hidden psychic thing or object behind it. I have argued elsewhere,[16] that an interpreter, who concludes from Wittgenstein's repudiation of a mentalistic explanation of the 'meaning' of our linguistic expressions that Wittgenstein was therefore a behaviourist and was claiming that all 'psychological expressions' report (or ascribe) behaviour, may have in effect fallen under the spell of an illusion or general fiction about the nature of language and 'meaning' that Wittgenstein was bent on dispelling.

If behaviourism were Wittgenstein's position, his view would be quite consistent with the taxonomic view of psychological concepts, because he would simply be replacing the mentalistic taxonomy which language allegedly conjures, by a behaviouristic taxonomy. However, that was not his position at all, at least not in his later philosophy. Indeed at the very outset of the manuscripts from which *Remarks on the Philosophy of Psychology* was compiled Wittgenstein at one point expressly indicated: "I am attacking the pneumatic [mentalistic] view. . . The antithesis of the pneumatic view is the behaviouristic one, and they are both bad [*schlecht*]".[17]

It may seem odd to group together such diverse standpoints as mentalism and behaviourism, but this is possible because both mentalism and behaviourism, and for that matter also materialism and phenomenalism, are all instances of a taxonomic approach to psychological concepts, and it is the taxonomic view itself which Wittgenstein would consider fundamentally flawed. The affinity between mentalism and behaviourism is clearly illustrated, for instance, in the philosophy of Russell. During the course of his career, Russell seemed to shift with relative ease from a Humean, phenomenalist account of 'meaning' to a behaviouristic account. The ease of that transition was indicative of the affinity lying in not only the causal but also the *taxonomic* nature of both the mentalistic and behavioural explanatory paradigms (to both of which explanatory paradigms Wittgenstein was critically reacting). Russell's mentalistic explanation of meaning was encapsulated fairly early in his career in the associationist principle he called the "law of mnemic causation":

> If a complex stimulus A has caused a complex reaction B in an organism, the occurrence of a part of A on a future occasion tends to cause the whole reaction B.[18]

Unpacking this explanatory principle as a *mentalistic* explanation of 'meaning' is quite straightforward.[19] This is accomplished simply by interpreting the 'complex reaction B' in mentalistic terms. If one is repeatedly subjected to a complex stimulus A which is composed of a verbal stimulus (a word) and an 'object' or thing along with the word, and one's complex mental reaction (B) to the complex stimulus is to have the roughly simultaneous sensations of a specific 'word' and a specific 'object', then, according to the above explanatory law, if at a later time one were to be subjected to only part of complex stimulus A (e.g. only the word), one's reaction should be the whole of complex reaction B, namely, both a sensation of the word and a copy or mental image of the sensation of the object associated with the word. Thus the 'meaning' of a word would be the effect the word has on our mind, namely the image or 'mental object' that it conjures, and the latter would be a causal by-product of our history of conditioning. Such an explanation gives not only a causal, but also a *taxonomic* account of the notion of meaning.

But the very same sort of causal and taxonomic explanatory paradigm can be adapted to a purely behavioural account of meaning. In fact, Russell did precisely that when later in his career he shifted from mentalism to behaviourism:

> People used to speak of 'association of *ideas*', but now-a-days association is rather between bodily movements. The essential phenomenon is what Dr. Watson calls a 'learned reaction'. Two stimuli A and B occur together, and B causes a bodily movement C. Later on, A may cause C, though it previously had no tendency to do so. All words are 'learned reactions' in this sense. There is no need to postulate a 'mental' intermediary between the stimulus and the reaction.[20]

Russell simply replaced '*mental reaction*' in his earlier account of 'meaning' by '*behavioural reaction*'. One's taxonomy of 'things' has simply shifted from a mentalistic taxonomy to a behaviouristic one. The causal character of the explanation has of course remained constant, but so has the essential taxonomic nature of the explanation of 'meaning' also remained unchanged. Nor would the taxonomic nature of the explanatory paradigm be essentially altered by including under the rubric 'behavioural reaction' the neuro-physiological effects of verbal stimuli rather than solely the gross behavioural effects (the 'actions' elicited).

The taxonomic view is characterized not so much by a theory about the nature of our 'inner life', but rather by a fundamental supposition about the use of our vocabulary of so-called 'psychological concepts'. The fundamental supposition is that our ordinary psychological vocabulary functions as a taxonomy of things or processes. The various taxonomic standpoints differ only in *what* they claim the taxonomy is or *should* be, not in *whether* the primary function of our psychological concepts is to present us with a taxonomy of things or processes in the first place, whether mental, physical or phenomenal. Viewed in this light, mentalism, phenomenalism, behaviourism and materialism are all birds of a feather (if you will pardon the taxonomic idiom), in that they *all* constitute taxonomic standpoints vis-à-vis an interpretation of the function of our psychological vocabulary. It is for this reason that it was possible for Russell's overall taxonomic outlook to remain essentially undisturbed when he shifted between materialism, phenomenalism and behaviourism. And likewise the basic taxonomic outlook remains similarly undisturbed whether one simply advocates that the old so-called mentalistic taxonomy can be *translated* into a new materialistic, or phenomenalistic or behaviouristic taxonomy, or whether one instead advocates the more extreme view that we ought completely to *eliminate* the old allegedly mentalistic taxonomy and start afresh with a new neuro-psychological taxonomic vocabulary.

If mentalism stems from a misunderstanding of our vocabulary *as* taxonomic, then by rejecting our psychological vocabulary because it allegedly presents us with a *false* mentalistic mythology, and advocating replacing that vocabulary with a neuro-psychological vocabulary, one commits two errors (according to Wittgenstein): 1) the error of misconstruing our psychological vocabulary as functioning as a taxonomy of mental events or processes, and 2) the error of perpetuating, by means of a materialistic substitute, the underlying taxonomic myth or confusion that gave rise to mentalism in the first place. A materialistic or behaviouristic taxonomic outlook and vocabulary would not be a cure for the so-called mentalistic mythology of language, but rather, since mentalism is simply an instance of the taxonomic outlook, behaviourism and materialism would simply be perpetuations of the taxonomic myth which underlies mentalism – namely, the simplistic taxonomic view that the primary function of our psychological vocabulary is to name or report things or processes, whether of a behavioural, mental, phenomenal or material nature.

Although Wittgenstein did not use the expression "taxonomic myth", it can safely be said that one of the central critical thrusts of his later philosophy was the attempt to debunk such a myth about the function of our psychological vocabulary. Soon after his return to philosophy in the late 1920s, he wrote the following illuminating remark which bears on the nature of the mistake underlying the taxonomic myth:

> If it is charged that language can express everything by means of substantives, adjectives and verbs, then we must say that it is then in any case necessary to differentiate between wholly different kinds of substantives (etc.), for different grammatical rules apply for them. . . Thereby it is evident that their substantive character is only a superficiality [*Äußerlichkeit*] and that we are actually dealing with wholly diverse families of words. The word-family is (first/now) determined by *all* grammatical rules which apply to a word. And thus viewed our language has a countless number of different kinds of words.[21]

The last two decades of Wittgenstein's philosophical activity can without distortion be characterized broadly as an exploration of the rich diversity and multiplicity of grammars that constitute our language, but which diversity and multiplicity tend to be concealed by the misleading simplicity and superficiality of the surface grammar of our language. And it is a failure to grasp the grammatical diversity and richness of our language that is at the root of the 'mythology' of language against which Wittgenstein was struggling. Indeed in his chapter on "Philosophy" in the 'Big Typescript', almost immediately after identifying the mythology of language as a fundamental problem, Wittgenstein went on to explain the way in which language gives rise to such mythology: "The primitive forms of our language – substantive, adjective and verb – show a simple picture to which it tries to make everything conform".[22]

The mistaken, simplistic picture to which the primitive forms of our language give rise is perhaps most clearly characterised in the 'Blue Book', where Wittgenstein indicates that:

> The mistake we are liable to make could be expressed thus: We are looking for the use of a sign, but we look for it as though it were an object *co-existing* with the sign. (One of the reasons for this mistake is again that we are looking for a 'thing corresponding to a substantive.')[23]

Furthermore, the way in which this fundamental mistake manifests itself more specifically in the domain of the philosophy of psychology is, as Wittgenstein explained a few years later in one of his notebooks:

> The philosophical confusion concerning the idea of thinking, together with the problems of psychology, is presented to us by the picture of a hidden (invisible) mechanism. . .
>
> It seems: 'believing' describes something which occurs with the sentence – as 'digesting,' something which occurs with the meal.
>
> One could then understand the belief if one knew what really went on at the time. One would then have analysed the 'process of belief.'
>
> . . . It is absolutely not unexplored processes of belief that interest us, but rather the use of our well-known processes of belief, e.g. of the expression of the sentence 'I believe. . .'[24]

The assumption that our psychological concepts *function* as names of hidden processes or events, rests on the simplistic assumption that the meaning of a word is an object, thing or process to which it refers. From Wittgenstein's perspective, the issue of whether the thing named is a mental process, a physical process or a behavioural process is really a derivative controversy, and one which is played out by philosophers from within the taxonomic spell that the surface grammar of language has allegedly thrown over us. To use Wittgenstein's own illustration, the surface grammar would suggest that believing is something which occurs with the sentence, just as digesting is something which occurs with the meal; and similarly for a host of other psychological concepts.

The taxonomic view seems to fall prey to just this sort of surface grammatical analogy. Let me illustrate this in more detail with respect to the concept of 'expecting'. A mentalistic taxonomic view might explain the meaning of expressions of the form 'I expect. . .' as that of *reporting* the occurrence of a hidden mental process accompanying such an utterance (for example, perhaps the occurrence of hidden phenomenal pictures or images in one's 'mind'). On the other hand, a behaviouristic taxonomic view might explain an expression of the form 'I expect. . .' as a report about how one is behaving or how one is disposed to behave (perhaps one's tendency to pace, fidget, make tea, carry an umbrella, and so on) when one utters such an expression. (One could if one wanted also throw in here neuro-physiological behaviour.) Wittgenstein denies such explanations, but in doing so he is neither denying 'the mental' (for example that we may have mental images that accompany speech), nor denying that we may pace, fidget and make tea (or be disposed to doing so) when saying we 'expect' something to happen. Rather Wittgenstein is denying that, in the context of the conventional language games in which the expression 'I expect. . .' is embedded, the function (the use, the logic, the grammar) of the expression is to *report* one's own mental or behavioural processes or states.

As for what specific sort of function such an expression *does* serve, Wittgenstein hints:

> If I say "I expect . . ." – am I remarking that the situation, my actions, thoughts etc. are those of expectancy of this event; or are the words "I expect . . ." part of the process of expecting?
> In certain circumstances these words will mean (will be replaceable by) "I believe such-and-such will occur." Sometimes also: "Be prepared for this to happen . . . ."[25]

I have indicated elsewhere,[26] that Wittgenstein can be taken in this remark to be suggesting that the expression 'I expect. . .' is itself the expecting, and we might ascertain this by noticing that the utterance 'I expect. . .' in certain circumstances does the same sort of job as (means, is replaceable by) the utterance 'I believe such-and-such will occur' or 'Be prepared for this to happen.' That an 'expression of expectation' might in certain circumstances involve the same type of linguistic move as an 'expression of belief' does not of course help us very much, if one is inclined to view an 'expression of belief' as a report about one's own mental events or behaviour. But to recognize that in certain circumstances the expression 'Be prepared for this to happen' plays the same role, serves the same duty, as 'I expect. . .' is more directly helpful in that it is an indication that the linguistic job performed by 'I expect. . .' is not that of reporting about the occurrence of mental or behavioural processes but rather a matter of offering some assurance of the likelihood of a predicted state of affairs (and perhaps a matter of offering a warning to that effect). To offer assurance or to forewarn that something will happen obviously has a quite different function from that of reporting what one is picturing in one's mind's eye, one's nervous fidgeting, one's blood pressure, or one's disposition to do any of these. Thus the claim is that in some circumstances, although the expression 'I expect. . .' may seem to be a report about a psychological or behavioural accompaniment to one's utterance, it in fact offers assurance or forewarns – and in this sense Wittgenstein suggests that the expression of expectation is the expectation, for, if the language game actually being played were that of offering assurance or forewarning, clearly we would call an 'expression of assurance' an 'assurance' and an 'expression of warning' a 'warning'.

Just what the meaning (the logic or grammar) of any given linguistic expression is will depend on the particular language game or system of communication in which it is embedded. It was Wittgenstein's view that this is the case for any linguistic expression, even those that might be called 'psychological expressions' (e.g. of the form 'I expect. . .', 'I believe. . .', 'I hope. . .', and so on). Upon investigating the logic of psychological expressions in the context of the multiplicity of language games which are constitutive of their sense, he was led to the conclusion that it is an erroneous and simplistic view of language to assume that the function of such expressions is that of naming or *reporting* psychological or behavioural mechanisms, processes or events.

The taxonomic myth is precisely the simplistic view of the grammar or function of our psychological concepts as solely a matter of *naming* or reporting the occurrence of processes and events. Mentalism, behaviourism, materialism and phenomenalism bicker about the *nature* of what is reported, but they all share this grammatical presumption.

The train of thought that 'if the meaning of a linguistic expression is not an 'inner' mental thing associated with the expression, it must be an inner physical thing or perhaps an associated 'outer' behavioural thing', is a form of reasoning that *remains* under and is held captive by the grammatical illusion, the taxonomic myth that all words serve (all language functions) to name or report something (whether object, process or state, whether mental or behavioural, inner or outer). An essential part of Wittgenstein's criticism of, for example, both mentalism and behaviourism is that the grammar or 'meaning' (the function) of our so-called 'psychological expressions' in the language games in which they are embedded may not be that of *reporting* (or naming) an associated something at all (whether a hidden 'inner' something, or an 'outer' behavioural one).

Wittgenstein's remarks in the philosophy of psychology during the last years of his life are a sustained effort to debunk this taxonomic myth by probing the rich and diverse grammars of our psychological concepts. As he explained in a revealing note that did not find its way into the published versions of his last writings:

> In order to be able to get a surview of these concepts, you must compare them with something other than what their surface grammar [*Oberflächengrammatik*] suggests to you. . . .
> The concepts are disguised.
> . . .
> The verbs and substantives of psychology (seeing, hearing, fearing, willing, hoping, believing; dread, will, fear, belief, wish) are so *very* misleading. One must again and again return to the concepts to describe their conduct and its occasions, in order not to be misled about the nature of the language-games [*Sprachspiele*] by the surface of grammar.[27]

The terrain of language which he explored so extensively during his last years is certainly very complex, and his published remarks in the philosophy of psychology are far from finished products, but this may not necessarily detract from what he perhaps would have wished to be one of the effects of his last writings in the philosophy of psychology, namely, that the more closely scholars scrutinize his exploration of the grammatical terrain of our psychological concepts, the more they would come to appreciate the richness and diversity of that terrain. And with that appreciation, perhaps would come the abandonment of the simplistic though bewitching taxonomic categories and mind-sets that have traditionally vied for dominance on the philosophical scene.

I indicated at the outset that there is the potential for a positive relationship between Wittgenstein's last writings and the tasks of empirical psychology. Science's task is an empirical one, yet its questions, particularly in the area of empirical psychology, are to a great extent spawned by and couched in terms of those very psychological concepts whose confusing grammar was the subject of Wittgenstein's investigations. Consequently, science itself is to some extent vulnerable to the conceptual confusions which our psychological concepts engender, and thus a clear surview of the logical contours of our psychological concepts may be useful to the empirical psychologist as well as to the philosopher.

To this extent, Wittgenstein's efforts to clarify the grammar of those concepts in order to debunk the grammatical myths they engender create at least the possibility for constructively construing the potential relationship between his last writings in the philosophy of psychology and the tasks of empirical psychology – in spite of the methodological chasm which seems to separate his grammatical investigations from the task of empirically investigating man's psycho-physical nature.

Austin, Texas

NOTES

[1] B. Russell, Symposium on 'The Meaning of Meaning', *Mind*, 29 (Oct. 1920), p. 401.
[2] *Ibid.*, p. 400.
[3] B. Russell, 'On Propositions: What They Are and How They Mean', *Proceedings of the Aristotelian Society*, 1919, Suppl. Vol. II, pp. 27–28.
[4] P. M. Churchland, *Matter and Consciousness: A Contemporary Introduction to the Philosophy of Mind*, MIT Press, Cambridge Mass. 1988, pp. 179–180.
[5] TS 213, p. 434. All references to Wittgenstein's papers are to the Cornell microfilm version, 'The Wittgenstein Papers', Cornell University Microfilms, Ithaca, N.Y. The system of reference is that of G. H. v. Wright, 'The Wittgenstein Papers', in *Wittgenstein*, Oxford, B. Blackwell 1982. Within quoted material, double obliques mark off terms or phrases that Wittgenstein entered as possible alternate wording. Broken or wavy underlines (reproduced here simply as broken underlines) were used by Wittgenstein to indicate that he was not sure about the appropriateness of a word or phrase.
[6] TS 213, p. 145; see *Zettel* (Z), ed. by G. E. M. Anscombe & G. H. von Wright. Trans. by G. E. M. Anscombe, B. Blackwell, Oxford 1967, § 211.
[7] MS 160, pp. 43–44; see my discussion in S. Hilmy, *The Later Wittgenstein*, B. Blackwell, Oxford 1987, nn. 174–5 and pp. 55 ff.
[8] L. Wittgenstein, *Philosophische Untersuchungen. Philosophical Investigations* (PI), ed. by G. E. M. Anscombe & R. Rhees. Trans. by G. E. M. Anscombe, B. Blackwell, Oxford 1953.
[9] MS 137, p. 29; see L. Wittgenstein, *Vermischte Bemerkungen. Culture and Value* (CV), ed. by G. H. von Wright in collaboration with H. Nyman. Trans. by P. Winch, University of Chicago Press, Chicago 1980, p. 44.
[10] MS 137, p. 30; published in *Bemerkungen über die Philosophie der Psychologie. Remarks on the Philosophy of Psychology*, Vol. II (RPP II), ed. by G. H. von Wright & H. Nyman. Trans. C. G. Luckhardt & M. A. E. Aue, B. Blackwell, Oxford 1980, § 453.
[11] PI I, § 109.
[12] MS 152, p. 94.
[13] MS 158, p. 54.
[14] See, for example, G. Bergmann, 'The Glory and the Misery of Ludwig Wittgenstein', in *Essays on Wittgenstein's Tractatus*, ed. by I. Copi & R. Beard, Hafner Press, New York 1973, pp. 354–355; C. Mundle, *A Critique of Linguistic Philosophy*, Glover & Blair Limited, London 1979, p. 199; G. Bowie, 'Wittgenstein's "Behaviorism" and Psycho-physical Laws', in *Wittgenstein and his Impact on Contemporary Thought: Proceedings of the Second International Wittgenstein Symposium*, ed. by E. Leinfellner *et al.*, Hölder-Pichler-Tempsky, Vienna 1978, p. 445.
[15] MS 124, pp. 5–6. See MS 161, pp. 79–80.
[16] S. S. Hilmy, 'Wittgenstein and Behaviourism', *Grazer Philosophische Studien: Wittgenstein in Focus – Im Brennpunkt: Wittgenstein*, ed. by B. McGuinness & R. Haller, Rodopi, Amsterdam-Atlanta,Ga, 33–34, 1989, pp. 335–352.
[17] MS 130, p. 3.
[18] B. Russell, *The Analysis of Mind*, G. Allen & Unwin, London 1921, p. 86.
[19] See my discussion in S. S. Hilmy, *The Later Wittgenstein*, cit., p. 116.
[20] B. Russell, 'The Meaning of Meaning', *The Dial* (August 1926), p. 116. See Carnap's adoption of a form of this causal explanatory principle, R. Carnap, 'Psychology in Physical Language', in *Logical Positivism*, ed. by A. J. Ayer, The Free Press, London 1963, pp. 196–197.
[21] MS 107, p. 210.

[22] TS 213, p. 434.
[23] L. Wittgenstein, *The Blue and Brown Books*, ed. and with a Preface by R. Rhees, B. Blackwell, Oxford 1958; Harper & Row, New York 1965^2, p. 5.
[24] MS 117, pp. 127–129. For a more extensive discussion of this passage in the broader context of Wittgenstein's later view of the nature of language, see S. S. Hilmy, *The Later Wittgenstein*, cit., pp. 163 ff.
[25] Z, § 65.
[26] S. S. Hilmy, *The Later Wittgenstein*, cit., pp. 146 ff.
[27] MS 134, pp. 126–129.

JOACHIM SCHULTE

EMOTION: REMARKS ON WITTGENSTEIN AND WILLIAM JAMES

In a remarkable passage of his *Principles of Psychology* William James writes:

> My theory [. . .] is that *bodily changes follow directly the perception of the exciting fact, and that our feeling of the same changes as they occur* IS *the emotion*. Common-sense says, we lose our fortune, are sorry and weep; we meet a bear, are frightened and run; we are insulted by a rival, are angry and strike. The hypothesis here to be defended says that this order of sequence is incorrect, that the one mental state is not immediately induced by the other, that the bodily manifestations must first be interposed between, and that the more rational statement is that we feel sorry because we cry, angry because we strike, afraid because we tremble, and not that we cry, strike, or tremble because we are sorry, angry, or fearful, as the case may be.[1]

Then James comes to what he calls 'the vital point' of his whole theory and mentions what has since come to be called a thought-experiment:

> *If we fancy some strong emotion, and then try to abstract from our consciousness of it all the feelings of its bodily symptoms, we find we have nothing left behind*, no 'mind-stuff' out of which the emotion can be constituted and that a cold and neutral state of intellectual perception is all that remains. It is true that, although most people when asked say that their introspection verifies this statement, some persist in saying theirs does not. Many cannot be made to understand the question. When you beg them to imagine away every feeling of laughter and of tendency to laugh from their consciousness of the ludicrousness of an object, and then to tell you what the feeling of its ludicrousness would be like, whether it be anything more than the perception that the object belongs to the class 'funny', they persist in replying that the thing proposed is a physical impossibility, and that they always *must* laugh if they see a funny object. Of course the task proposed is not the practical one of seeing a ludicrous object and annihilating one's tencency to laugh. It is the purely speculative one of subtracting certain elements of feeling from an emotional state supposed to exist in its fulness, and saying what the residual elements are. [. . .] What kind of an emotion of fear would be left if the feeling neither of quickened heart-beats nor of shallow breathing, neither of trembling lips nor of weakened limbs, neither of goose-flesh nor of visceral stirrings, were present, it is quite impossible for me to think. Can one fancy the state of rage and picture no ebullition in the chest, no flushing of the face, no dilatation of the nostrils, no clenching of the teeth, no impulse to vigorous action, but in their stead limp muscles, calm breathing, and a placid face? The present writer, for one, certainly cannot. [. . .] Every passion in turn tells the same story. A purely disembodied human emotion is a nonentity. I do not say that it is a contradiction in the nature of things, or that pure spirits are necessarily condemned to cold intellectual lives; but I say that for us, emotion dissociated from all bodily feeling is inconceivable.[2]

Wittgenstein mentions this theory and this thought-experiment from William James more than once in his later writings on the philosophy of psychology. Of particular interest, however, is an earlier passage from the 'Brown Book',

R. Egidi (ed.), Wittgenstein: Mind and Language, 249–262.
© *All Rights Reserved.*
1995 *Kluwer Academic Publishers. Printed in the Netherlands.*

which was dictated in the academic year 1934–35. There, in the context of discussing the question of the correctness of our translations of foreign, or radically foreign, words, Wittgenstein writes:

> You will find that the justifications for calling something an expression of doubt, conviction, etc., largely, though of course not wholly, consist in descriptions of gestures, the play of facial expressions, and even the tone of voice. Remember at this point that the personal experiences of an emotion must in part be strictly localized experiences; for if I frown in anger I feel the muscular tension of the frown in my forehead, and if I weep, the sensations around my eyes are obviously part, and an important part, of what I feel. This is, I think, what William James meant when he said that a man doesn't cry because he is sad but that he is sad because he cries. The reason why this point is often not understood, is that we think of the utterance of an emotion as though it were some artificial device to let others know that we have it. Now there is no sharp line between such 'artificial devices' and what one might call the natural expressions of emotion. Cf. in this respect: a) weeping, b) raising one's voice when one is angry, c) writing an angry letter, d) ringing the bell for a servant you wish to scold.[3]

This passage contains several themes which keep cropping up in Wittgenstein's later philosophy of psychology and, interestingly enough, even the combination of themes is typical of his later thought. The first point concerns the fact that we justify our ascriptions of psychological states to people by referring to facial expressions, gestures, and the tone of voice. This is important, not merely because these expressions often are our *only* grounds for attributing certain states to others, but also because often they are more *reliable* than what people say. After all, many of these expressions are what we call involuntary ones: they are difficult or impossible to suppress or imitate, and frequently our attempts at suppressing or imitating them are not successful, so that others can perfectly easily and with confidence tell how we feel.

The second point is a very involved one. Here Wittgenstein talks about our experiences of emotions and the questions arising in this context are largely coloured by problems about observing one's own state of mind. But here Wittgenstein seems to intend something narrower than what is suggested by the word 'observe'. When he speaks of experiencing a certain emotion he seems to mean that the occurrence of some emotions tends to go with certain typical feelings, and that this way of experiencing our emotions is possible only if these feelings concern a well-circumscribed part of our bodies. Wittgenstein mentions the frown which is felt in my forehead and the weeping felt in the area around my eyes. But this does not seem to apply to all kinds of emotions. Grief, for example, is not felt in a particular spot of one's body. Certain more specific manifestations, on the other hand, are felt in this way. Thus it may be the case that, being very sad, I weep; and this of course *is* closely related to my sadness and at the same time something that I do feel.

Tears, however, can also be shed from joy, so the mere fact of crying is not very indicative of the emotion in question. But in the case of joy, at any rate, there normally are other typical feelings in certain parts of our face which tend to correspond to natural and thus reliable outer expressions of this emotion.

The third point mentioned by Wittgenstein again regards the question of expressing an emotion. He says that the relation between feeling or experiencing an emotion and a certain place at which it, or part of it, is felt is often overlooked because 'we think of the utterance of an emotion as though it were some artificial device to let others know that we have it'. Now this connection may not appear altogether obvious. It consists, I suppose, in this: that the typical localized feeling going with a certain emotion is at the same time and through the good offices of our gestures, weepings, groans, and stammerings related to that which other people perceive as expressions or utterances of that emotion. And Wittgenstein wants to stress that these expressions or utterances tend to be natural, not artificial, ones. What appears as a way of communicating a certain emotion normally is not primarily intended as such. When I sit at the breakfast-table and notice that my father's nose is twitching, his fingers jerking, his brows knit, and his face purple, I literally see that he is angry. These unmistakable signs of his anger are not artificial devices he uses to tell me that he is angry; they are his natural, even instinctive, expressions of anger which however do tell me more clearly than words could that he is angry. Of course, he may, on another occasion, try to use these signs or symptoms as devices for giving me the impression that he is angry. This may work, especially if he manages to 'feel' part of what he is attempting to convey. But it is nevertheless a very different situation.

However, the roles of these expressions of our emotions are not always clearly separable; they tend to shade into each other. This is why Wittgenstein asks us to consider four different cases where our expressions are more or less natural or, as the case may be, more or less artificial. The first case is that of weeping which, I take it, is a paradigmatic natural expression. The second case is that of raising one's voice when one is angry. Normally this would also be a fairly natural expression which, however, can relatively easily be used, that is, artificially employed, to give the impression that one is angry. The third case is that of writing an angry letter. In this case it is the medium of writing which by itself takes away a lot of the naturalness of the expression. Writing may come fairly natural to some people, but even they have to look for the *mot juste* and to adjust their grammar, and these activities tend to take one's mind off the emotion that originally inspired the letter. The fourth case, in which you are ringing the bell for a servant you wish to scold, is a curious one. Here the means used, viz. to press a buzzer or to pull a string, is decidedly an artificial one but it may still be a natural way of expressing your anger. The servant, however, will not normally be able to tell from the way the bell sounds whether you are angry or not.

We can, as we have seen, use what normally are natural expressions of emotions as more or less artificial means of giving the impression that one feels a certain emotion. But it is also possible to rely on our capacity to feign such expressions in order to cause oneself really to feel that emotion. This possibility is clearly suggested by James's words, and Wittgenstein mentions it in his later remarks on the philosophy of psychology. He writes:

> And how does it come about that – as James says – I have a feeling of joy if I merely make a joyful face; a feeling of sadness, if I make a sad one? That, therefore, I can produce these feelings by imitating their expression? Does that show that muscular sensations are sadness, or part of sadness?[4]

Wittgenstein's last question concerns the problem of whether there is a conceptual or merely an empirical connection between statements about emotions and statements about the expressions of emotions. If there is such a conceptual connection, then it will be admissible to say that having certain feelings in your facial muscles is part of sadness. If, on the other hand, there is no such connection, that is, if statements about the expressions of emotions only say something about empirical relations between those expressions and the emotions themselves, then it will be incorrect to claim that those muscular sensations are part of sadness. Thus Wittgenstein writes:

> Suppose someone were to say: 'Raise your arm, and you will feel that you are raising your arm.' Is that an empirical proposition? And is it one if it is said: 'Make a sad face, and you will feel sad'?
>
> Or was that meant to say: 'Feel that you are making a sorrowful face, and you will feel sorrow'? and is that a pleonasm? (RPP I, § 452)

Now what about the first question: Is 'Raise your arm, and you will feel that you are raising your arm' an empirical statement? One thing we must remember is that this statement can be false. This is the case, for instance, when the person we are talking to has taken a drug which renders his limbs numb, so that he won't feel anything. Another possibility is that he is bombarded with so many exceptionally strong stimuli that the feeling of raising his arm won't register at all. The possible falsity of the statement seems to speak in favour of thinking that it is an empirical one. On the other hand, it will be false only under very special circumstances whereas in normal situations it will be true. And we have the feeling that normally somehow it cannot help being true, that feeling that one raises one's arm is simply part of raising one's arm. Yet it surely is not a purely conceptual truth, that is, it is not what we may wish to call an analytic statement. Is the same true of the statement 'Make a sad face, and you will feel sad'? Now of course, when we are really sad we do not normally bother to find out about what it feels like to be sad. But in many cases of, as it were, minor sadness we do observe ourselves, trace our feelings and may even compare them with our feelings on other occasions. In such situations you may realize that you are making a sad face and literally feel that you are sad. But if you are not sad but attempt to make a sad face and succeed in making a face which looks like the face of a sad person, does that really mean that you feel sad? Wittgenstein reformulates the question in order to make it sound even more pleonastic and says, 'Feel that you are making a sad face, and you will feel sad'. But even this formulation, I think, is not fully pleonastic; it is not, if true, a mere conceptual truth. After all, it seems quite possible to make a sad face, so to speak savouring the various aspects of the sensation, and still be quite aware that you are not really sad. In that case the statement in question would be false and hence not a conceptually true one. This impression is reinforced by Wittgenstein's next remark:

> Does one say: 'Now I feel much better: the feeling in my facial muscles and round about the corners of my mouth is good'? And why does that sound laughable, except, say, when one had felt pain in these parts before? (RPP I, § 454)

Of course one does not normally say things like that. And it is also clear that such a statement sounds silly because when talking about myself feeling well or better I do not intend to talk about what it looks like to feel well or better. That would be treating myself as if I were a different person, which of course is a possible attitude to take but not in talking to others.

Wittgenstein does not really come to a conclusion at this point, but he does throw doubt on the claim that certain sensations or feelings are to count as constituents of our emotions. He writes:

> Now granted – although it is extremely doubtful – that the muscular feeling of a smile is a constituent part of feeling glad; – where are the other components? Well, in the breast and belly etc.! – But do you really feel them, or do you merely conclude that they *must* be there? Are you really conscious of these localized feelings? –And if not – why are they supposed to be there at all? Why are you supposed to mean *them*, when you say you feel happy?
>
> Something that could only be established through an act of *looking* – that's at any rate not what you meant.
>
> For 'sorrow', 'joy' etc. just are not used like that. (RPP I, §§ 456-457)

Our words for emotions like sorrow, joy, etc. are used for states which we need not find out about. And whatever is part of these emotions is something we are immediately aware of, not through making an introspective effort or an inference.

In considering the problems concerning our concepts of the various emotions as well as their relations to sensations, muscular feelings, etc. special attention must be paid to the possibilities of arousing emotions artificially and of pretending to be in a certain emotional state. Looking at these non-natural situations, as we may call them, might help us to see what is really essential to such states.

An example which, I think, springs to mind in this connection is that of the actor who tries to play his dramatic part as convincingly as possible. William James quotes a number of actors who have given accounts of their experiences. Thus one actress says:

> I often turn pale in scenes of terror or great excitement. I have been told this many times, and I can feel myself getting very cold and shivering and pale in thrilling situations. (James, op. cit., p. 464)

And another actress gives the following answer:

> Playing with the brain is far less fatiguing than playing with the heart. An adventuress taxes the physique far less than a sympathetic heroine. Muscular exertion has comparatively little to do with it.

The example of the actor is used by Wittgenstein too in his attempt to get clearer about what an emotion consists of and whether it is legitimate to describe it in terms of various components. He writes in his *Remarks on the Philosophy of Psychology*:

> But wouldn't I say that the actor does experience something like real longing? For isn't there something in what James says: that the emotion consists in the bodily feelings, and hence can be at least partially reproduced by voluntary movements?
>
> Is it so disagreeable, so sad, to draw down the corners of one's mouth, and so pleasant to pull them up? What is it that is so frightful about fear? The trembling, the quick breathing, the feeling in the facial muscles? – When you say: 'This fear, this uncertainty, is frightful!' – might you go on 'If only I didn't have this feeling in my stomach!'? (I, §§ 727-728)

I take it that it is unproblematic to say that an actor who is acting his emotional part convincingly really experiences something like the relevant emotion. I suppose it is very much a question of degree. On the other hand, there is something puzzling in saying that a man who has *no reason* to yearn for anything feels something like yearning; or that a person who has absolutely no reason to be angry feels something like anger. But still, this kind of experience does exist, and actors are not the only people who know it; we all know it from occasions like reading thrilling books, watching romantic films, or hearing ghastly stories which arouse emotional feelings in us, although in our real lives there may be nothing to cause them. And in such situations of what one may call vicarious emotional feelings we can easily observe the palpitations of our hearts, the trembling of our hands, the biting of our lips, etc. But the relations between these occurrences and our emotions is still far from clear. We have seen that it is difficult or impossible to imagine certain emotions without their typical physical accompaniments. Thus it seems for example quite meaningless to imagine fear without fear-feelings in your heart and your stomach. When one is in the grip of a strong emotion one does say things like, 'I'm so afraid; I cannot stand this feeling in my stomach any more', or, 'If this state of uncertainty will continue much longer, I shall die of heart-failure', etc. And yet, when Wittgenstein asks whether it is this kind of feeling which is so frightful when one fears something, we tend to answer, 'Of course not. The frightful thing is that which arouses the feeling'. A similar point is made by Wittgenstein:

> The expression 'This anxiety is frightful!' is like a groan, a cry. Asked 'Why do you cry out?', however – we wouldn't point to the stomach or the chest etc. as in the case of pain; rather, perhaps, at what gives us our fear.
>
> When anxiety is frightful, and when in anxiety I am conscious of my breathing and of a tension in the muscles of my face – does that mean that I find *these feelings* frightful? Might they not even signify an alleviation? (RPP I, §§ 729-730)

There can be little doubt that it is not the feelings accompanying one's state of anxiety which are frightful. Or, to change our example, think of the case of exhilaration. It's not the tension caused by the upward moving corners of your mouth which you find exhilarating, nor the extraordinary lightness you feel in your chest. Rather, it may be the good weather, or the good news, or the fact that you slept marvellously which fills you with high spirits. In such cases we are dealing with the causes of our emotion, and they are easy to indicate. Other states have no straightforward causes in this sense. If I am worried about the future in general, it would be foolish to suggest that the uncertain state of the future is the cause of my worrying. It may be possible that the future is in a certain sense the object of my emotion in a similar way in which the cook who has spoiled my dinner is the object of my wrath. But that does not get us very far because there obviously are many emotional states of excitement, nervousness, languor, etc. which have neither a clear cause nor anything one might wish to call an object. But in these cases too we should not find it natural to say that the feelings of nervousness are that which we find enervating, or that the palpitating heart is that which we find exciting.

The distinction between emotions with a clear cause or object on the one hand and emotions without any identifiable cause or object on the other is, I think, what Wittgenstein has in mind when he asks us to compare fear and anxiety with care. The original German word translated as care is *Sorge*. And Wittgenstein continues by asking himself what sort of description is given by words like 'Ewiges Düstre steigt herunter', translated as 'Perpetual cloud descends'. These words are spoken by the figure of *Sorge* in the last act of Goethe's *Faust*. *Sorge*, or care, represents an emotion which is typically causeless and objectless. It is a state you can be in in spite of material and physical well-being; it is directed at nothing. It does not go with very strong typical accompanying muscular or other physical feelings. And still you can describe it. That is the point of Wittgenstein's quotation from Goethe. He wants to bring out that we can describe an emotion or, if you like, the *content* of an emotion by using words in a certain way, even though you cannot refer to causes or objects or accompaniments of that emotion. Now it would of course be interesting to use this idea in looking at the other emotions, that is, those which do have causes, objects, and accompanying feelings. And this is what I propose to do after I have gone a bit more thoroughly into some further questions arising from the theory suggested by William James.

In his most extended discussion of James' theory Wittgenstein writes:

> James says it is impossible to imagine an emotion or a mood without the corresponding bodily sensations (of which it is composed). If you imagine the latter absent then you can see that you are thereby abolishing the very existence of the emotion. This might happen in the following way: I imagine myself sorrowing, and now in the imagination I try to picture and to feel myself rejoicing at the same time. To do that I might take a deep breath and imitate a beaming face. And now indeed I have trouble forming an image of sorrow; for forming an image of it would mean play-acting it. But it does not follow from this that our bodily feeling at that point is sorrow, or even something like it. – To be sure, a person who is sorrowful cannot laugh and rejoice convincingly, and if he could, what we call the expression of sorrow would not really be that, and rejoicing would not be the expression of a different emotion. – If the death of a friend and the recovery of a friend equally caused us to rejoice or – judging by our behaviour – both caused us sorrow, then these forms of behaviour would not be what we call the expressions of joy or sorrow. Is it clear *a priori* that whoever imitates joy will feel it? Couldn't the mere attempt to laugh while one was feeling grief bring about an enormous sharpening of the grief?[5]

The passage in James to which Wittgenstein here alludes is evidently the quoted one in which James claims that it is impossible to abstract from our consciousness of an imagined emotion the feelings of its bodily symptoms. But while James argues his case by simply pointing out that if we subtract those physical feelings from our image of a given emotion nothing will remain of the original emotion, Wittgenstein introduces a new element into the discussion. He describes a case in which a person tries to imagine being in a certain emotional state and in that state trying to imagine to be in a different emotional state. His example is that of a man who imagines being sad and then tries to form a mental picture of himself being in a good mood. It does not work. And that it does not work is at least partly due to our being unable to combine the imagined physical symptoms and feelings going with one emotion with the symptoms and feelings of the other. But as the whole experiment is not a real one but entirely taking place in the imagination, it is not owing to any really felt physical sensations that the combination of those two states appears impossible. It looks more like a formal or Gestalt incompatibility and reminds one of our incapacity to see a picture or hear a tune in two different ways at the same time.

This part of Wittgenstein's remark concerns imagination. But then he goes on to real cases and observes that someone who is in a state of sadness cannot convincingly laugh, even if what he is confronted with is exceedingly funny. If he were capable to laugh, we should think that something was wrong with him, that his mind had suffered from the terrible event that brought on his sadness. This, however, has nothing to do with the physical feelings which James calls symptoms of our emotions. It is connected with what we regard as our natural ways of expressing our emotions. And some of these expressions exhibit a pattern in time which is incompatible with interruptions of a certain type. Of course, a great many emotions are quite compatible with other ones. Thus there is nothing strange in laughing about a joke when one is in a generally hopeful mood. But there are other emotions, notably sadness, melancholy, or grief, which form a pattern that is not compatible with interruptions by outbursts of good humour.

A person who is terribly sad can nonetheless try to show joy or pleasure, for instance because he does not want to disturb others by showing his true feelings. In that case he may, as Wittgenstein points out, feel his sadness even more acutely than before. And that again speaks against the existence of a special logical relation between statements about emotions and statements about the feelings or sensations normally accompanying these emotions. For in cases like that of the sad man trying to laugh about a joke the attempt at imitating the typical behaviour of a person in high spirits can, even though the muscular sensations normally accompanying such behaviour are felt, have the effect that he senses his grief even more sharply than before. Thus it is, to use Wittgenstein's words, by no means 'clear a priori that whoever imitates joy will feel it'.

However, all these considerations should not make us forget that the point from which we started – viz. the closeness of the connection between emotion and certain bodily symptoms – is incontrovertible. It is, although we cannot really find anything in the nature of an a priori relation, confirmed by the incompatibility of what Wittgenstein in the *Philosophical Investigations* calls the *patterns* of our emotions with certain other expressions of emotions, sensations, or feelings. This state of things is emphasized by Wittgenstein when he writes the following remark:

> Yet still I mustn't forget that joy goes along with physical well-being, and sadness, or at least depression, often with being physically out of sorts. – If I go for a walk and take pleasure in everything, then it is surely true that this would not happen if I were feeling unwell. But if I now express my joy, saying, e. g., 'How marvellous all of this is!' – did I mean to say that all of these things were producing pleasant physical feelings in me?
>
> In the very case where I'd express my joy like this: 'The trees and the sky and the birds make me feel good all over' – still what's in question here is not causation, nor empirical concomitance, etc. etc. (RPP II, § 322)

The first part of this remark is aimed at making clear that in talking about, or verbally expressing, a certain emotion we do not *mean* to speak of the physical feelings accompanying the emotion in question, although these feelings are in a certain sense inseparable from the emotion. And it is a sort of corollary of this that an expression of a given emotion mentioning those physical feelings is not meant to be a statement about 'causation' or 'empirical concomitance', as Wittgenstein says. This is naturally not intended as a denial of the existence of causal relations between that which makes me feel a certain emotion and the bodily symptoms accompanying that emotion. It should bring out that in expressing our emotions by way of mentioning what may possibly be their causes we still do not make a causal claim; at most we can be said to be trying to make our expressive behaviour specific.

The role of causes is made clearer by the following passage:

> [. . .] Possibly one could be sad because one is crying, but of course one is not sad *that* one is crying. It would after all be possible that people made to cry by application of onions would become sad; that they would either become generally depressed, or would start thinking about certain events, and then grieve over them. But then the *sensations* of crying would not thereby have turned into a part of the 'feeling' of grief. (RPP II, § 323)

The causal story told by Wittgenstein is this: People are made to cry by some artificial means; the crying causes them to become depressed or to think of sad events and thus come to feel sad. The sensation of crying, which presumably is indistinguishable from that felt in normal situations of sadness, is here clearly separated from the emotion of feeling sad. And it becomes increasingly clearer that what James calls the bodily symptoms of an emotion are in a certain sense not part of the emotion itself. They are closely related to the emotion but not an essential ingredient of it.

William James was led to his position, according to which the physical feelings accompanying an emotion count as its essential component, by starting, so to speak, from the inside. His most convincing argument is his thought-experiment, in which we try in our imagination to subtract the physical feelings from the emotion and do not succeed. But that still does not show that these feelings are particularly characteristic of our emotions. That which is really characteristic of them may become clearer by changing our perspective and looking at the emotions not so much from the inside but from the outside, that is, from the point of view of a man who tries to understand how another person feels by observing his behaviour, his expressions, his gestures, his utterances.

Some things are obvious. We judge that somebody is in a certain emotional state because his behaviour, or part of his behaviour, is a natural expression of a certain emotion. These natural expressions are often directly connected with the bodily symptoms William James talks about: the crying, the trembling, the raised voice, and the facial expression. But oddly enough, these phenomena now seem to be of much greater relevance when looked at from the outside. After all, they often are our only criteria for telling what state another person is in. They are the basis on which we judge whether someone is lying or telling the truth about himself.

It is as seen from the outside that these natural expressions can be useful for understanding and predicting other people's behaviour. Now it cannot be excluded that I can look at my own behaviour as it were from the outside and on the basis of that predict my future states of mind. But this would still be an exception. And equally rare or even rarer would be the type of case in which through noticing certain physical feelings I get clearer about my emotional state.

But the real point I want to make is this: Natural expressions like groaning, crying, trembling and so on are still only very rough indications of what another person may feel. Taken by itself, weeping does not tell you whether the person concerned is crying from joy or grief, jealousy or relief. And the same goes for the other natural expressions, like groaning, trembling, blushing, etc. In order to understand what they express we need to know more, either about the history of the person in question or about his present state.

Knowing more about his history, telling a story on the basis of that and the natural expression observed would amount to telling a plausible causal story. But if this type of knowledge is unavailable or insufficient we shall have to know more about the other person's present state. One way of finding out about that is by looking at the finer shades of his behaviour, at additional gestures, and especially at the peculiar *tone* in which he utters his expressions.

This point is of the greatest importance, for only now are we approaching that which is really specific and characteristic of certain emotions. It is the nuances, the shades and colourings of my expressions which make it possible for others to tell what I feel. An inkling of this is given by the following passage from the *Philosophical Investigations*[6]:

> I say 'I am afraid'; someone else asks me: 'What was that? A cry of fear; or do you want to tell me how you feel; or is it a reflection on your present state?' – Could I always give a clear answer? Could I never give him one?
>
> We can imagine all sorts of things here, for example:
> 'No, no! I am afraid!'
> 'I am afraid. I am sorry to have to confess it.'
> 'I am still a bit afraid, though I won't confess it to myself.'
> 'I torment myself with all sorts of fears.'
> 'Now, just when I should be fearless, I am afraid!'
>
> To each of these sentences a special tone of voice is appropriate, and a different context. (PI II, ix, pp. 187-188)

It is the specific tone of our expressions which allows other people to understand how we really feel. And it is this tone which is so difficult to imitate when you try to feign or to play-act a certain emotion. It is the existence of a certain tone of expression which enables us to make relatively fine distinctions between types of emotions and feelings, and it is the specificity of the tone which accounts for cultural variations in the expression of emotion. Many feelings can be distinguished from others only because there is a tone going with them. As an example Wittgenstein mentions the feeling of conviction and writes:

> One speaks of a feeling of conviction because there is a *tone* of conviction. For the characteristic mark of all 'feelings' is that there is expression of them, i.e. facial expression, gestures, of feeling. (RPP II, § 320)

I think, however, that there are at least two further outstanding features of our expressions of emotions which render them specific and help us to understand them. The first of these features is connected with what Wittgenstein likes to call the pattern of a certain emotion, for instance when he writes in the *Philosophical Investigations*:

'Grief' describes a pattern which recurs, with different variations, in the weave of our life. If a man's bodily expression of sorrow and of joy alternated, say with the ticking of a clock, here we should not have the characteristic formation of the pattern of sorrow or of the pattern of joy. (PI II, i, p.174)

The expressions of our emotions follow a certain pattern in the sense of having a kind of form or rhythm which can sometimes be imitated in another medium. An example would be music where we sometimes speak of sad or joyful, melancholic or boisterous tunes, or movements, or even entire pieces. And this possibility is surely due to the fact that something of the rhythm or pattern – and certainly also something of the tone – of these emotions or feelings can be expressed by musical means.

The third and last feature of the expression of emotions I want to mention is related to what was said about the quotation from Goethe's *Faust* concerning *Sorge*, or care. Here the point is that there are words or phrases by means of which one can contrive to describe exactly the characteristics of a certain emotion, although these expressions are descriptive neither of a cause nor of an object of these emotions. Goethe's poetic evocation of *Sorge* is a particularly accomplished description; but there are many more conventional, even hackneyed, phrases which serve to bring out the specific characteristics of our emotions. Thus we often use colour words to describe what we feel, for instance blue, grey, black, and yellow. But there are many more expressions – and sometimes quotations from poetry or song – which serve the same purpose.

These three features – tone, pattern, descriptive phrase – are not merely means of telling how another person feels. They have a lot to do with the perception of our own feelings. The rhythm or pattern, the expressive tone, and the visual or tactile impression corresponding to certain descriptive phrases are *felt* by the person who is in a certain emotional state. I think one could repeat William James' thought-experiment and ask whether it would be possible to subtract the typical greyness, the subdued tone and the slow rhythm from our emotion of melancholy and still retain an idea of what the feeling was like. And the answer, I suspect, would again have to be that an emotion without those ingredients would be a 'nonentity'.[7]

Bologna, Italy

NOTES

[1] W. James, *The Principles of Psychology*, Volume Two, H. Holt, New York 1890, p. 499.
[2] *Ibid.*, pp. 451 f.
[3] L. Wittgenstein, *The Blue and Brown Books*, ed. and with a Preface by R. Rhees, B. Blackwell, Oxford 1958, p. 103.
[4] L. Wittgenstein, *Bemerkungen über die Philosophie der Psychologie. Remarks on the Philosophy of Psychology*, Volume I (RPP I), ed. by G. E. M. Anscombe & G. H. von Wright. Trans. by G. E. M. Anscombe, B. Blackwell, Oxford 1980, § 451.
[5] L. Wittgenstein, *Bemerkungen über die Philosophie der Psychologie. Remarks on the Philosophy of Psychology*, Volume II (RPP II), ed. by G. H. von Wright & H. Nyman. Trans. by C. G. Luckhardt & M. A. E. Aue, B. Blackwell, Oxford 1980, § 321.
[6] *Philosophische Untersuchungen. Philosophical Investigations* (PI), ed. by G. E. M. Anscombe & R. Rhees. Trans. by G. E. M. Anscombe, B. Blackwell, Oxford 1953.
[7] This paper is an earlier version of Chapter 8 of J. Schulte, *Experience and Expression: Wittgenstein's Philosophy of Psychology,* Claredon Press, Oxford 1993.

GUIDO FRONGIA

WITTGENSTEIN AND MEMORY

1. One of the most significant explicit references to the philosophical problem of memory in *Philosophical Investigations*[1] is contained in § 265 which occupies a very central position in the discussion of private language. Because of its position, that section has been quoted and paraphrased innumerable times in the context of the wide debate provoked by Wittgenstein's thesis concerning the latter issue, and in this way it has become quite widely known. On the other hand, it has been far less considered in itself as containing certain notions which, while they are connected with other more general aspects of Wittgenstein's theory, put forward a viewpoint on memory which has its own intrinsic interest. The reason for this probably lies in the fact that in *Philosophical Investigations* the other references to the specific philosophical problem connected with memory are somewhat rare and unsystematic. The recent edition of the previously unpublished works of his late years provides new material of considerable interest to this question for the first time, and this leads us to attempt a summary of what Wittgenstein thought on this specific issue.

I shall limit myself here to considering certain aspects of the problem of memory which may have broader implications, useful for understanding his general theory on language and the bases of knowledge. Although I shall begin with a brief comment from PI, § 265, I shall seek to leave out as far as possible questions directly connected with private language. There has been so much written and said on this subject that it would prove not merely impossible but also in a way inappropriate to reintroduce this question directly here; if for no other reason than the complexity of the twists and turns of debate which it has recently produced.

2. The most important theoretical indication contained in PI, § 265 is that the deep grammar which regulates the uses of language connected with memory is not in fact homogeneous, as it might seem to be. A close analysis shows rather that there exist substantial differences within it, which make it possible to define two quite different notions of "memory", each characterised by its own rules which cannot be brought into line with those of the other.

In order to demonstrate this theory, Wittgenstein attacks what is probably

R. Egidi (ed.), Wittgenstein: Mind and Language, 263–277.
© 1995 *Kluwer Academic Publishers. Printed in the Netherlands.*

the most commonly accepted presupposition (though this acceptance may well be unconscious) of the opposite position, i.e. the idea that it is possible in each case, and following the same procedure, "to appeal from one memory to another", or even to "justify" the assumption that both refer to a "same" event or object. Naturally, the critical notion is not necessarily restricted to two memories only, but can be extended to a greater number of them, and may present complex ramifications. In any case, I shall limit myself here to the simplest case, and for reasons of clarity I will introduce an elementary formalisation.

Let us call M1 and M2 two successive memories which, we shall presume, come from the same event or object. The idea which Wittgenstein intends to criticise is the one which maintains that in the case of uncertainty about the reliability of the more recent record (M2), it is possible to have recourse to a memory which has arisen previously (M1), and thus to proceed with a sort of comparison, and from that to a choice between the two options. This operation should make it possible in some way to confirm or deny M2 in authoritative manner by use of M1. The example which Wittgenstein puts in the mouth of his ideal objector is that of someone who has to take a train and being in some doubt as to its departure-time recalls to mind an image of a railway timetable which he presumably consulted some time in the past. The second memory (M2) – the one over which there is some doubt – would thus be confronted with a memorisation (M1) which has been previously acquired and which deals with the same source of information, the physical railway timetable. If the two should coincide, then M1 would provide a confirmation of the correctness of M2.

Wittgenstein's counter-argument consists in observing that the example of his ideal opponent is certainly plausible enough in itself, but it is not pertinent to the purpose of demonstrating that one can speak of "justification" of the memory M2 *through* memory M1. If there is any "justification" here, it depends uniquely on the fact that the example envisages the possibility "of appealing to something independent" of both the memories (the physical railway timetable), which makes it possible to go on to a conclusive test of the correctness and thus eliminate all further doubts.

In general, then, Wittgenstein seems to be stating, recalling to mind a mental image (M1) with the aim of "justifying" a memory of the same event (M2) has meaning if, and only if, M1 can be considered "a memory which is actually *correct*." But this obviously does not always happen: indeed, it probably does not happen in the majority of cases.

3. As we can see, in this discussion Wittgenstein is assuming a particularly strong acceptance of the term "justification" *(Rechtfertigung)*. This provides the outline of a confirmation needing the support of a proof which, at least in the specific context in which it is produced, can be held to be certain and conclusive. The paradigm of certainty taken here as the measure of proof is that offered by the perception of a physical object, observed in normal conditions. Thus any further objections which may arise are not considered on this occasion; Wittgenstein discusses these elsewhere, at length, in his later writings. These may be put forward by a radical sceptic, who maintains that, in the example cited above, the further verification by M1, which refers back to an independent text (the page of the railway timetable), would not be definitive either. The reasons for this would probably be those which are generally adopted in this type of discussion: the person who proceeds with the verification may still be in doubt as to whether he is not dreaming, or having hallucinations, etc.

But if we exclude this type of reservation, Wittgenstein's conclusions seem to be fairly clear and convincing. Once the type of confirmation which can be provided by an "independent" physical object (i.e. an object which has an existence not dependent on the memory which one possesses of it) has been assumed as a paradigm of "justification", then no memory can itself justify another.

Naturally, if we consider the great variety of situations to which the function of memory must be adapted, we can easily imagine an infinity of intermediate grades between the definitive "justification" and the insoluble doubt. Wittgenstein, as is well known, does not deal thoroughly and in unified fashion with this variety of cases, though here and there he provides interesting footnotes which may concern (sometimes indirectly) different aspects of their grammar. It would certainly be very useful to reconstruct these rather fragmentary and technical references of his analysis in a systematic way.

Simply in order to provide examples of these grades of certainty which can be discerned in the use of memory in various effective contexts, we may refer here to certain consequences which can be extrapolated from the case discussed in PI, § 265. Even before reaching a "final justification", it is certainly "legitimate" in Wittgenstein's sense to compare two different memories with each other. And this may lead to varying results. Two possibilities can be provided for. The first is that M2 coincides with M1. Then it can be said that M1 provides a confirmation of M2 (even if not in a definitive sense); but it may equally be said that this coincidence provides a confirmation of M1. The second possibility is that M1 and M2 do not coincide. One could then say that M1 contradicts M2, as one could also state the opposite, that M2 contradicts M1. In each case, such a contrast does not provide *in itself* any conclusive indication on which of the two memories is to be considered more reliable. By what can be deduced solely on the basis of this comparison, M1 and M2 might both (totally or in part) be inaccurate or false, and be revealed as such in the light of subsequent verifications. In general, the simple fact that M1 is nearer in time than M2 to the event or object which is (presumably) at the origin of both, cannot be held in itself to be a criterion for greater reliability. Many other interesting considerations about this and other aspects of the grammar of memory could be entered into, and significant philosophical implications could be drawn from them. However, it is not these which we propose to discuss here.

4. From the standpoint we are taking, on the other hand, it is a basic requirement to return to the observation that in PI, § 265 Wittgenstein lays the conceptual basis for a fairly clear distinction within the grammar of memory. And this distinction makes it possible to define two major categories of cases: a first made up of the uses of memorising which are fully justifiable, and a second made up of those uses for which such a justification is not possible, even in principle.

The error of *reducing* the first to the second is likened in PI, § 265 to the meaningless claim of a person who buys "several copies of the morning paper to assure himself that what it said was true." There are two implications, one fairly clear and the other somewhat problematic of the meaning which this well-known image can assume in relation to the specific problem of memory. The first is fully in agreement with what we have observed so far. The absurdity of someone who compares two copies of the same newspaper is similar to that of the person who claims to compare two memories of the "same" event or object which cannot be "justified" conclusively by recourse to a definitive text, in order to find out which of them is true. In both cases, this operation is useless because it makes a comparison between entities which are pre-supposed to be homogeneous, and for this reason have the effect of proving nothing in relation to each other.

On the more general plane, (and this is certainly the most important philosophical element, which above all links our theme directly to that of private language), the *reductio ad absurdum*, within which the image of the newspaper which we have already mentioned appears, tends to show, as we noted previously, how wrong it is to reduce one area of the grammar of memory to another, for example that related to the cases for which it is not possible to refer to objective justifications to that which concerns cases for which such a possibility does in fact exist. It can be observed that a similar reductionist tendency may also be found in the opposite direction, that is, from the second category of cases to the first. And this naturally produces quite a few paradoxes, even though they be of reverse tendency.

Taking account of this, we shall use the term "reductionism" from here onwards (without any further specification) to denote the theory which tends to explain the rules of one of the two areas of the grammar of memory in terms of the other, independently of the direction in which such an attempt is headed. Conversely, we shall use further specification in cases where we wish to refer to one of the two possible variants of this type of theoretical approach.

5. It may in any case be useful to point out that, however brilliant on the plane of the argument and effective in literary terms it may be, the analogy between memory of the "same" event on the one hand, and copies of the same newspaper on the other, may well be quite inappropriate in other ways, and can create a number of uncertainties, as has in fact been the case.

The first is easier to locate, and for that reason theoretically less dangerous. In contrast to what happens with copies of the same newspaper, (always assuming that it only appeared in one edition), different memories of a "same" event or object (which are bound to occur at different times), may also be very different to each other, as Wittgenstein in fact admits on a number of occasion. Hence, at least from this standpoint, the metaphor of the newspaper does not work.

The second possible observation concerns grammatical aspect which is more difficult to grasp, but which – if it remains unclarified – may prove to be somewhat underhand. The use of the analogy of the copies of the same newspaper may be inappropriate in that it could easily provide water for the mill of the reductionist, in this case the one who seeks to reduce the whole grammar of memory to the objectivity of comparison. The reference to the copies of the newspaper could in fact suggest that, at least in the case in which it makes sense "to appeal from one memory to another" (because a particular link in the chain of memories can be checked against objective evidence), this takes place in the same way in which two physical objects are compared, for example, two editions of the "same" document written at different times, or even two copies taken from *two differing editions* of the same newspaper.

Naturally, Wittgenstein has shown, through innumerable analyses of linguistic cases, that he does not in fact believe in this second type of reduction of the psychological functions to the grammatical state which is proper to observable physical processes or events. Indeed, he always pointed out the serious conceptual difficulties to which such an attempt would lead. But this image of the newspaper could effectively induce some distracted commentator to forget this evidence.

6. The issues considered up to this point have concerned directly only one of the two major areas of the grammar of memory – that constituted by the memories for which it is possible to provide objective "justifications". So far, the other area (the other half of the apple) has only been considered for purposes of confrontation, as having certain grammatical characteristics opposed to the former area. We must therefore ask ourselves at this point, in more positive terms, what happens to those memories which are not "justifiable" in the sense defined above. Could they be considered (at least in a weaker sense than that implied by a "justification") more or less "reliable" – according to each individual case? And if two memories of the "same" event or object do not agree, how is one to decide between them? On what grammatical basis could such a comparison be made?

In order to deal with such a complex set of questions more closely and systematically, we would need to take into consideration a very broad series of linguistic cases discussed by Wittgenstein, which all seem to fall within the grammatical area of memory which we have now moved on to consider. To mention just a few examples: memories of thoughts which we might have had in the past, memories of particular sensations and states of mind, of mental images, etc. Moreover, there is a whole complex of psychological phenomena to which the problem of memorising is intrinsically connected. And apart from offering, certain more general grammatical aspects common to other cases, this has a specific interest in itself. This is the subject of dreams.

Dreams are the result of a psychic activity which, inasmuch as it takes place in the state of sleep, is not stimulated in direct, unequivocal and regular fashion by objects or phenomena of the external world. For this reason, the memories which we have of dreams when we pass into the state of wakefulness cannot be in any way "justified", in the sense in which Wittgenstein uses the term in PI, § 265. Therefore, we are here faced with the problem of the memory of dreams as an extremely significant case which makes it possible for us to gain a better understanding of some more general aspects of that very broad area of the grammar of memory which we are now considering.

7. As is well-known, the notes by Rhees dated 1942–46, and published under the title *Conversations with Freud* (CF)[2] are dedicated to the specific theme of dreams, as well as a large number of scattered notes contained principally

in Wittgenstein's late unpublished writings. Although not systematic in the conventional sense, this material nevertheless enables us to glean certain indications which are consistent in themselves and which can effectively help us to respond to the questions which we are now facing.

In *Conversations with Freud*, reference is often made to dream activity as characterised by a certain regularity; for example images which occur according to certain constant sequences, or which seem to follow particular internal logics, or which are arranged according to plots with recurrent schemes which sometimes make it possible even to make predictions about their future recurrence, etc. All this would seem justify, as Wittgenstein sometimes acknowledges, the commonly accepted idea that dream activity is "a way of saying something or a way of symbolising something" (CF, p.48); that there is, in fact, "something like a dream language" (CF, p.43).

However, to what degree is it justified to use the term "language" to refer to these regular formal and structural aspects of dream-perception? Does this use reflect a genuine analogy with the ordinary concept of "language", or is it only a way of showing some extrinsic similarity, but one which tends to ignore or hide substantial differences? These questions obviously have very complex implications, which cannot be considered here. Certainly, in support of the notion of a language of dreams there is the undeniable evidence, even historical evidence, of the contributions made by a large areas of contemporary psychology, which, like psychoanalysis itself, have made this notion a key factor in the elaboration of theory and analysis.

Wittgenstein does not seem to show any preconceived hostility to these trends of contemporary psychology, despite what has often been maintained. He certainly sees certain way of understanding and using their results as potentially "dangerous"; just as he sees danger in the fact that they tend to affirm the existence of a "powerful mythology" in contemporary culture (CF, p.52). But in overall terms his remarks do not have the aim of rejecting the bases of Freudian theory and psycho-analysis *en bloc*; they are tended rather to provoke critical reflection directed mainly at forestalling certain easily predictable misunderstandings, and at combating the mechanical, repetitive and dogmatic applications which can be made of certain of their theoretical principles.

In any case, the subjects dealt with in the conversations with Rhees have a fairly wide theoretical and philosophical range, which goes beyond the more circumscribed polemical concerns and particular historical points of reference of contemporary psychology.

8. Wittgenstein directs a particularly significant critique to the tendency to

see the exercise of memory, when this is applied to past dream-material, as a function aimed at explaining or describing in the most faithful possible way what has been dreamed. The reason for this criticism is that, in his opinion, this way of understanding attributes an exclusively passive, reproductive, figurative task to memory.

Naturally, Wittgenstein is well aware of the degree to which this aspect has been studied in the context of depth psychology. Almost all these studies, independently of their specific orientation, have revealed the various kinds of interferences which we meet with in the correct memorising of dreams. And this image of a distortion which memory imposes on its own objects is commonly aimed in this context not only at localising the nature of such interference, but also – and above all – at reconstructing the "real" development of the dream processes. The objective of this attempt to go beyond the appearances is obviously not aimed exclusively at the observance of a disinterested and abstract principle of truth. Such an exercise in this case has rather the aim of intercepting, as far as is possible in their immediate and integral state, the messages which the more profound layers of the Ego send to the conscious mind through the "dream language", to be deciphered later through "interpretation".

The most important aspect of Wittgenstein's reflection on this subject is therefore aimed at a critical review of the various moments of this theoretical and procedural scheme. His polemic is aimed above all at the idea of a supposed "objective" existence, with an identity in itself, of a dream-material which the memory must tend to reproduce faithfully. To this notion he counterposes the thesis that "the criteria for the truth" of an account of a private event, such as a thought which has arisen in the past, or a dream, cannot be "the criteria for a true description of a process"; and that the importance of such truth "does not reside in its being a correct and certain report of a process".[3]

As can also be seen quite clearly from *Conversations with Freud*, Wittgenstein sees the attempt to assume such a criterion for a "faithful" reproduction of a past and irrecoverable reality, such as a dream, as more a reflection of the aspiration of psychology to adapt itself to a model of knowledge and scientific explanation which is completely extraneous to it. This model, in his opinion, is moreover derived from a nineteenth century positivist conception of physical sciences (cf. CF, p.48), and hence not even capable of adaptation to many of the physical phenomena which are more frequently studied today. A double misunderstanding, this, which makes such an interpretation of the problem of memorisation of dreams quite impracticable. We should note that, if we wish to simplify matters, the model criticised here could fairly well be adapted to the case of a memory which is objectively "justifiable", as considered in PI, § 265, and discussed above.

9. The alternative way indicated by Wittgenstein is in line with his general theoretical approach, aimed at demonstrating the great variety of linguistic games, and at pointing out the impossibility of reducing certain areas of the deep grammar of the language to other areas which present apparent similarities with the first.

In this particular case, he induces us to recognise the specific nature of the rules which govern the various linguistic games concerning dreams:for example, teaching the use of the expression "I dreamed that. . .", recounting one's own dreams to others, recounting them to one's self in a low voice, describing them in writing, listening to the accounts of others, recalling to one's own mind visible images which have appeared in dreams, etc., etc.

The exercise of memory, inasmuch as it is aimed at reconstructing the content of the dreams, must be considered in all this variety of contexts of use. It could be said that in general memory is not here subject to limitations of justifiability of the same type as those to which it is subject when it refers to physical events or to objects of perception. The reason for this is that, in the case of a dream, in contrast to what applies in the latter cases, the memory itself *necessarily* (and hence not in some occasional fashion) contributes to the *identity* of what has been dreamed. Thus if, after return to the state of wakefulness, there are doubts as to the contents of a dream, the possibility of recourse to any incontestable protocol which has an identity independent of the memory itself is precluded. Similarly, in the case in which two differing memories (and thus two memories which have arisen at different moments) of the "same" dream differ from each other, it will be impossible to "justify" conclusively the validity of one or the other. The actual question: "How do I know if I *really* dreamed that?" (RPP I, §363) thus loses the meaning which an apparently similar demand might otherwise have, if it were aimed at reference to the objective and external source of a fact which is presumed to be reliable: for example: "How do I know that the train leaves at 7.30?" (cf. PI I, §265). The reply, in the case of the dreams, is necessarily univocal, and internal to the question itself: "This is what I remember". In short, memory seems here to be playing a productive and constitutive role in its own object – a role which has no parallel in other areas of the grammar of memory.

Now, must we hold that such a productive function, which memory exercises in these contexts, is deprived of restrictions? That it has unlimited power to give an identity to its own object? For instance, am I obliged to take everything I remember of a certain dream as equally valid? And if someone recounts his dream to me, and I assume his total sincerity, must I accept as definitive everything which he tells me? And what happens to any doubts I may have about certain particulars of my dream? And how should I decide, in the case of discordant memories, some of them recorded with great care in different moments? In short, are we forced to admit that memory becomes anarchic when it is applied to dreams? But then, how would it be possible to distinguish the memories of dreams from illusions or the creations of fantasy? (cf. RPP I, §201).

10. The replies that Wittgenstein provides for these and other similar questions are often not very clear, but show a certain indecision in assuming a precise theoretical position. Even considering the growing frequency of references to the theme of dreams contained in his later works, it is not impossible that he himself, as he progressively matured a more systematic response to this issue, did not then have the time to develop it completely. Certainly if there was evolution on this specific theme, it is fairly difficult to follow it with continuity through the labyrinth of the writings left unpublished. But if we exclude concerns of a genetic type, at least two rather different solutions can be discerned in his later work.

According to the first of these two solutions, in a certain sense, memory becomes a "*criterion* for the past" (cf LRPP I, §837), and thus the source of a sort of "subjective truth" (LRPP I, §897). While it is true that these expressions have a rather uncertain meaning, and recur only infrequently in his later writings, it is perhaps possible to hypothesise what Wittgenstein intended to state by them. For dreams, and for other analogous psychic phenomena, memory is the true source of information grammatically permitted to act as a reference for their content. And if it does not succeed in resolving any "doubts" which may arise, or fails to cover its own lacunae, the uncertainties which all this may produce will be destined to remain such, for the most part.

However, it is interesting to observe that both the expression here mentioned disappear in the corresponding passages in *Philosophical Investigations* – passages which probably reflect a more advanced stage of his work. On the contrary, there is a section in *Last Writings on the Philosophy of Psychology* which provides a rather different, and quite indicative, point of reference for our subject:

> Assuming that peoples' dreams can yield important information about a dreamer, what yielded the information would be truthful accounts of dreams. The question whether the dreamer's memory sometimes, often, or always deceives him cannot even arise, unless indeed we introduce a completely new criterion for the 'correctness' of the account of a dream (LWPP I, § 898)

As we can see, Wittgenstein seems here to be quite ready to admit a certain regulatory limitation in the reporting of dreams. In *Philosophical Investigations* (II, p.222) this indication receives further reinforcement. At the end of the paragraph corresponding to the one cited above, the "completely new criterion" which is required in order to decide about the "report's 'agreeing' with the dream", is given the specific title of *Wahrhaftigkeit*. So, something more than a genuine appeal to "correctness" is implied here. And yet, Wittgenstein does not provide us with any specific indication of the significance which should be attributed to such a criterion. He only says that it is quite distinct from that of the "truth" which must be valid for those accounts of memory which, according to his contention in PI, §265, are in proper sense "justifiable".

11. We may gather some indications, even if only indirect ones, which might help to clarify this crucial point from the *Conversations on Freud*, which will perhaps enable us to venture on a hypothetical response, tendential and schematic though it may be. Here in general, as previously observed, Wittgenstein does not appear to be preconceptually resistant to the idea that dreams can provide important information on the person who is dreaming. As we have seen, his reservations about the theory of Freud and psycho-analysis do not exclude the possibility that the instrument of dreams could be used for a better knowledge of the human mind. More often, his doubt concerns *the way* in which this possibility is frequently interpreted and translated into action.

In this way, we may feel that it is justified to remove the reservation of the conditional at the beginning of the passage cited earlier. Thus Wittgenstein would now really be searching for a "special" criterion of "truthfulness" for the memorising of dreams. But since the memory, when it is applied to this field of experience, remains the only effective source of information, the desired criterion of correctness must in some way be held to be *within* the actual *construction* which memory imposes on its own objects. And this would introduce an important reason for agreement with a fundamental principle of the theory and practice of psycho-analysis: in its productive function of dream-material, the memory does not proceed in chaotic and unpredictable fashion, but follows certain regular procedures. It is these latter which have a primary interest for the analysis, even if they have to be "interpreted" appropriately.

This very concept of "interpretation" provides a further indication of where a criterion of "correctness" might be found for the use of memory as applied to dreams. In the light of Rhees' notes, Wittgenstein does not actually seem disposed to set aside the problem of the interpretation of dreams from that of the reconstruction carried out in the light of the memory which one has of them. In other words, the relationship between these two moments is not one of succession. It is not a matter of *first* the memory and *then* the interpretation of what is remembered of a certain dream. Instead, it is the actual record which constitutes an interpretation (CF, p.46). Or, if we wish to put it in another way, interpretations and memories follow similar procedures.

12. We still need to clarify the meaning to be attributed to such a substantial identification between interpretation and memory. In *Conversations on Freud*, at least one significant pointer is given about this question. Dream material is presented to the memory in a way not different from the way one might present a complex mass of fragments of a figure which is indecipherable in itself (CF, pp.45-46). The memory tends to unify these fragments according to a coherent design, for example by attributing a unified plot to images and fragments of episodes which otherwise would seem incomprehensible in themselves. Naturally, this unifying function can be called into action by the need to make the dream known to others by recounting it, or may be influenced by the interaction of the person recounting it with the listener (the analyst, for example) (cf. CF, pp.43-44).

What Wittgenstein seems to be seeking to indicate here is the need not to consider these "plots" which the memory uses to unify the material which it has succeeded in retaining, as a "strange disturbance" (cf. LWPP I, §§656, 292). Instead, a "plot" is an integral part of this function of memory, aimed at gathering and mingling things remembered which may be derived from the most varied sources, and "turning them into the memory of an event which took place while a person was sleeping" (LWPP I, §656). It is this type of constructive intervention which confers singularity the linguistic game constituted by remembering dreams.

It seems clear that here Wittgenstein is thinking of certain characteristics of remembering dreams which are not dissimilar to that of a literary construction suited to certain canonical models of a structural or aesthetic kind – for instance, a particular pattern of recounting, widely accepted and easily discernible logical connections and oppositions, particular stylistic choices, and so on. Both these two activities, just because they adopt formal procedures and structurally regular typologies would succeed in transferring on to the plane of language contents which otherwise would remain relegated to the private world of the conscience, in the amorphous and non-communicable form of unique and unrepeatable impressions. In other words, here the "plot", far from being "a strange disturbance of memory", in its function of expounding dreams constitutes the essential go-between which gives them their identity. Without the medium of the "plot", it could be said, a dream would not exist, since memory could not confer any identity on it.

As can be seen, from this point of view, the problem of the correctness of the memory (its truthfulness, or *Wahrhaftigkeit*, PI II, p.223) comes to be identified with the problem of the correctness of the procedure of interpretation, such as is realised for instance in the reconstruction of a dream on the basis of a plot. And it is highly significant that in *Conversations with Freud* the most severe criticism which Wittgenstein directs at psychoanalysis concerns this very point. From these annotations some very illuminating indications not to be found elsewhere can be gathered.

The idea that Wittgenstein views with the greatest suspicion and hostility is that the interpretation of dream constitutes a kind of mechanical and univocal translation from one "symbolism" (specifically that of dreams) to another (that represented by our language of discourse), according to certain rigid rules of correspondence (cf. CF, p.42). The origin of this mistake is sited by Wittgenstein in the claim of Freud (and perhaps by psychoanalysis in general) to "be scientific" (CF, p. 44) and in this way to protect the interpretation (and the memory) of dreams from arbitrary interference. The intention of this is to confer objective security, uniformity of results, and hence reliability and prestige, on analysis. In fact the "scientific" model seems to require that the interpretation must either be "complete" or otherwise have no authentic value at all; and in this second case it must be substituted by a more "complete" interpretation (CF, p.49).

13. What we have seen so far provides clear indications on the critical plane, and on the objective polemics which inspire the analysis of Wittgestein. These aspects have been generally accepted by the critics, both for their historical interest and for their effectiveness in discussion. But it has been far less often noted that Wittgestein's observations also provide positive theoretical indications, though often not explicit ones. The interpretation of a dream, he seems to be saying, must be treated as such, i.e. as an *attempt* to pin down a fleeting object which cannot do other than escape. Such an attempt is made by assuming a particular standpoint or hermeneutic attitude which may change for very varied reason: "What is done in interpreting dreams is not all of one

sort" (CF, p.46). Thus, to summarise Rhees' testimony, reflecting on a dream and recording certain details of it "will give it a different aspect"; and remembering in different moments and context *will produce* different dreams (CF, p.46). Given the analogy between recording and interpreting: "there is a work of interpretation which, so to speak, still belongs to the dream itself"; and remembering the dream is a little like "dreaming the dream over again" (*ibid.*)

Therefore, what characterises an interpretation (for instance, in contrast to the certainty which can be attributed to the observation of a physical object or to the "justifiable" memory of it) is the fact of not being conclusive. An interpretation, in a certain sense, tends to produce a "myth" (CF, p.51). A negative connotation need not necessarily be given to this term. It is conferred on it by a comparison of an emulatory type with a (nineteenth century) image of science, tied to the concept of "exactitude" and to the principle of univocality of explanations and solutions. But there are extremely significant pages in Wittgenstein's last works[4] where the latter term does not have such a meaning. Here a "mythology" constitutes "the inherited background against which we distinguish between true and false" (OC, § 94) – a knowledge, then, that we accept, in more or less provisional fashion, with the aim of allowing other sectors of enquiry (interpretations) to proceed in relatively freer and speedier fashion with their proper exploratory tasks. Thus a "mythology" may itself become a "norm of description" (OC, §167) and hence a paradigm of interpretation.

If (as may be assumed) a critical intention is present in the observation that psychoanalysis tends to produce a "new myth", this is solely because Wittgenstein (rightly or wrongly) sees in it a certain propensity to undervalue the unavoidable complexity and indeterminacy of the examination of dreams and similar psychological phenomena – a propensity which he believes has its origin in a misconceived intention of "being scientific", which leads that discipline to aspire to an identity which is both alien and inconvenient to it. It is, then, against this unauthentic attitude rather than against particular theoretical and therapeutic advances that Wittgenstein's critique is directed. An attitude which may make for reluctance to accept what probably today many analytical schools would be ready to recognise: that a dream is destined to remain "an enigma", which "doesn't *have* to have a solution. It intrigues us" (LWPP I, §195).

Università di Roma 'Tor Vergata'

NOTES

[1] *Philosophische Untersuchungen. Philosophical Investigations* (PI), ed. by G. E. M. Anscombe & R. Rhees. Trans. by G. E. M. Anscombe, B. Blackwell, Oxford 1953.
[2] See in L. Wittgenstein, *Lectures & Conversations on Aesthetics, Psychology and Religious Belief*, ed. by C. Barrett, B. Blackwell, Oxford 1966, pp. 41–52.
[3] See *Letzte Schriften über die Philosophie der Psychologie. Last Writings on the Philosophy of Psychology*, Vol. I (LWPP I), ed. G. H. von Wright & H. Nyman. Trans. by C. G. Luckhardt & M. A. E. Aue, B. Blackwell, Oxford 1982, § 897; PI, II, p. 222.
[4] Cf. *Über Gewissheit. On Certainty* (OC), ed. by G. E. M. Anscombe & G. H. von Wright. Trans. by D. Paul & G. E. M. Anscombe, B. Blackwell, Oxford 1969, §§ 94 ff.

FRITZ G. WALLNER
ASSISTED BY MARKUS COSTAZZA

HOW LUDWIG WITTGENSTEIN WOULD HAVE REACTED TO RECENT CHANGES IN PSYCHOLOGY

In this paper we want to examine whether L. Wittgenstein's ideas can be made fruitful for basic discussion in cognitive psychology, as the development of methodological discussion in the field of psychology is not as remarkable as in physics.

The first main point in this discussion is the confrontation between psychology of experience and cognitive psychology. After this confrontation the perspectives in psychology changed drastically; this was due to the so called cognitive turn. At this point the contrast between behaviorism and psychology of experience is not only overcome, but abolished (in the philosophical sense that we owe to Hegel). I.e., it is solved at a higher level, without sacrifice of the claims underlying those two approaches. One could characterize those claims as the primacy of immediateness and the primacy of probability. We shall see that both play an important role in Wittgenstein's confrontation with psychology, especially when he distances himself from Carnap's approach.

Cognitive psychology, however, in contrast to psychology of experience and behaviorism already assumes a concept of "self" adapted to psychology. The "self" is included in the system's structure in this case. We shall find that in this case a deficiency appearing in Wittgenstein's writings is solved by a single discipline.

Wittgenstein's position concerning behaviorism is often misunderstood and seems to be strange; he was often seen as a behaviorist. For example, Rorty (1977, p. 169) points out that Wittgensteinians have entered into an alliance with traditional behaviorism. As everybody knows, there are indeed some places in Wittgenstein's works, which could be interpreted in a behavioristic way. This becomes evident in particular if we cite the distinction of the first and third person, as it occurs in *Zettel*, § 472:

> Plan for the treatment of psychological concepts. Psychological verbs characterized by the fact that the third person of the present is to be verified by observation, the first person not.

Statements like the last quotation seduced Robert Fogelin to talk of a form of behaviorism in the third person in Wittgenstein's work, which makes him accuse Wittgenstein of a basic weakness.[1]

For Wittgenstein the point is that one can apply verification to psychological

R. Egidi (ed.), Wittgenstein: Mind and Language, 279–288.
© 1995 *Kluwer Academic Publishers. Printed in the Netherlands.*

sentences in the third person; applying it to sentences in the first person, however, would be senseless. He draws, however, another conclusion than R. Carnap.

In our opinion Luckhardt (1983) is right in interpreting Wittgenstein's concept of psychological sentences as reaction to Carnap's approach in this context. We can also find one of the few places in which Wittgenstein explicitly quotes Carnap:

> One may also say: 'He made *this* face' or 'His face altered like *this*', imitating it – and again one can't describe it in any other way. ((There just are many more language-games that are dreamt of in the philosophy of Carnap and others.))[2]

Carnap was influenced by the principle of verification when solving the problem of the psychological sentence in the first person. He rejected introspection, since, given the lack of verification, it cannot lead to sentences being intersubjectively meaningful. He offers another explanation for psychological sentences in the first person; it runs as follows: if A learns to use psychological sentences about himself, then he is learning an intersubjective physical language which not only describes the relevant state of his body, but also has to describe him in such a way as another person could describe it in principle.[3]

It can be recognized that Carnap is defending the classical concept of behaviorism, as it was formulated by Watson (1913). It seems that the following is valid for Carnap as well as for Watson: Although consciousness cannot be topic of psychology, it has the "not unimportant function of being the instrument of scientists".[4] This implies however a strange ambivalence: it is a metaphysical fiction and yet has scientific function.[5]

In contrast, Wittgenstein follows another path. He interprets only psychological sentences in the third person as sentences having representative function; sentences in the first person are expressive.

> . . . The word 'pain' is the name of a sensation is equivalent to 'I've got a pain' is an expression of sensation.[6]

Having these ideas in mind, Wittgenstein rejects introspection as well as traditional behaviorism. As a consequence we have to ask, if Wittgenstein creates a new kind of behaviorism: behaviorism of linguistic behavior which we could call the "linguistic behaviorism".

Wittgenstein considers this in his *Philosophical Investigations* I, §307:

> 'Are you not really a behaviourist in disguise? Aren't you at bottom really saying that everything except human behaviour is a fiction?' – If I do speak of a fiction, then it is of a *grammatical* fiction.

It becomes clear that Wittgenstein is not concerned with empirical statements in such as occur in particular discipline. We shall realize the point of this position only if we look at the behaviorism in psychology.

E. C. Tolman (1932) already attacked Watson's position. His argument against Watson's physiological reductionism was that referring to physiological knowledge which has not yet been attained means running the risk of a pseudoscientific behaviorism. One should not refer to Watson's "muscle-twitching" – but to behavioral acts as conceptual units. This means that intentions become more important for psychological investigations while the intended states of affairs also are determined in regard to the relevant behavior.[7]

As in phenomenological psychology the facts have a different meaning, depending on whether the agent is an American, a child or an animal.[8]

The logic of this idea led Tolman to his propositionalism (Tolman 1959). By propositional behavior he understands behavior cognitively directed towards a goal. The psychological analysis founded on these ideas is not aimend at describing experience, but rather behavior. Following Tolman's example a chair could have the meaning of "sit upon ableness", of "climb upon ableness" as well as of "throw about ableness" (according to the behavior of the user). Things may be comprehended intentionally in a means-end-relation without having to make use of introspective data of experience. By these distinctions Tolman (1959) also introduces distinctions within manifestations which have been normally attributed to subjective soul life. He differentiates between means-end readiness, beliefs and expectancies. On this basis he determines intentional phenomena by behavioral description (Graumann 1965, pp. 248f). In this way he anticipates a position labeled by Miller, Galanter and Pribram (1960, p.211ff) as subjective behaviorism. Using this approach Tolman could hope to fulfill the tasks of psychology, as conceived, for instance, by phenomenological psychology, without having to recourse to descriptions of experience.

Applying these arguments we have made a big development compared to Watson. Watson's reductionism expects too much from the concept of organism. One cannot explain the differentiated structures of behavior, as for instance the bearing in mind of a certain goal, from the organism alone, unless one makes pseudoscientific assumptions and constructions. Subjective behaviorism overcomes this deficiency without having to make the assumption of the hypothesis of two domains of reality as psychology of experience does.

Let us return to the different views of Wittgenstein and Carnap. Carnap conforms with Watson. He succumbs to a bewitchment of language in his desire to make language a language of science. This bewitchment lies in the demand that even psychological sentences in the first person should be a descriptive language game. This leads to an ontological change in his approach in order to fulfill the claims of the principle of verification. The change consists in the banishment of fictions mental objects out of the body; Carnap succumbs to a misunderstanding of language (as Heidegger did in his ontologisation of "the nothing"). While Heidegger postulated meaning for a word, Carnap undertakes the effort of deciding between possibility and impossibility of meaning of words. It was Wittgenstein's achievement to unmask introspectionism as well as behaviorism as a grammatical fiction. His position is quite near to Miller, Galanter and Pribram's approach of subjective behaviorism, but, unlike theirs, his approach permits the overcoming of the dualistic gap.

Due to its ontological undecidedness subjective behaviorism runs the risk of being subject to metaphysical fictions, as the literature on its phenomenological interpretation shows.[9] In contrast Wittgenstein does not succeed in accommodating subjectivity and sense in psychology. If cognitive psychology follows him, it has to face an undecidedness in principle. Ontologizing subjectivity represents a similar danger for cognitive psychology as previously for introspectionism and physiologism. The accusation that Wittgenstein forfeits subjectivity represents a similar misunderstanding to the accusation of his being a behaviorist.

The ontological undecidedness of modern psychology causes its uncertainty about its object of research. This is shown in Wittgenstein's *Zettel,* § 468:

> 'Man thinks, is afraid etc. etc.': that is the reply one might give to someone who asked what chapters a book on psychology should contain.

This implies the special position of psychology and its different character compared to exact natural sciences. Wittgenstein writes in his *Remarks on Philosophy of Psychology*:

> Psychological concepts are just everyday concepts. They are not concepts newly fashioned by science for its own purpose, as are the concepts of physics and chemistry. Psychological concepts are related to those of the exact sciences as the concepts of the science of medicine are to those of old women who spend their time nursing the sick.[10]

Would it be a possible suggestion to use the method of explicative definition in order to state these concepts more precisely or to delimit them from their everyday use. We will see that Wittgenstein's insight into epistemological interrelations keeps him away from this naive cul-de-sac.

Wittgenstein's investigations of psychological terms can, I believe, be used for the development of a new approach to psychology. In this respect he has gone beyond the dispute between natural scientific and humanistic psychology. In this respect Stromberg (1987, p. 28) is correct when claiming that Wittgenstein's critical remarks do not play an important role from the point of view of present-day trends in psychology. To my mind, however, one cannot accept the self-understanding (i.e. the unification of application and theory) of some of these trends. Humanistic psychology is an example of this.

The joint development of theory and application is represented in psychology in those approaches which suppose that the creation of psychological terms and specific human selfinterpretation go hand in hand. One could talk of a pragmatic turn in psychology, when its motto is the "self-making" of the human being. We have to differentiate this from the action-theoretical turn in psychology. It can be understood as the reaction to behaviorism, in which the human being was seen as a stimulus-response mechanism.

Wittgenstein's ideas must also be correlated with the approach of action theory. This relation is quite similar to the relations between the late Wittgenstein philosophy of language and Searle's speech act theory (cf. Wallner 1981). There is a twofold relationship: in contrast to approaches in theory of action Wittgenstein is not in search of an explanation of human behavior. The aim, again, is to gain clarity. Wittgenstein's view of behavior represents an essential difference to the theory of action. The representation of mental processes against the background of behavior reveals structures which are not taken into consideration in theory of action.

> My own behaviour is sometimes, but *rarely* the object of my own observation. And this hangs together with the fact that I intend my behaviour. Even if an actor observes the expressions of his own face in a glass, or the musician attends closely to every note in playing, and judges it, this happens after all so that he shall direct his action accordingly. [*Zettel*, § 591][11]

As Wittgenstein sees it, two aspects of behaving are of special interest: its deliberateness and the question of the role the observer plays in this context. Because of its deliberateness it is different from mental life (this is also the argument against the misinterpretation of Wittgenstein as a behaviorist). This determines the role of behavior in the argument: undoubtedly it represents the first-person point of view; hence its expressivity, but its deliberateness makes behavior an object of observation. This happens quite rarely, usually when it serves some purpose of action. This becomes clearer when Wittgenstein talks about the role of self-observation, as in § 839 of *Remarks on philosophy of psychology*. Two aspects become clear: on the one hand the strange role of self-observation which does in "first person" what is done to the other by means of the "third person" and, on the other hand, the possibility of self-observation is made clear.

We have to focus on the conceptual relation between action and behavior. This relation is not easy to realize and we cannot make use of traditional categories to comprehend it. The criterion of immediate planning represents an important distinctive feature. Action is immediately determined by planning, whereas behavior is planned seldom and only indirectly.

> After all I know that when I am angry, I simply don't need to learn this from my behaviour. – But do I draw a conclusion from my anger to my probable action? One might also put the matter, I think, like this: my relation to my *actions* is not one of observation.[12]

An argument for this approach to the concept of action is evident in RPP I, § 561

> The expression, the behaviour, of considering. Of what do we say: It is considering something? Of a human being, sometimes of a beast. (Not of a tree or a stone.) *One* sign of considering is hesitating in what you do (Köhler). (Not just *any* hesitation.)

Acting is planned immediately if reflection disturbs it. The function of observation is different in the context of behavior and in that of action. Action is not observed because it is planned, behavior is observed in order to be planned. The relation between anger and angry actions could be observed in principle; it would not correspond, however, to normal human behavior. Hence the methodological argument against such an observation is not lack of clearity in the observed object.

Let us turn to RPP I, § 705

> Why do I never conclude from my words to my probable actions? For the same reason as I don't conclude from my facial expression to my probable behaviour, – for the interesting thing isn't that I don't conclude from my expression of emotion to my emotion, but rather than I don't conclude from that expression to my later behaviour either, as others do, who observe me. [*Zettel*, § 576]

The conceptual relation between actions and words corresponds to the relation between behavior and facial expression, it is that of specification. The relation between the two forms of behavior is not a relation of causality: as regards behavior we rather have to differentiate the first-person point of view from the third-person point of view. If one infers from behavior (1) to behavior (2), one looks at behavior (1) from a third-person point of view and interprets it as action. Language clarifies the situation in the phrase "expression of emotion". Behavior (1) is an expression of emotions in so far as I make my emotions public. When I am expressing my emotions, the conclusion from this to my emotions is trivial. The conclusion from such expression to future behavior is not trivial, however.

In my opinion, this is the crucial point of modern cognitive psychology. Turning to the investigation of human action-plans implied turning away from the incorrect anthropological concept of behaviorism. But a category mistake was committed. The unsatisfactoriness of interchanging the first-person point of view with the third-person observation was seen; it was realized that this unsatisfactoriness could not be rectified by addition a data from introspective psychology. It was been seen that in the concept of action-plans there arises a similar problem. One could describe it as follows: If action-plans are made from the perspective of the third person, they will become relatively unimportant and trivial for the individual, and for his emotional life.[13] One can try to draw up action-plans from the first-person point of view. However, these would not be action-plans in the same sense as discussed above. The problems may be covered up by a concept of "participating observation", but they will not be surmounted. Neisser (1976, p.7) describes the problems in the context of history of psychology:

> Indeed, the assumptions that underlie most contemporary work on information processing are surprisingly like those of nineteenth-century introspective psychology, though without introspection itself.

His suggestions for improvement, however, do not go beyond methodological corrections:

> First, cognitive psychologists must make a greater effort to understand cognition as it occurs in the ordinary environment and in the context of natural purposeful activity. This would not mean an end to laboratory experiments, but a commitment to the study of variables that are ecologically important rather than those that are easily manageable. Second, it will be necessary to pay more attention to the details of the real world in which perceivers and thinkers live, and the fine structure of information which that world makes available to them.[14]

A similar judgment underlies Wittgenstein's thesis concerning the role of experiments in psychology, however without hope of improvement:

> Rather, in psychology there is what is problematic and there are experiments which are regarded as methods of solving the problems, even though they quite by-pass the thing that is worrying us.[15]

In the view of the misunderstanding of the conceptual role of "action" in cognitive psychology, turning to experiments, leads, only to apparent pseudo-solutions.[16]

Wittgenstein's thoughts, as mentioned and reconstructed above, show us the reasons why cognitive psychology was not capable of fulfilling the hopes which had been placed in it. We must ask whether Wittgenstein can offer us suggestions for finding a way out of the present crisis in cognitive psychology.

This crisis can be briefly characterized as follows: one group of approaches does not reach our emotional life, the other group overrides it. The computational approach of simulating cognitive processes is based on a formal assumption, i.e. that emotional life follows strict and explicable rules, and on the assumption that psychological performances can be understood criterially; i.e. they have to be interpreted as given results. Artificial Intelligence (the symbolic paradigm) abandons all claim to explain emotional life, but rather tries to reproduce cognitive processes under the general aspect of information processing but with results drawn from artificial models.[17]

Finally we must estimate in what respect we can learn from Wittgenstein in this situation. It seems that Allan Turing's successful approach in Artificial Intelligence research narrows the horizon of Wittgenstein's thoughts interpretation by criteria. It seems that Allan Turing's approach, so succesful in Artificial Intelligence, narrows the horizons of Wittgenstein's thought by a purely criterial interpretation. It puts in the shade the "psychological propositions" whose conceptual elucidation is so ticklish, but which could be important even for present-day psychology. We observed this, for instance, in differentiating between acting and behaving.

Universität Wien

NOTES

[1] J. R. Fogelin, *Wittgenstein*, p. 176.
[2] Wittgenstein (1980), § 920.
[3] Carnap (1959), p. 192.
[4] Graumann (1965), p. 244.
[5] *Ibid.*
[6] RPP I, § 313.
[7] Tolman poses 1922 in his m paper the following question: "By what single common set of concepts can we possibly take care both of the facts of gross behavior and of those of consciousness and imagery?" (p. 3). And he opines that a "non-physiological behaviorism", which nevertheless does also investigate behavior, has to distinguish at least four aspects: "stimulating agency, behavior cue, behavior object and behavior act" (p. 4). "Behavior object" represents in this case the meaning (depending on the situation) of a bundle of possibilities of behaving, "because of his particular training and past experience and state of behavior readiness . . . arouse in him a very specific group of behavior tendencies; e.g., those of sitting upon, getting up from, kneeling on, moving up to

the table, etc. This group of aroused tendencies defines his behavior object." (p. 6).
[8] Tolman (1959), p. 6.
[9] Cf. e.g. Wann (1964).
[10] RPP II, § 62.
[11] RPP I, § 838.
[12] RPP I, § 712.
[13] Zitterbarth and Werbik (1987) already locate this methodological weakness in the beginnings of cognitivistic explanations of behavior: "Es kommt also, wenn Miller, Galanter & Pribram von 'Bildern' und 'Plänen' reden, nicht auf den 'subjektiven Sinn' an, den Handelnde mit ihrem Tun verbinden . . . 'Pläne' und 'Bilder' meinen also nicht die faktischen Sinngehalte, mit denen ein Subjekt operiert, sondern 'theoretische' Konstrukte, die an die Stelle des physiologischen Reflexbogens im Behaviorismus treten." (When Miller, Galanter and Pribram speak of "images" and "plans" they are not referring to the "subjective meaning", which acting people connect with their actions . . . therefore "plans" and "images" don't mean the actual meaning content present to an acting subject, but "theoretical constructs" replacing the physiological reflex arc assumed by behaviorism.) (p. 231).
[14] Neisser (1976), pp. 7–8.
[15] Wittgenstein (1967), § 1039.
[16] This opinion, saying that shortcomings arise when unclear basic concepts coincide with unproved empirical methodology, is shared by Neisser (1976) too, who writes: "Perhaps as a result, the actual development of cognitive psychology in the last few years has been disappointingly narrow, focusing inward on the analysis of specific experimental situations rather than outward toward the world beyound the laboratory." (Preface, p. XI).
[17] Compare Schiffner (1989), especially p. 23 ff.

REFERENCES

Carnap R.: 1959, 'Psychology in Physical Language'. In: *Logical Positivism*, ed. by A. J. Ayer, New York, p. 165 ff.

Fogelin J. R.: 1976, *Wittgenstein*, Routledge & Kegan Paul, London.

Graumann C. -F.: 1965, 'Subjektiver Behaviorismus?', *Archiv für die gesamte Psychologie*, Bd. 117, pp. 240–251

Luckhardt C. G.: 1983, 'Wittgenstein and Behaviorism', *Synthese*, 56, pp. 319–338.

Miller G. A., Galanter E. & Pribram K. H.: 1960, *Plans and the Structure of Behavior*, Holt, Rinehart & Winston, New York.

Neisser U.: 1976, *Cognition and Reality. Principles and Implications of Cognitive Psychology*, W. H. Freeman, New York.

Rorty R.: 1977, 'Wittgensteinian Philosophy and Empirical Psychology', *Philosophical Studies*, 31, pp. 151–172.

Schiffner E.: 1989, *Verstehende Maschinen versus Verstehensmaschinen. Eine philosophische Untersuchung über Bedingungen der Möglichkeit von Maschinenintentionalität*, Phil. Diss., Wien.

Stromberg W.: 1987, 'Wittgenstein. Theoretical Psychology and the Classification of Psychological Concepts', *Philosophical Investigations*, 10, pp. 11–30.

Tolman E. C.: 1966, 'A New Formula for Behaviorism', in E. C.Tolman, *Behavior and Psychological Man. Essays in Motivation and Learning*, Univ. of California Press, Berkeley,

Los Angeles, pp. 1–8.
Tolman E. C.: 1932, *Purposive Behavior in Animals and Men*, Century, New York.
Tolman E. C.: 1959, 'Principles of Purposive Behavior', in *Psychology: A Study of a Science*, ed. by S. Koch, Vol. 2., McGraw-Hill, New York, pp. 92–157.
Wallner F.: 1983, 'Läßt sich die Sprechakttheorie als eine präzisierende Fortführung und systematische Ausarbeitung von Wittgensteins späterer Philosophie verstehen?', *Der Mensch und die Wissenschaft vom Menschen. Die Beiträge des XII. Deutschen Kongresses für Philosophie*, ed. by G. Frey, Innsbruck, Vol. 2, p. 1033 ff.
Wann, T. W. (ed.): 1964, *Behaviorism and Phenomenology. Contrasting Bases for Modern Psychology*, Univ. of Chicago Press, Chicago.
Watson J. B.: 1913, 'Psychology as the Behaviorist views it', *Psychological Review*, 20, pp. 158–177.
Wittgenstein L.: 1953, *Philosophische Untersuchungen. Philosophical Investigations*, trans. by G. E. M. Anscombe, B. Blackwell, Oxford.
Wittgenstein L.: 1980, *Bemerkungen über die Philosophie der Psychologie. Remarks on the Philosophy of Psychology*, Vol. I, ed. by G. E. M. Anscombe & G. H. v. Wright. Oxford: Basil Blackwell, Oxford (Quoted as RPP I).
Wittgenstein L.: 1967, *Zettel*, ed. by G. E. M. Anscombe & G. H. v. Wright. B. Blackwell, Oxford
Zitterbarth W., Werbik H.: 1987, 'Subjektivität als methodisches Prinzip. Argumente und Verfahrensweisen einer dialogisch-verstehenden Psychologie', in *Wissenschaftstheorie und Erkenntnistheorie. Eine Einführung für Psychologen und Humanwissenschaftler*, ed. by J. Kriz, H. E. Lück, H. Heidbrink, Leske & Budrich.

INDEX OF NAMES

Ambrose A., 81n, 136n, 183n, 231n
Anderson C., 169n
Anderson T., 56n
Angelelli I., 81n
Anscombe G. E., 35n, 36n, 46n, 56n, 81n, 91n, 105n, 114n, 135n, 168n, 182n, 183n, 205n, 210n, 222n, 231n, 247n, 262n, 277n
Aristotle, 33, 37, 42, 43, 45, 47, 121
Armstrong D., 203, 205n
Aue M.A.E., 81n, 182n, 205n, 247n, 262n, 277n
Augustine, 121
Austin J.L., 26
Ayer A.J., 247

Baker G.P., 93–95, 105n, 136
Beard R., 247n
Bell J., 90n
Bergmann G., 137, 144–146, 157, 167, 169n, 247
Beth E.W., 88
Black M., 11, 17
Blackmann L., 169n
Barrett C., 136n, 277
Boltzmann L., 4
Boričić B.R., 90n
Bosanquet R.G., 81n
Bouveresse J., 59
Bower G., 247n
Bradley H., 152, 154, 170n
Brandl J., 47
Brentano F., 3, 131, 136, 171, 181
Broch H., 9
Burnyeat M., 222n
Burkhardt H., 169n

Cantor G., 69, 70, 81
Carl W., 211
Carnap R., 2, 3, 4, 5, 7, 14, 16n, 17n, 175, 181, 183n, 221, 222n, 225, 247n, 279–282, 286n
Casati R., 185
Cellucci C., 83, 90n
Chisholm R.M., 53, 54, 56n
Chomsky N., 77
Church P.A., 247n
Claesges U., 192n
Clark R.W., 135n
Clarke T., 216
Coleridge S.T., 39
Comte A., 1
Copi I., 247n
Corsi G., 90n
Costazza M., 279

D'Agostino M., 90n
Darwin C., 1
Davidson D., 166–168, 170n, 221
Dell'Utri M., 117
Descartes R., 38, 42, 44, 118, 122n, 211–216, 218, 222
Diamond C., 81n, 91
Dostoyevsky F., 4, 6
Dummett M., 84, 90n

Eames E., 169n
Egidi R., 171, 231n
Einstein A., 3, 17, 225, 230
Elie H., 81n
Euclide., 68

Feyerabend P.K., 224, 231n
Fichte J.G., 7
Fodor J.A., 107–115, 117, 120, 121, 123n
Fogelin R., 279, 286n
Frank Ph., 225
Frazer J.G., 7
Frege G., 8, 11, 68, 69, 75, 81n, 170n, 171, 175, 181, 177, 182, 207, 208

Freud S., 9
Frongia G., 263

Gabbay D., 91n
Galanter E., 281, 282, 287n
Galilei G., 18n
Gargani A.G., 127
Geach P.T.164, 169n
Gentzen G., 84, 86–90
Gödel K., 71
Goethe J.W., 256, 261
Goodman N., 231n
Graumann C. F., 281, 286n
Grillparzer F., 6
Grimm J., 1
Guenther F., 91n

Hacker P.M.S., 93–95, 105n
Hahn H., 3, 17n, 225
Haller R., xi, 17n, 223, 247n
Hanson N.R., 231n
Hargreaves R., 135n, 183n, 191
Hegel G.W.F., 1, 4
Heidegger M., 282
Hilmy S.S., 7, 8, 17n, 18n, 230, 231n, 235, 247n, 248n
Hintikka J., 90n, 228, 231n
Hintikka M.B. 231n
Hochberg H., 137, 168n, 169n
Hollis M., 231n
Holtzman S.H., 136n
Hume D., 44, 49, 142, 240
Husserl E., 77–79, 81n, 192n

James W., 80, 249–262
Janik A., 6–8, 17n

Kaila E., 2
Kant I., 12, 218, 225
Kelkel L., 81n
Keim R., 56n
Kenny A.J.P., 37, 47, 56n, 91n, 135n, 169n, 183n, 222n
Kierkegaard S., 4
King J., 81n, 135n, 183n
Kneale M. & W., 90n
Köhler W., 284
Kraus K., 9, 10
Krausz M., 231n
Kripke S., 50, 93, 94, 97, 99, 100, 113, 225, 226
Kuhn T.S., 231n

Lee D., 81n, 135n, 183n
Leibniz G.W., 16, 66
Leich C.M., 136n
Leinfellner E., 247n
Lewis H.D., 90n
Loos A., 9, 10
Luckhardt C.G., 81n, 135n, 182n, 205n, 247n, 262n, 277n, 280n
Lukes S., 231n
Lyell C., 1

Macdonald M., 81n, 136n, 183n, 231n
Mach E., 3
Machover M., 90n
Malcolm N., 81n, 94, 192n, 195, 207–210
Marconi D., 107, 117–122
Martinich A.P., 105n
Marty A., 184n
Mauthner F., 10
McAlister L.L., 136n, 183n, 231n
McGinn C., 228, 231n
McGinn M., 219–220, 222n
McGuinness B.F., xi, 35n, 81n, 91n,135n, 183n, 192n, 247n
Meiland J., 231n
Miller G.A., 281, 283, 287n
Monk R., 170n
Mondadori M., 88, 89, 90n, 91n
Moore G.E., xi, 18n, 26, 72, 137–147, 156, 168n, 169n, 174, 183n, 195–205, 211, 213, 215, 219
Mulligan K. 170n
Mundle C., 247n
Musil R., 9

Nagel T., 101, 213
Neisser U., 285, 287n
Neurath O., 3, 7, 225
Nietzsche F., 4, 6

Nyíri C. 5, 6, 17n–19n
Nyman H., 18n, 35n, 81n, 136n, 182n, 205n, 247n, 262n, 277n

Paul D., 183n, 222n, 231n, 277n
Pears D., 1, 23, 35n, 135n, 148, 155–164, 170n, 183n
Picardi E., 207
Pitcher G., 122n
Plato., 33, 34, 39,160
Prawitz D., 84, 85, 87, 88, 91n
Pribram K.H., 281, 282, 287n
Prior A., 55, 56n
Protagoras., 223
Putnam H., 117–119, 121, 122n, 123n, 231n

Quine W.O.V., 56n, 231n

Rask R.K., 1
Reichenbach H., 225
Rhees R., 35n, 36n, 46n, 56n, 81n, 105n, 114n, 122n, 123n, 136n, 182n, 183n, 191n, 211n, 222n, 223n, 231n, 247n, 248n, 262n, 268n, 277n, 279n
Russell B., 2, 3, 8–11, 17n, 83, 127–129, 135n, 137, 139–170, 173, 182, 184n, 191, 235, 236, 240, 241, 247n

Sambin G., 90n
Sauer W., 231n
Savage C. 169n
Schättle M. 136n, 183n, 231n
Schérer R., 81n
Schiffner E., 287n
Schilpp A.N., 205n
Schlick M., 3, 191, 225
Schönberg A., 9
Schubert F., 68
Schulte J., 249, 135n
Searle J., 283
Shanker S., 192n
Shoesmith D.J., 91n
Šikiš Z., 91n
Slater J.G., 169n
Smiley T.J., 91n
Smith B., 56n, 169n
Smullyan R.M., 91n
Smythies Y., 81n
Spencer H., 1
Spengler O., 4, 5
Spiegelberg H., 192n
Spinoza B., 16, 42
Sraffa P., 4
Stroll A., 93
Stromberg W., 383
Stroud B., 212, 216–218, 222n
Sundholm G., 91n
Szabo M.E., 90n

Tarski A., 167
Tolman E.C., 281, 286n
Tolstoy L., 8–10
Toulmin S., 231n
Travis C., 56n
Turing A., 286

Waismann F., 91n, 174
Wallner F., 279, 283
Wann T.W., 287n
Watson J.B., 280, 281
Werbik H., 287n
White R., 135n. 183n, 191n
Whitehead A.N., 169n
Williams B., 222n
Winch P., 35n, 136n, 247n
Wright C., 231n
Wright G.H. (von), 1, 16, 35, 46, 81n, 90n, 91n, 135n, 136, 168n, 182n, 205n, 222n, 231n, 247n, 262n, 277n

Zermelo E., 81n
Zitterbach W., 287

INDEX OF SUBJECTS

Abilities, 43, 44, 47
Absolute concepts, 230
Acquaintance, 162
Action/s, 54, 201, 202, 209
  and certainty, 215, 216
  and facts, 176–178
  individual, 179
  intentional/voluntary, 175, 176,178
  rule-following 22. *See also* Explanation, Intentional explanation
Action theory, 283–286, 288
Actualisations, 43
Actuality, 37, 43
  and potentiality, 47. *See also* Potentiality
Analytic truth, 214
Analyticity, 169
Artificial intelligence, 286
Aspects, theory of, xi
Assertion, 207, 208, 210
  and supposition, 197–199, 203
  indirect, 200
  tentative or hesitant, 197, 202, 204. *See also* Assumption; Force indicator
Associating
  and thinking, 29
Associationist principle
  Russell's, 130
Asimmetry, *See* First/third person
Assumption ('Annahme'), 208, 209
'Attente', 60, 61. *See also* Expectation
Attitude/s, 207
  non epistemic, 210, 222, *See* Dispositional states, Judgment/s
'Bedeutung', 182
'Bedeutungskörper', *See* 'Corps de signification'
Behavior, 179, 180, 279–287
  cognitivistic explanations of, 287
Behavioral data, 224
Behaviorism, 37, 44, 45, 239, 240, 241, 245, 279–287
  linguistic, 280
  non-physiological, 286
  subjective, 281, 282
Believe/belief, 137–144, 168, 176–178, 195–205, 207–210, 280
'Bewegungsgefühlraum', 190
Bodily states, 48. *See also* Mental states
Body, 48
Brain, 39
  'brain in the vat', 117–119, 212

Care ('Sorge'), 256–261
Cartesian ego, 9
Cartesianism, 37, 47
Causal explanation, *See* Explanation
Causal laws, 29
Causality 284. *See also* Intentionality
Causation
  and empirical concomitance, 228
  mnemic, *See* Associationist principle
Causes, *See* Motives
Certainty/ties, xi, 211–222
Chemistry, 282
'Classes', 'théorie de', 71
Cognitive psychology, 279, 282, 285, 286
  computational approach to, 286
Colours, 172, 185, 186
  phenomenological theory of, 174
Complexes, 156, 157
'Compositionalité, 'principe de', 75
Computer, 119, 212. *See also* Machine language
'Conceptual analysis', 173
Conditional/s, 208, 209
Consequentialist theory of meaning, *See* Meaning
Conservatism, 6
'Contradiction'

logical, 200
'loi de', 80
Conventionalism, 24, 224, 225
'Corps de signification', 74, 75,79, 80
'Couleur', 72, 73. *See also* Colours
Criteria, 177, 181
pragmatic, 176
semantic, 218
Custom, 227

Deduction
system of natural, 85–87, 90
Definition
ostensive, 38
'Démonstration', 65, 66, 68, 71, 76
'Dénotation', 75
Descriptions
definite, 139, 140, 162, 165
'Désir', 60, 61
Desires, 176
Dispositional states/dispositions, 43, 203
Doubt, xi, 177, 212–214
and certainty, 221
cartesian, 211, 212, 215
existential, 14
internal, 215–217
sceptical, 211, 213, 214
Dream language, 278–276
Dreaming, argument from, 212, 217, 218
Dreams, interpretation of, 270, 274–276

Emotions, xi, 249–262, 285
and bodily sensations, 249, 250, 252–256
and care ('Sorge'), 256, 261
and imagination, 257
and one's own mind, 250
causes of, 256–261
expressions of, 252–260, 285
symptons of, 257–259
'Energeia', 43
'Ensembles', 'théorie de', 69–71
'Entités'/'êtres'
géométriques, 68, 69
intermédiaires/fantômatiques, 59–81
Epistemic preference, 53–55
Epistemic states, 8
Error, possibility of, 216, 217
'Espace'
'physique', 73
'visuel', 73
Essentialism, 11, 55
Excluded middle, principle of, 80, 90
Expectation/expecting, expect, 129–133, 176, 178, 243–245
Explanation, 29, 30
Carnap's causal, 247
causal/intentional, 175, 178, 180
mentalistic, 329, 240
non–causal, 177
of actions, 175, 178
of behavior, 283. *See* Intentional explanation
Extensionality, 47–49, 54
External relations, *See* Relation/s

Fact/s
atomic, 170n
empirical, 214
existential, 153
existentially general, 153, 160, 161, 163, 164
grammatical, 208
intentional, 153
non–existent, 138, 139, 141, 145, 151
possible, 144, 145, 166
potential, 145
relational, 149. *See also* Possibility/ies
'Fait'
'negatif', 64
'possible', 61
Falsity/false, *See* Truth/true
First/third person, 49, 51, 98–1O1, 196, 202, 205, 209, 279, 280, 282–285
Force indicator, 207, 208
Forms of life, 228, 229
Freudian theory, *See* Psychoanalysis

'Gedankeninhalte'. *See* Sense/s
'Généralité', 71
Generality
empirical/inductive, 172, 181

high, 172, 180, 181
'Géométrie', 68
'Grammaire'
'règles de', 63, 72, 73, 75, 76
'philosophique', 62, 63, 72
'pure logique', 77, 79
Grammar
and psychology, 171–184
philosophical, 171–184, 225
surface/depth grammar, 208, 242, 243, 246. *See also* Rules

'Hexis', 43, 44
'Homunculus fallacy', 39
Hopes, 176, 178

I (pronoun), 48, 98–100, 104
Idealism
ontological, 221
Imagination, 257
Immediate experience, 173, 180
Implication, 88
'Impossibilité'/és', *See* 'Possibilité'/és'/'Impossibilité'/és'
Impossibility, degrees of, 59, 61
Incommensurability, thesis of, 224
Incomplete symbols, *See* Descriptions
Indeterminacy of translation, 224
Individuals, 188, 189, 225
Induction/inductive procedures, 178
'Infini'
'potentiel'/'actuel', 69–72
Infinity axiom, 70
Inner, *See* Mental
Inner eye, 47
Inner experience, *See* Internal experience
Inner life, xi, 117, 236
Inner/outer, 145, 246
Intensions, 75
Intention/s, 176–178, 182
sense and content of, 146
Intentionality, xi, 127–136, 153, 175
Bergmann's and Moore's analysis of, 144
and causality, 178. *See also* Relational analysis
Intentional
action, *See* Action/s
contexts/propositions, 158, 159, 163, 175, 177–179, 181
element of language, 182
acts, 128, 129, 131, 132
Intentional explanation, 178
Intentional reference, 137–170
Internal experience/s, 174, 175, 182
phenomenological analysis of, 171. *See also* Mental processes
Internal properties, 143–145, 152, 155
Internal relations, *See* Relation/s
Internal statements, 221
Introspection/introspectionism, 249, 254, 280, 282
Intuition
in Husserl, 78
Irrationalism, xi

Judgment/s
agreement in, 227, 228
concept of, 219
relational analysis of, 149
synthetic a prior, 225
value, 178

'Kinesis', 43
Knowledge
and action, 215, 216
and certainty, 221
empirical, 214
factual, 176
foundations of, 220
holistic, 218
objective, 213
theory of, 171
trascendental conditions of, 218

'Langage des valences', 67
Language, 24, 25, 185
and mind, 235–248
and world, 26
as a calculus, 129
as a system of internal relations, 128–129
critique of, 10, 12

ideal, 10
intentional use of, 181
mental, 31
mythology of, *See also* Mythology
natural state of , 11, 12
'never–never', 11
non–public, 114
of data, 173
of thought, 107–114, 117, 120, 121
philosophy of, xi, 10, 28, 181
pragmatic nature of, 181
primary, *See* Physical language
public, 122
secondary, *See* Phenomenological language
social dimension of, 117, 123
thing, 221. *See also* Dream language, Machine language, Private language
Language–game/s, 191, 216, 218, 219, 228
verbal fallacy of, 228, 229
Law of mnemic causation, *See* Associationist principle
Laws, 174
'Lebensformen', *See* Forms of life
'Lingua mentalis', *See* Private language
Linguistics, 181
Logic, 171–173
modal, 55
philosophy of, 10
truth–functional/extensional, 172, 177, 181
Logical
constants, 128
form/s, 127
product, 127
relation, *See* Intentional reference
Logical atomism, 24
Logical positivism, 2, 15, 191
Logical symbols, meaning of, 83–91
Logical relation, *See* Intentional reference

'Mathématiques', 65, 66, 68–71
Machine-language, 112, 113, 133
computer's and programmer's, 112, 113. *See also* Computer
Materialism, 44, 45, 240, 241, 245
Materialist theory of mind, 205
Mathematical
and physical language, 225
concepts, xi, 177, 179
formulas, *See* Rules
Mathematics, 180, 224
foundations of, 171
Meaning
as use, 128
causal theory of, 118
concept of, 235
consequentialist theory of, 84, 89, 90
criteria of, 176
non-Fregean theory of, 171
of the word, 214
theories of, 171, 175
'Means that', 167
Medicine, 282
Memory, xi, 263–277
and knowledge, 263, 270
and language, 263, 270
as 'criterion for the past', 271, 272
concept of, 29
justification of, 264–268
two notions of, 263, 264
Mental
and bodily states, 48
and physical distinction, 211
causes, 30
concepts, 179
images, 244
involvement, 53–55
objects, 181, 282
phenomena, 30, 175, 176,
process/es, 39, 181, 243
states, 177, 199–201, 203, 207, 209, 244. *See also* Psychological/mental
Mental life, grammar of, 127
Mentalese/mentalesians, 117–123
Mentalism, 119, 121, 122, 237, 238, 240, 245
Metaphysical claims
three types of, 47–53
Metaphysics, 14–16, 37–38
dynamic, 43, 46
foundationalist, 42, 45

of mind, 47–56
speculative, 54
spiritualistic, 38, 41, 42, 44–46
trascendental, 52, 53
Mind, 44, 45
and brain, 28, 29, 44,
metaphysical concept of, 41
metaphysical theory of, 37
metaphysics of, 37, 47–56
philosophy of, 15, 43, 44, 171, 175
Russell's theory of, 235
Modal involvement, 55
Modernity, 1–7, 16
Moore's paradox, *See* Paradox
Motives
and causes, 178, 278. *See also* Reasons
'Myth'/s', 238, 246, 276
Mythology, 235–248
of mental processes, 237
of symbolism, 237

Natural history, knowledge of, 213
Necessary propositions
and non–necessary, 225
and probabilistic, 173
Necessity, 55
nomological, 110, 115
'Négation', 64
Neural explanation, 31
Neurology, 28
Neurophysiology, 29
'Nombres'
'entiers', 70
'transfinis', 69

'Ombre', *See* 'Possibilité'/és'/'Impossibilité'/és'
Objectivity, 213, 221–224
Objects
material, 186
visual, 187
Optimism
rationalist, 7
Other minds/people/persons, 44–49, 54, 55, 201, 209, 221, 259, 260

Pain, 49, 52, 255, 280
Paradox
Kripke's, 226
Moore's, xi, 195–205, 207–210
Russell's, 169n
Parts
visual, 187. *See also* Whole/parts
Perception, problem of, 192
Person/s, 48, 54, 55
"personal experiences", 175. *See also* First/third person, Other minds
Persuasion, 229
Phenomenalism, 240, 241, 245
Phenomenalistic sensualism, 3
Phenomenological-not-phenomenological, 172, 178
Phenomenological/phenomenal
analysis, 173–175
language, 191
propositions/sentences, 174
problems/"problems", 173, 192
Phenomenology, 171, 172–178, 183n, 185–192
Philosophical psychology, xi, 174, 175, 181, 235–288
Philosophy
'first', 42
of psychology, *See* Philosophical psychology
Physical
language, 280
objects, 174
phenomena, 270
sentences/propositions, 174. *See also* Phenomenological
Physics, 173, 174, 224, 235, 282
Physiologism, 281, 282
Platonism, 31–33
'Possibilité'/és'/'Impossibilité'/és', 61, 64, 66, 67, 70, 72, 73, 75–79
'chimique', 63, 66, 67
'et existence', 66
'et réalité'/'réel', 59–81
'et sens', 59–81
'géométrique', 67
'logique'/'grammaticale', 62

Possibility/ies
  and facts, 25–28
  in Russell, 143, 145
  logical, 195, 197, 198, 202
Possible/'Possible'
  See Fact/s, 'Possibilité'/és'/'Impossibilité'/és'/Possibility/ties
Possible worlds, 56n, 175
Potentiality, 37, 43, 44
  and actuality, 47
Powers
  active and passive, 43, 44
Prediction, 177, 178
Present/past tense, 196, 197, 199, 205, 207, 209
Primary phenomena, *See* Phenomenological language
Private language, xi, 51, 52, 53, 93, 263
  argument, 107–115, 117–123
  as 'lingua mentalis', 230
  as non public language, 109, 110
Proof, 225
  method of, 180
Propositional
  attitudes, 173, 175, 177
  attributes, 144–147
  entities, 137, 140, 149, 150–152, 158
Propositions/sentences/statements
  external existential, 221
  'grammaticales'/'mathématiques', 67, 68, 70
  normative, 239
  not–extensional, 182
  systematic/empirical, 178, 179, 225
Propositionalism
  Tolman's, 281
Psychoanalysis, 269, 273, 275
Psychological/mental
  concepts/propositions, xi, 171, 172, 175, 180, 181, 202, 236, 238–243, 245, 246, 279, 280, 282, 286
  events, 178
  expressions, 245
  phenomena/processes, 177
  taxonomical/causal view of, 236, 240
  verbs, 173
Psychologism, xi, 174–177, 182
Psychology, 137, 171–184, 237, 279–287
  and grammar, 171–184
  and philosophy, 29
  and physics, 279
  cognitive, *See* Cognitive psychology
  contemporary, 269
  depth, 270
  empirical, 236, 238, 246
  history of, 285
  humanistic, 283
  introspective, 285
  of mental states, 185
  phenomenological, 281
  philosophical, *See* Philosophical psychology

Quantification/quantifier
  existential, 86
  universal, 83, 84, 88
Quantities and measures, problem of, 185, 186

Realism, 11, 213,
'Realité', *See* 'Possibilité'/és'/'Impossibilité'/és'
Reasons, 216, 219
Reductionism, 44
  in the theory of memory, 267
  physiological, 281
  vehicle, 44
  Watson's, 281
'Réel'
  'et son ombre', 59–81. *See also* 'Possibilité'/és'/'Impossibilité'/és'
Reference, 119, 120. *See also* Intentional reference
Relation/s
  external, 128, 182
  internal, 128, 184n
Relational analysis
  Russell's, 139, 141, 142, 145, 147, 148, 151, 156,164–166
Relational contexts
  non–symmetrical, 158, 159, 163
Relativism, xi, 180, 223–231

epistemological, 223–231
ontological, 224
Protagoras thesis, 223
Relativity, theory of, 230
Representation/s, 201, 212
internal, 107, 110, 111
mental, 117, 119, 121
physical, 119
'Représentation', 'systeme de', 62, 63
Representational relation
Bergmann's, 145
Representational system, 122
Rule following, 28, 31–35, 93–105, 226
Kripke's argument of, 50, 51
Rule/s, 178, 179, 181, 182, 218
and agreement, 227, 228
and mathematical formulas, 95,104
and postulates, 225, 226
and trespass, 103, 104
as edicts, decrees, etc., 94–97
constative, 96, 97, 99–103
disjunction, 86
elimination, 84–87, 89, 90
inference, 83, 85
introduction, 84, 88, 90
of a game, 162
of grammar/grammatical, 162, 225
of induction, 87
system of, 176, 178–180
universal, 83

Scepticism, xi, 26, 47, 49, 211–222
Science/s, 173
and philosophy, 23–36
empirical, 172
mental, 179
'naturelle', 70
of science, 23
Scientific theories, 235, 236
Seeing
concept of, 237
Self, concept of, 279
Self-contradictory sentence, 196, 202, 203
Self-interpretation/self-understanding, 283
Self-observation, 202, 282-284
Semantics of possible worlds, 56n
Sensation/s, xi
terms, 117
expression of, 280
'Sens', 66, 73–75, 80
'détermination de' ('Sinnbestimmung'), 68
et 'contresens', 77
et 'non-sens', 59, 61–66, 70, 73, 74, 79, 80
'remplissant', 76
'signification', 76–78
'utilisation de' ('Sinnverwendung'), 68
Sense/s, 174–176
as 'Gedankeninhalte', 181
as 'Sinne', 181, 182
Sense datum, language of, 132
Sense–impressions, 174, 209
Sense–perception, 217
Set theory 180. *See also* 'Ensembles', 'théorie des'
Shadow
of the expected fact, 131
Shadowy beings, *See* 'Entités'/'êtres'
'Signification', 76–80
Similarity, notion of, 130–132
'Sinn'/'Sinne', *See* Sense/s, 'Sens'
Solipsism, 25–28
Soul–body, 39, 46. *See also* Body
Space, 189
absolute, 186–188, 191
'Gesichtsraum', 192
orientation ('Gefühlsraum'), 189–191, 192n
visual 185–192. *See also* 'Espace'
Speech act theory, 283
Subjectivism, xi, 180
Supposition, 197–199

T-sentences, 167
Tautologies, 225
Thinking
without looking, 203
Thought, 208
language of, *See* Assumption, Language of thought
Thought-experiment

James', 249, 259, 251
True
'for me', 224
'per se', 223, 224
Truth-conditions, 166
Truth/true, 176, 223, 228–230
analytica/conceptual, 253
subjective, 272
Truth-functions, 11
Truth-makers, 151, 170n
Truth values, 174

Understanding, 39, 40, 41, 127, 182, 214, 218, 228
'Übersichtliche Darstellung' ('perspicuous presentation'), 12

Validity, 85, 87, 88
Value judgments, *See* Judgment/s
Verification, principle of, 280, 282
Verificationism, 107, 108
'View from nowhere', 101
Vienna Circle/'Wiener Kreis', 3, 8
Visual coordinates
system of 188–190. *See also* Space

Whole/parts, 186, 187, 190
Will, xi
'Wissenschaftliche Weltauffassung', 3
World
and representation, 212
external, 211, 217
World picture/'Weltbild', 214, 218–221

SYNTHESE LIBRARY

Studies in Epistemology, Logic, Methodology, and Philosophy of Science

1. J. M. Bochénski, *A Precis of Mathematical Logic.* Translated from French and German by O. Bird. 1959 ISBN 90-277-0073-7
2. P. Guiraud, *Problèmes et méthodes de la statistique linguistique.* 1959 ISBN 90-277-0025-7
3. H. Freudenthal (ed.), *The Concept and the Role of the Model in Mathematics and Natural and Social Sciences.* 1961 ISBN 90-277-0017-6
4. E. W. Beth, *Formal Methods.* An Introduction to Symbolic Logic and to the Study of Effective Operations in Arithmetic and Logic. 1962 ISBN 90-277-0069-9
5. B. H. Kazemier and D. Vuysje (eds.), *Logic and Language.* Studies dedicated to Professor Rudolf Carnap on the Occasion of His 70th Birthday. 1962 ISBN 90-277-0019-2
6. M. W. Wartofsky (ed.), *Proceedings of the Boston Colloquium for the Philosophy of Science, 1961–1962.* [Boston Studies in the Philosophy of Science, Vol. I] 1963 ISBN 90-277-0021-4
7. A. A. Zinov'ev, *Philosophical Problems of Many-valued Logic.* A revised edition, edited and translated (from Russian) by G. Küng and D.D. Comey. 1963 ISBN 90-277-0091-5
8. G. Gurvitch, *The Spectrum of Social Time.* Translated from French and edited by M. Korenbaum and P. Bosserman. 1964 ISBN 90-277-0006-0
9. P. Lorenzen, *Formal Logic.* Translated from German by F.J. Crosson. 1965 ISBN 90-277-0080-X
10. R. S. Cohen and M. W. Wartofsky (eds.), *Proceedings of the Boston Colloquium for the Philosophy of Science, 1962–1964.* In Honor of Philipp Frank. [Boston Studies in the Philosophy of Science, Vol. II] 1965 ISBN 90-277-9004-0
11. E. W. Beth, *Mathematical Thought.* An Introduction to the Philosophy of Mathematics. 1965 ISBN 90-277-0070-2
12. E. W. Beth and J. Piaget, *Mathematical Epistemology and Psychology.* Translated from French by W. Mays. 1966 ISBN 90-277-0071-0
13. G. Küng, *Ontology and the Logistic Analysis of Language.* An Enquiry into the Contemporary Views on Universals. Revised ed., translated from German. 1967 ISBN 90-277-0028-1
14. R. S. Cohen and M. W. Wartofsky (eds.), *Proceedings of the Boston Colloquium for the Philosophy of Sciences, 1964–1966.* In Memory of Norwood Russell Hanson. [Boston Studies in the Philosophy of Science, Vol. III] 1967 ISBN 90-277-0013-3
15. C. D. Broad, *Induction, Probability, and Causation.* Selected Papers. 1968 ISBN 90-277-0012-5
16. G. Patzig, *Aristotle's Theory of the Syllogism.* A Logical-philosophical Study of *Book A* of the *Prior Analytics.* Translated from German by J. Barnes. 1968 ISBN 90-277-0030-3
17. N. Rescher, *Topics in Philosophical Logic.* 1968 ISBN 90-277-0084-2
18. R. S. Cohen and M. W. Wartofsky (eds.), *Proceedings of the Boston Colloquium for the Philosophy of Science, 1966–1968, Part I.* [Boston Studies in the Philosophy of Science, Vol. IV] 1969 ISBN 90-277-0014-1
19. R. S. Cohen and M. W. Wartofsky (eds.), *Proceedings of the Boston Colloquium for the Philosophy of Science, 1966–1968, Part II.* [Boston Studies in the Philosophy of Science, Vol. V] 1969 ISBN 90-277-0015-X

20. J. W. Davis, D. J. Hockney and W. K. Wilson (eds.), *Philosophical Logic.* 1969 ISBN 90-277-0075-3
21. D. Davidson and J. Hintikka (eds.), *Words and Objections.* Essays on the Work of W. V. Quine. 1969, rev. ed. 1975 ISBN 90-277-0074-5; Pb 90-277-0602-6
22. P. Suppes, *Studies in the Methodology and Foundations of Science. Selected Papers from 1951 to 1969.* 1969 ISBN 90-277-0020-6
23. J. Hintikka, *Models for Modalities.* Selected Essays. 1969 ISBN 90-277-0078-8; Pb 90-277-0598-4
24. N. Rescher *et al.* (eds.), *Essays in Honor of Carl G. Hempel.* A Tribute on the Occasion of His 65th Birthday. 1969 ISBN 90-277-0085-0
25. P. V. Tavanec (ed.), *Problems of the Logic of Scientific Knowledge.* Translated from Russian. 1970 ISBN 90-277-0087-7
26. M. Swain (ed.), *Induction, Acceptance, and Rational Belief.* 1970 ISBN 90-277-0086-9
27. R. S. Cohen and R. J. Seeger (eds.), *Ernst Mach: Physicist and Philosopher.* [Boston Studies in the Philosophy of Science, Vol. VI]. 1970 ISBN 90-277-0016-8
28. J. Hintikka and P. Suppes, *Information and Inference.* 1970 ISBN 90-277-0155-5
29. K. Lambert, *Philosophical Problems in Logic.* Some Recent Developments. 1970 ISBN 90-277-0079-6
30. R. A. Eberle, *Nominalistic Systems.* 1970 ISBN 90-277-0161-X
31. P. Weingartner and G. Zecha (eds.), *Induction, Physics, and Ethics.* 1970 ISBN 90-277-0158-X
32. E. W. Beth, *Aspects of Modern Logic.* Translated from Dutch. 1970 ISBN 90-277-0173-3
33. R. Hilpinen (ed.), *Deontic Logic.* Introductory and Systematic Readings. 1971 *See also* No. 152. ISBN Pb (1981 rev.) 90-277-1302-2
34. J.-L. Krivine, *Introduction to Axiomatic Set Theory.* Translated from French. 1971 ISBN 90-277-0169-5; Pb 90-277-0411-2
35. J. D. Sneed, *The Logical Structure of Mathematical Physics.* 2nd rev. ed., 1979 ISBN 90-277-1056-2; Pb 90-277-1059-7
36. C. R. Kordig, *The Justification of Scientific Change.* 1971 ISBN 90-277-0181-4; Pb 90-277-0475-9
37. M. Čapek, *Bergson and Modern Physics.* A Reinterpretation and Re-evaluation. [Boston Studies in the Philosophy of Science, Vol. VII] 1971 ISBN 90-277-0186-5
38. N. R. Hanson, *What I Do Not Believe, and Other Essays.* Ed. by S. Toulmin and H. Woolf. 1971 ISBN 90-277-0191-1
39. R. C. Buck and R. S. Cohen (eds.), *PSA 1970.* Proceedings of the Second Biennial Meeting of the Philosophy of Science Association, Boston, Fall 1970. In Memory of Rudolf Carnap. [Boston Studies in the Philosophy of Science, Vol. VIII] 1971 ISBN 90-277-0187-3; Pb 90-277-0309-4
40. D. Davidson and G. Harman (eds.), *Semantics of Natural Language.* 1972 ISBN 90-277-0304-3; Pb 90-277-0310-8
41. Y. Bar-Hillel (ed.), *Pragmatics of Natural Languages.* 1971 ISBN 90-277-0194-6; Pb 90-277-0599-2
42. S. Stenlund, *Combinators, λ-Terms and Proof Theory.* 1972 ISBN 90-277-0305-1
43. M. Strauss, *Modern Physics and Its Philosophy.* Selected Paper in the Logic, History, and Philosophy of Science. 1972 ISBN 90-277-0230-6
44. M. Bunge, *Method, Model and Matter.* 1973 ISBN 90-277-0252-7
45. M. Bunge, *Philosophy of Physics.* 1973 ISBN 90-277-0253-5

46. A. A. Zinov'ev, *Foundations of the Logical Theory of Scientific Knowledge (Complex Logic).* Revised and enlarged English edition with an appendix by G. A. Smirnov, E. A. Sidorenka, A. M. Fedina and L. A. Bobrova. [Boston Studies in the Philosophy of Science, Vol. IX] 1973 ISBN 90-277-0193-8; Pb 90-277-0324-8
47. L. Tondl, *Scientific Procedures.* A Contribution concerning the Methodological Problems of Scientific Concepts and Scientific Explanation. Translated from Czech by D. Short. Edited by R.S. Cohen and M.W. Wartofsky. [Boston Studies in the Philosophy of Science, Vol. X] 1973 ISBN 90-277-0147-4; Pb 90-277-0323-X
48. N. R. Hanson, *Constellations and Conjectures.* 1973 ISBN 90-277-0192-X
49. K. J. J. Hintikka, J. M. E. Moravcsik and P. Suppes (eds.), *Approaches to Natural Language.* 1973 ISBN 90-277-0220-9; Pb 90-277-0233-0
50. M. Bunge (ed.), *Exact Philosophy.* Problems, Tools and Goals. 1973 ISBN 90-277-0251-9
51. R. J. Bogdan and I. Niiniluoto (eds.), *Logic, Language and Probability.* 1973 ISBN 90-277-0312-4
52. G. Pearce and P. Maynard (eds.), *Conceptual Change.* 1973 ISBN 90-277-0287-X; Pb 90-277-0339-6
53. I. Niiniluoto and R. Tuomela, *Theoretical Concepts and Hypothetico-inductive Inference.* 1973 ISBN 90-277-0343-4
54. R. Fraissé, *Course of Mathematical Logic* – Volume 1: *Relation and Logical Formula.* Translated from French. 1973 ISBN 90-277-0268-3; Pb 90-277-0403-1 (For *Volume 2* see under No. 69).
55. A. Grünbaum, *Philosophical Problems of Space and Time.* Edited by R.S. Cohen and M.W. Wartofsky. 2nd enlarged ed. [Boston Studies in the Philosophy of Science, Vol. XII] 1973 ISBN 90-277-0357-4; Pb 90-277-0358-2
56. P. Suppes (ed.), *Space, Time and Geometry.* 1973 ISBN 90-277-0386-8; Pb 90-277-0442-2
57. H. Kelsen, *Essays in Legal and Moral Philosophy.* Selected and introduced by O. Weinberger. Translated from German by P. Heath. 1973 ISBN 90-277-0388-4
58. R. J. Seeger and R. S. Cohen (eds.), *Philosophical Foundations of Science.* [Boston Studies in the Philosophy of Science, Vol. XI] 1974 ISBN 90-277-0390-6; Pb 90-277-0376-0
59. R. S. Cohen and M. W. Wartofsky (eds.), *Logical and Epistemological Studies in Contemporary Physics.* [Boston Studies in the Philosophy of Science, Vol. XIII] 1973 ISBN 90-277-0391-4; Pb 90-277-0377-9
60. R. S. Cohen and M. W. Wartofsky (eds.), *Methodological and Historical Essays in the Natural and Social Sciences. Proceedings of the Boston Colloquium for the Philosophy of Science, 1969–1972.* [Boston Studies in the Philosophy of Science, Vol. XIV] 1974 ISBN 90-277-0392-2; Pb 90-277-0378-7
61. R. S. Cohen, J. J. Stachel and M. W. Wartofsky (eds.), *For Dirk Struik. Scientific, Historical and Political Essays.* [Boston Studies in the Philosophy of Science, Vol. XV] 1974 ISBN 90-277-0393-0; Pb 90-277-0379-5
62. K. Ajdukiewicz, *Pragmatic Logic.* Translated from Polish by O. Wojtasiewicz. 1974 ISBN 90-277-0326-4
63. S. Stenlund (ed.), *Logical Theory and Semantic Analysis.* Essays dedicated to Stig Kanger on His 50th Birthday. 1974 ISBN 90-277-0438-4
64. K. F. Schaffner and R. S. Cohen (eds.), *PSA 1972. Proceedings of the Third Biennial Meeting of the Philosophy of Science Association.* [Boston Studies in the Philosophy of Science, Vol. XX] 1974 ISBN 90-277-0408-2; Pb 90-277-0409-0

65. H. E. Kyburg, Jr., *The Logical Foundations of Statistical Inference*. 1974
ISBN 90-277-0330-2; Pb 90-277-0430-9
66. M. Grene, *The Understanding of Nature*. Essays in the Philosophy of Biology. [Boston Studies in the Philosophy of Science, Vol. XXIII] 1974
ISBN 90-277-0462-7; Pb 90-277-0463-5
67. J. M. Broekman, *Structuralism: Moscow, Prague, Paris*. Translated from German. 1974 ISBN 90-277-0478-3
68. N. Geschwind, *Selected Papers on Language and the Brain*. [Boston Studies in the Philosophy of Science, Vol. XVI] 1974 ISBN 90-277-0262-4; Pb 90-277-0263-2
69. R. Fraissé, *Course of Mathematical Logic* – Volume 2: *Model Theory*. Translated from French. 1974 ISBN 90-277-0269-1; Pb 90-277-0510-0
(For *Volume 1* see under No. 54)
70. A. Grzegorczyk, *An Outline of Mathematical Logic*. Fundamental Results and Notions explained with all Details. Translated from Polish. 1974
ISBN 90-277-0359-0; Pb 90-277-0447-3
71. F. von Kutschera, *Philosophy of Language*. 1975 ISBN 90-277-0591-7
72. J. Manninen and R. Tuomela (eds.), *Essays on Explanation and Understanding*. Studies in the Foundations of Humanities and Social Sciences. 1976
ISBN 90-277-0592-5
73. J. Hintikka (ed.), *Rudolf Carnap, Logical Empiricist*. Materials and Perspectives. 1975 ISBN 90-277-0583-6
74. M. Čapek (ed.), *The Concepts of Space and Time*. Their Structure and Their Development. [Boston Studies in the Philosophy of Science, Vol. XXII] 1976
ISBN 90-277-0355-8; Pb 90-277-0375-2
75. J. Hintikka and U. Remes, *The Method of Analysis*. Its Geometrical Origin and Its General Significance. [Boston Studies in the Philosophy of Science, Vol. XXV] 1974 ISBN 90-277-0532-1; Pb 90-277-0543-7
76. J. E. Murdoch and E. D. Sylla (eds.), *The Cultural Context of Medieval Learning*. [Boston Studies in the Philosophy of Science, Vol. XXVI] 1975
ISBN 90-277-0560-7; Pb 90-277-0587-9
77. S. Amsterdamski, *Between Experience and Metaphysics*. Philosophical Problems of the Evolution of Science. [Boston Studies in the Philosophy of Science, Vol. XXXV] 1975 ISBN 90-277-0568-2; Pb 90-277-0580-1
78. P. Suppes (ed.), *Logic and Probability in Quantum Mechanics*. 1976
ISBN 90-277-0570-4; Pb 90-277-1200-X
79. H. von Helmholtz: *Epistemological Writings. The Paul Hertz / Moritz Schlick Centenary Edition of 1921 with Notes and Commentary by the Editors*. Newly translated from German by M. F. Lowe. Edited, with an Introduction and Bibliography, by R. S. Cohen and Y. Elkana. [Boston Studies in the Philosophy of Science, Vol. XXXVII] 1975 ISBN 90-277-0290-X; Pb 90-277-0582-8
80. J. Agassi, *Science in Flux*. [Boston Studies in the Philosophy of Science, Vol. XXVIII] 1975 ISBN 90-277-0584-4; Pb 90-277-0612-2
81. S. G. Harding (ed.), *Can Theories Be Refuted?* Essays on the Duhem-Quine Thesis. 1976 ISBN 90-277-0629-8; Pb 90-277-0630-1
82. S. Nowak, *Methodology of Sociological Research*. General Problems. 1977
ISBN 90-277-0486-4
83. J. Piaget, J.-B. Grize, A. Szemińska and V. Bang, *Epistemology and Psychology of Functions*. Translated from French. 1977 ISBN 90-277-0804-5

84. M. Grene and E. Mendelsohn (eds.), *Topics in the Philosophy of Biology.* [Boston Studies in the Philosophy of Science, Vol. XXVII] 1976 ISBN 90-277-0595-X; Pb 90-277-0596-8
85. E. Fischbein, *The Intuitive Sources of Probabilistic Thinking in Children.* 1975 ISBN 90-277-0626-3; Pb 90-277-1190-9
86. E. W. Adams, *The Logic of Conditionals.* An Application of Probability to Deductive Logic. 1975 ISBN 90-277-0631-X
87. M. Przełęcki and R. Wójcicki (eds.), *Twenty-Five Years of Logical Methodology in Poland.* Translated from Polish. 1976 ISBN 90-277-0601-8
88. J. Topolski, *The Methodology of History.* Translated from Polish by O. Wojtasiewicz. 1976 ISBN 90-277-0550-X
89. A. Kasher (ed.), *Language in Focus: Foundations, Methods and Systems.* Essays dedicated to Yehoshua Bar-Hillel. [Boston Studies in the Philosophy of Science, Vol. XLIII] 1976 ISBN 90-277-0644-1; Pb 90-277-0645-X
90. J. Hintikka, *The Intentions of Intentionality and Other New Models for Modalities.* 1975 ISBN 90-277-0633-6; Pb 90-277-0634-4
91. W. Stegmüller, *Collected Papers on Epistemology, Philosophy of Science and History of Philosophy.* 2 Volumes. 1977 Set ISBN 90-277-0767-7
92. D. M. Gabbay, *Investigations in Modal and Tense Logics with Applications to Problems in Philosophy and Linguistics.* 1976 ISBN 90-277-0656-5
93. R. J. Bogdan, *Local Induction.* 1976 ISBN 90-277-0649-2
94. S. Nowak, *Understanding and Prediction.* Essays in the Methodology of Social and Behavioral Theories. 1976 ISBN 90-277-0558-5; Pb 90-277-1199-2
95. P. Mittelstaedt, *Philosophical Problems of Modern Physics.* [Boston Studies in the Philosophy of Science, Vol. XVIII] 1976 ISBN 90-277-0285-3; Pb 90-277-0506-2
96. G. Holton and W. A. Blanpied (eds.), *Science and Its Public: The Changing Relationship.* [Boston Studies in the Philosophy of Science, Vol. XXXIII] 1976 ISBN 90-277-0657-3; Pb 90-277-0658-1
97. M. Brand and D. Walton (eds.), *Action Theory.* 1976 ISBN 90-277-0671-9
98. P. Gochet, *Outline of a Nominalist Theory of Propositions.* An Essay in the Theory of Meaning and in the Philosophy of Logic. 1980 ISBN 90-277-1031-7
99. R. S. Cohen, P. K. Feyerabend, and M. W. Wartofsky (eds.), *Essays in Memory of Imre Lakatos.* [Boston Studies in the Philosophy of Science, Vol. XXXIX] 1976 ISBN 90-277-0654-9; Pb 90-277-0655-7
100. R. S. Cohen and J. J. Stachel (eds.), *Selected Papers of Léon Rosenfield.* [Boston Studies in the Philosophy of Science, Vol. XXI] 1979 ISBN 90-277-0651-4; Pb 90-277-0652-2
101. R. S. Cohen, C. A. Hooker, A. C. Michalos and J. W. van Evra (eds.), *PSA 1974. Proceedings of the 1974 Biennial Meeting of the Philosophy of Science Association.* [Boston Studies in the Philosophy of Science, Vol. XXXII] 1976 ISBN 90-277-0647-6; Pb 90-277-0648-4
102. Y. Fried and J. Agassi, *Paranoia.* A Study in Diagnosis. [Boston Studies in the Philosophy of Science, Vol. L] 1976 ISBN 90-277-0704-9; Pb 90-277-0705-7
103. M. Przełęcki, K. Szaniawski and R. Wójcicki (eds.), *Formal Methods in the Methodology of Empirical Sciences.* 1976 ISBN 90-277-0698-0
104. J. M. Vickers, *Belief and Probability.* 1976 ISBN 90-277-0744-8
105. K. H. Wolff, *Surrender and Catch.* Experience and Inquiry Today. [Boston Studies in the Philosophy of Science, Vol. LI] 1976 ISBN 90-277-0758-8; Pb 90-277-0765-0

106. K. Kosík, *Dialectics of the Concrete.* A Study on Problems of Man and World. [Boston Studies in the Philosophy of Science, Vol. LII] 1976
ISBN 90-277-0761-8; Pb 90-277-0764-2
107. N. Goodman, *The Structure of Appearance.* 3rd ed. with an Introduction by G. Hellman. [Boston Studies in the Philosophy of Science, Vol. LIII] 1977
ISBN 90-277-0773-1; Pb 90-277-0774-X
108. K. Ajdukiewicz, *The Scientific World-Perspective and Other Essays, 1931-1963.* Translated from Polish. Edited and with an Introduction by J. Giedymin. 1978
ISBN 90-277-0527-5
109. R. L. Causey, *Unity of Science.* 1977 ISBN 90-277-0779-0
110. R. E. Grandy, *Advanced Logic for Applications.* 1977 ISBN 90-277-0781-2
111. R. P. McArthur, *Tense Logic.* 1976 ISBN 90-277-0697-2
112. L. Lindahl, *Position and Change.* A Study in Law and Logic. Translated from Swedish by P. Needham. 1977 ISBN 90-277-0787-1
113. R. Tuomela, *Dispositions.* 1978 ISBN 90-277-0810-X
114. H. A. Simon, *Models of Discovery and Other Topics in the Methods of Science.* [Boston Studies in the Philosophy of Science, Vol. LIV] 1977
ISBN 90-277-0812-6; Pb 90-277-0858-4
115. R. D. Rosenkrantz, *Inference, Method and Decision.* Towards a Bayesian Philosophy of Science. 1977 ISBN 90-277-0817-7; Pb 90-277-0818-5
116. R. Tuomela, *Human Action and Its Explanation.* A Study on the Philosophical Foundations of Psychology. 1977 ISBN 90-277-0824-X
117. M. Lazerowitz, *The Language of Philosophy.* Freud and Wittgenstein. [Boston Studies in the Philosophy of Science, Vol. LV] 1977
ISBN 90-277-0826-6; Pb 90-277-0862-2
118. Not published
119. J. Pelc (ed.), *Semiotics in Poland, 1894–1969.* Translated from Polish. 1979
ISBN 90-277-0811-8
120. I. Pörn, *Action Theory and Social Science.* Some Formal Models. 1977
ISBN 90-277-0846-0
121. J. Margolis, *Persons and Mind.* The Prospects of Nonreductive Materialism. [Boston Studies in the Philosophy of Science, Vol. LVII] 1977
ISBN 90-277-0854-1; Pb 90-277-0863-0
122. J. Hintikka, I. Niiniluoto, and E. Saarinen (eds.), *Essays on Mathematical and Philosophical Logic.* 1979 ISBN 90-277-0879-7
123. T. A. F. Kuipers, *Studies in Inductive Probability and Rational Expectation.* 1978
ISBN 90-277-0882-7
124. E. Saarinen, R. Hilpinen, I. Niiniluoto and M. P. Hintikka (eds.), *Essays in Honour of Jaakko Hintikka on the Occasion of His 50th Birthday.* 1979
ISBN 90-277-0916-5
125. G. Radnitzky and G. Andersson (eds.), *Progress and Rationality in Science.* [Boston Studies in the Philosophy of Science, Vol. LVIII] 1978
ISBN 90-277-0921-1; Pb 90-277-0922-X
126. P. Mittelstaedt, *Quantum Logic.* 1978 ISBN 90-277-0925-4
127. K. A. Bowen, *Model Theory for Modal Logic.* Kripke Models for Modal Predicate Calculi. 1979 ISBN 90-277-0929-7
128. H. A. Bursen, *Dismantling the Memory Machine.* A Philosophical Investigation of Machine Theories of Memory. 1978 ISBN 90-277-0933-5

129. M. W. Wartofsky, *Models.* Representation and the Scientific Understanding. [Boston Studies in the Philosophy of Science, Vol. XLVIII] 1979
ISBN 90-277-0736-7; Pb 90-277-0947-5
130. D. Ihde, *Technics and Praxis.* A Philosophy of Technology. [Boston Studies in the Philosophy of Science, Vol. XXIV] 1979 ISBN 90-277-0953-X; Pb 90-277-0954-8
131. J. J. Wiatr (ed.), *Polish Essays in the Methodology of the Social Sciences.* [Boston Studies in the Philosophy of Science, Vol. XXIX] 1979
ISBN 90-277-0723-5; Pb 90-277-0956-4
132. W. C. Salmon (ed.), *Hans Reichenbach: Logical Empiricist.* 1979
ISBN 90-277-0958-0
133. P. Bieri, R.-P. Horstmann and L. Krüger (eds.), *Transcendental Arguments in Science.* Essays in Epistemology. 1979 ISBN 90-277-0963-7; Pb 90-277-0964-5
134. M. Marković and G. Petrović (eds.), *Praxis.* Yugoslav Essays in the Philosophy and Methodology of the Social Sciences. [Boston Studies in the Philosophy of Science, Vol. XXXVI] 1979 ISBN 90-277-0727-8; Pb 90-277-0968-8
135. R. Wójcicki, *Topics in the Formal Methodology of Empirical Sciences.* Translated from Polish. 1979 ISBN 90-277-1004-X
136. G. Radnitzky and G. Andersson (eds.), *The Structure and Development of Science.* [Boston Studies in the Philosophy of Science, Vol. LIX] 1979
ISBN 90-277-0994-7; Pb 90-277-0995-5
137. J. C. Webb, *Mechanism, Mentalism and Metamathematics.* An Essay on Finitism. 1980 ISBN 90-277-1046-5
138. D. F. Gustafson and B. L. Tapscott (eds.), *Body, Mind and Method.* Essays in Honor of Virgil C. Aldrich. 1979 ISBN 90-277-1013-9
139. L. Nowak, *The Structure of Idealization.* Towards a Systematic Interpretation of the Marxian Idea of Science. 1980 ISBN 90-277-1014-7
140. C. Perelman, *The New Rhetoric and the Humanities.* Essays on Rhetoric and Its Applications. Translated from French and German. With an Introduction by H. Zyskind. 1979 ISBN 90-277-1018-X; Pb 90-277-1019-8
141. W. Rabinowicz, *Universalizability.* A Study in Morals and Metaphysics. 1979
ISBN 90-277-1020-2
142. C. Perelman, *Justice, Law and Argument.* Essays on Moral and Legal Reasoning. Translated from French and German. With an Introduction by H.J. Berman. 1980
ISBN 90-277-1089-9; Pb 90-277-1090-2
143. S. Kanger and S. Öhman (eds.), *Philosophy and Grammar.* Papers on the Occasion of the Quincentennial of Uppsala University. 1981 ISBN 90-277-1091-0
144. T. Pawlowski, *Concept Formation in the Humanities and the Social Sciences.* 1980
ISBN 90-277-1096-1
145. J. Hintikka, D. Gruender and E. Agazzi (eds.), *Theory Change, Ancient Axiomatics and Galileo's Methodology.* Proceedings of the 1978 Pisa Conference on the History and Philosophy of Science, Volume I. 1981 ISBN 90-277-1126-7
146. J. Hintikka, D. Gruender and E. Agazzi (eds.), *Probabilistic Thinking, Thermodynamics, and the Interaction of the History and Philosophy of Science.* Proceedings of the 1978 Pisa Conference on the History and Philosophy of Science, Volume II. 1981 ISBN 90-277-1127-5
147. U. Mönnich (ed.), *Aspects of Philosophical Logic.* Some Logical Forays into Central Notions of Linguistics and Philosophy. 1981 ISBN 90-277-1201-8
148. D. M. Gabbay, *Semantical Investigations in Heyting's Intuitionistic Logic.* 1981
ISBN 90-277-1202-6

149. E. Agazzi (ed.), *Modern Logic – A Survey.* Historical, Philosophical, and Mathematical Aspects of Modern Logic and Its Applications. 1981 ISBN 90-277-1137-2
150. A. F. Parker-Rhodes, *The Theory of Indistinguishables.* A Search for Explanatory Principles below the Level of Physics. 1981 ISBN 90-277-1214-X
151. J. C. Pitt, *Pictures, Images, and Conceptual Change.* An Analysis of Wilfrid Sellars' Philosophy of Science. 1981 ISBN 90-277-1276-X; Pb 90-277-1277-8
152. R. Hilpinen (ed.), *New Studies in Deontic Logic.* Norms, Actions, and the Foundations of Ethics. 1981 ISBN 90-277-1278-6; Pb 90-277-1346-4
153. C. Dilworth, *Scientific Progress.* A Study Concerning the Nature of the Relation between Successive Scientific Theories. 2nd, rev. and augmented ed., 1986. 3rd rev. ed., 1994 ISBN 0-7923-2487-0; Pallas Pb 0-7923-2488-9
154. D. Woodruff Smith and R. McIntyre, *Husserl and Intentionality.* A Study of Mind, Meaning, and Language. 1982 ISBN 90-277-1392-8; Pb 90-277-1730-3
155. R. J. Nelson, *The Logic of Mind.* 2nd. ed., 1989 ISBN 90-277-2819-4; Pb 90-277-2822-4
156. J. F. A. K. van Benthem, *The Logic of Time.* A Model-Theoretic Investigation into the Varieties of Temporal Ontology, and Temporal Discourse. 1983; 2nd ed., 1991 ISBN 0-7923-1081-0
157. R. Swinburne (ed.), *Space, Time and Causality.* 1983 ISBN 90-277-1437-1
158. E. T. Jaynes, *Papers on Probability, Statistics and Statistical Physics.* Ed. by R. D. Rozenkrantz. 1983 ISBN 90-277-1448-7; Pb (1989) 0-7923-0213-3
159. T. Chapman, *Time: A Philosophical Analysis.* 1982 ISBN 90-277-1465-7
160. E. N. Zalta, *Abstract Objects.* An Introduction to Axiomatic Metaphysics. 1983 ISBN 90-277-1474-6
161. S. Harding and M. B. Hintikka (eds.), *Discovering Reality.* Feminist Perspectives on Epistemology, Metaphysics, Methodology, and Philosophy of Science. 1983 ISBN 90-277-1496-7; Pb 90-277-1538-6
162. M. A. Stewart (ed.), *Law, Morality and Rights.* 1983 ISBN 90-277-1519-X
163. D. Mayr and G. Süssmann (eds.), *Space, Time, and Mechanics.* Basic Structures of a Physical Theory. 1983 ISBN 90-277-1525-4
164. D. Gabbay and F. Guenthner (eds.), *Handbook of Philosophical Logic.* Vol. I: Elements of Classical Logic. 1983 ISBN 90-277-1542-4
165. D. Gabbay and F. Guenthner (eds.), *Handbook of Philosophical Logic.* Vol. II: Extensions of Classical Logic. 1984 ISBN 90-277-1604-8
166. D. Gabbay and F. Guenthner (eds.), *Handbook of Philosophical Logic.* Vol. III: Alternative to Classical Logic. 1986 ISBN 90-277-1605-6
167. D. Gabbay and F. Guenthner (eds.), *Handbook of Philosophical Logic.* Vol. IV: Topics in the Philosophy of Language. 1989 ISBN 90-277-1606-4
168. A. J. I. Jones, *Communication and Meaning.* An Essay in Applied Modal Logic. 1983 ISBN 90-277-1543-2
169. M. Fitting, *Proof Methods for Modal and Intuitionistic Logics.* 1983 ISBN 90-277-1573-4
170. J. Margolis, *Culture and Cultural Entities.* Toward a New Unity of Science. 1984 ISBN 90-277-1574-2
171. R. Tuomela, *A Theory of Social Action.* 1984 ISBN 90-277-1703-6
172. J. J. E. Gracia, E. Rabossi, E. Villanueva and M. Dascal (eds.), *Philosophical Analysis in Latin America.* 1984 ISBN 90-277-1749-4
173. P. Ziff, *Epistemic Analysis.* A Coherence Theory of Knowledge. 1984

ISBN 90-277-1751-7

174. P. Ziff, *Antiaesthetics*. An Appreciation of the Cow with the Subtile Nose. 1984 ISBN 90-277-1773-7
175. W. Balzer, D. A. Pearce, and H.-J. Schmidt (eds.), *Reduction in Science*. Structure, Examples, Philosophical Problems. 1984 ISBN 90-277-1811-3
176. A. Peczenik, L. Lindahl and B. van Roermund (eds.), *Theory of Legal Science*. Proceedings of the Conference on Legal Theory and Philosophy of Science (Lund, Sweden, December 1983). 1984 ISBN 90-277-1834-2
177. I. Niiniluoto, *Is Science Progressive?* 1984 ISBN 90-277-1835-0
178. B. K. Matilal and J. L. Shaw (eds.), *Analytical Philosophy in Comparative Perspective*. Exploratory Essays in Current Theories and Classical Indian Theories of Meaning and Reference. 1985 ISBN 90-277-1870-9
179. P. Kroes, *Time: Its Structure and Role in Physical Theories*. 1985 ISBN 90-277-1894-6
180. J. H. Fetzer, *Sociobiology and Epistemology*. 1985 ISBN 90-277-2005-3; Pb 90-277-2006-1
181. L. Haaparanta and J. Hintikka (eds.), *Frege Synthesized*. Essays on the Philosophical and Foundational Work of Gottlob Frege. 1986 ISBN 90-277-2126-2
182. M. Detlefsen, *Hilbert's Program*. An Essay on Mathematical Instrumentalism. 1986 ISBN 90-277-2151-3
183. J. L. Golden and J. J. Pilotta (eds.), *Practical Reasoning in Human Affairs*. Studies in Honor of Chaim Perelman. 1986 ISBN 90-277-2255-2
184. H. Zandvoort, *Models of Scientific Development and the Case of Nuclear Magnetic Resonance*. 1986 ISBN 90-277-2351-6
185. I. Niiniluoto, *Truthlikeness*. 1987 ISBN 90-277-2354-0
186. W. Balzer, C. U. Moulines and J. D. Sneed, *An Architectonic for Science*. The Structuralist Program. 1987 ISBN 90-277-2403-2
187. D. Pearce, *Roads to Commensurability*. 1987 ISBN 90-277-2414-8
188. L. M. Vaina (ed.), *Matters of Intelligence*. Conceptual Structures in Cognitive Neuroscience. 1987 ISBN 90-277-2460-1
189. H. Siegel, *Relativism Refuted*. A Critique of Contemporary Epistemological Relativism. 1987 ISBN 90-277-2469-5
190. W. Callebaut and R. Pinxten, *Evolutionary Epistemology*. A Multiparadigm Program, with a Complete Evolutionary Epistemology Bibliograph. 1987 ISBN 90-277-2582-9
191. J. Kmita, *Problems in Historical Epistemology*. 1988 ISBN 90-277-2199-8
192. J. H. Fetzer (ed.), *Probability and Causality*. Essays in Honor of Wesley C. Salmon, with an Annotated Bibliography. 1988 ISBN 90-277-2607-8; Pb 1-5560-8052-2
193. A. Donovan, L. Laudan and R. Laudan (eds.), *Scrutinizing Science*. Empirical Studies of Scientific Change. 1988 ISBN 90-277-2608-6
194. H.R. Otto and J.A. Tuedio (eds.), *Perspectives on Mind*. 1988 ISBN 90-277-2640-X
195. D. Batens and J.P. van Bendegem (eds.), *Theory and Experiment*. Recent Insights and New Perspectives on Their Relation. 1988 ISBN 90-277-2645-0
196. J. Österberg, *Self and Others*. A Study of Ethical Egoism. 1988 ISBN 90-277-2648-5
197. D.H. Helman (ed.), *Analogical Reasoning*. Perspectives of Artificial Intelligence, Cognitive Science, and Philosophy. 1988 ISBN 90-277-2711-2

198. J. Wolenski, *Logic and Philosophy in the Lvov-Warsaw School.* 1989 ISBN 90-277-2749-X
199. R. Wójcicki, *Theory of Logical Calculi.* Basic Theory of Consequence Operations. 1988 ISBN 90-277-2785-6
200. J. Hintikka and M.B. Hintikka, *The Logic of Epistemology and the Epistemology of Logic.* Selected Essays. 1989 ISBN 0-7923-0040-8; Pb 0-7923-0041-6
201. E. Agazzi (ed.), *Probability in the Sciences.* 1988 ISBN 90-277-2808-9
202. M. Meyer (ed.), *From Metaphysics to Rhetoric.* 1989 ISBN 90-277-2814-3
203. R.L. Tieszen, *Mathematical Intuition.* Phenomenology and Mathematical Knowledge. 1989 ISBN 0-7923-0131-5
204. A. Melnick, *Space, Time, and Thought in Kant.* 1989 ISBN 0-7923-0135-8
205. D.W. Smith, *The Circle of Acquaintance.* Perception, Consciousness, and Empathy. 1989 ISBN 0-7923-0252-4
206. M.H. Salmon (ed.), *The Philosophy of Logical Mechanism.* Essays in Honor of Arthur W. Burks. With his Responses, and with a Bibliography of Burk's Work. 1990 ISBN 0-7923-0325-3
207. M. Kusch, *Language as Calculus vs. Language as Universal Medium.* A Study in Husserl, Heidegger, and Gadamer. 1989 ISBN 0-7923-0333-4
208. T.C. Meyering, *Historical Roots of Cognitive Science.* The Rise of a Cognitive Theory of Perception from Antiquity to the Nineteenth Century. 1989 ISBN 0-7923-0349-0
209. P. Kosso, *Observability and Observation in Physical Science.* 1989 ISBN 0-7923-0389-X
210. J. Kmita, *Essays on the Theory of Scientific Cognition.* 1990 ISBN 0-7923-0441-1
211. W. Sieg (ed.), *Acting and Reflecting.* The Interdisciplinary Turn in Philosophy. 1990 ISBN 0-7923-0512-4
212. J. Karpiński, *Causality in Sociological Research.* 1990 ISBN 0-7923-0546-9
213. H.A. Lewis (ed.), *Peter Geach: Philosophical Encounters.* 1991 ISBN 0-7923-0823-9
214. M. Ter Hark, *Beyond the Inner and the Outer.* Wittgenstein's Philosophy of Psychology. 1990 ISBN 0-7923-0850-6
215. M. Gosselin, *Nominalism and Contemporary Nominalism.* Ontological and Epistemological Implications of the Work of W.V.O. Quine and of N. Goodman. 1990 ISBN 0-7923-0904-9
216. J.H. Fetzer, D. Shatz and G. Schlesinger (eds.), *Definitions and Definability.* Philosophical Perspectives. 1991 ISBN 0-7923-1046-2
217. E. Agazzi and A. Cordero (eds.), *Philosophy and the Origin and Evolution of the Universe.* 1991 ISBN 0-7923-1322-4
218. M. Kusch, *Foucault's Strata and Fields.* An Investigation into Archaeological and Genealogical Science Studies. 1991 ISBN 0-7923-1462-X
219. C.J. Posy, *Kant's Philosophy of Mathematics.* Modern Essays. 1992 ISBN 0-7923-1495-6
220. G. Van de Vijver, *New Perspectives on Cybernetics.* Self-Organization, Autonomy and Connectionism. 1992 ISBN 0-7923-1519-7
221. J.C. Nyíri, *Tradition and Individuality.* Essays. 1992 ISBN 0-7923-1566-9
222. R. Howell, *Kant's Transcendental Deduction.* An Analysis of Main Themes in His Critical Philosophy. 1992 ISBN 0-7923-1571-5

223. A. García de la Sienra, *The Logical Foundations of the Marxian Theory of Value.* 1992 ISBN 0-7923-1778-5
224. D.S. Shwayder, *Statement and Referent.* An Inquiry into the Foundations of Our Conceptual Order. 1992 ISBN 0-7923-1803-X
225. M. Rosen, *Problems of the Hegelian Dialectic.* Dialectic Reconstructed as a Logic of Human Reality. 1993 ISBN 0-7923-2047-6
226. P. Suppes, *Models and Methods in the Philosophy of Science: Selected Essays.* 1993 ISBN 0-7923-2211-8
227. R. M. Dancy (ed.), *Kant and Critique: New Essays in Honor of W. H. Werkmeister.* 1993 ISBN 0-7923-2244-4
228. J. Woleński (ed.), *Philosophical Logic in Poland.* 1993 ISBN 0-7923-2293-2
229. M. De Rijke (ed.), *Diamonds and Defaults.* Studies in Pure and Applied Intensional Logic. 1993 ISBN 0-7923-2342-4
230. B.K. Matilal and A. Chakrabarti (eds.), *Knowing from Words.* Western and Indian Philosophical Analysis of Understanding and Testimony. 1994 ISBN 0-7923-2345-9
231. S.A. Kleiner, *The Logic of Discovery.* A Theory of the Rationality of Scientific Research. 1993 ISBN 0-7923-2371-8
232. R. Festa, *Optimum Inductive Methods.* A Study in Inductive Probability, Bayesian Statistics, and Verisimilitude. 1993 ISBN 0-7923-2460-9
233. P. Humphreys (ed.), *Patrick Suppes: Scientific Philosopher.* Vol. 1: Probability and Probabilistic Causality. 1994 ISBN 0-7923-2552-4; Set Vols. 1–3 ISBN 0-7923-2554-0
234. P. Humphreys (ed.), *Patrick Suppes: Scientific Philosopher.* Vol. 2: Philosophy of Physics, Theory Structure, and Measurement Theory. 1994 ISBN 0-7923-2553-2; Set Vols. 1–3 ISBN 0-7923-2554-0
235. P. Humphreys (ed.), *Patrick Suppes: Scientific Philosopher.* Vol. 3: Language, Logic, and Psychology. 1994 ISBN 0-7923-2862-0; Set Vols. 1–3 ISBN 0-7923-2554-0
236. D. Prawitz and D. Westerståhl (eds.), *Logic and Philosophy of Science in Uppsala.* Papers from the 9th International Congress of Logic, Methodology, and Philosophy of Science. 1994 ISBN 0-7923-2702-0
237. L. Haaparanta (ed.), *Mind, Meaning and Mathematics.* Essays on the Philosophical Views of Husserl and Frege. 1994 ISBN 0-7923-2703-9
238. J. Hintikka (ed.): *Aspects of Metaphor.* 1994 ISBN 0-7923-2786-1
239. B. McGuinness and G. Oliveri (eds.), *The Philosophy of Michael Dummett.* With Replies from Michael Dummett. 1994 ISBN 0-7923-2804-3
240. D. Jamieson (ed.), *Language, Mind, and Art.* Essays in Appreciation and Analysis, In Honor of Paul Ziff. 1994 ISBN 0-7923-2810-8
241. G. Preyer, F. Siebelt and A. Ulfig (eds.), *Language, Mind and Epistemology.* On Donald Davidson's Philosophy. 1994 ISBN 0-7923-2811-6
242. P. Ehrlich (ed.), *Real Numbers, Generalizations of the Reals, and Theories of Continua.* 1994 ISBN 0-7923-2689-X
243. G. Debrock and M. Hulswit (eds.): *Living Doubt.* Essays concerning the epistemology of Charles Sanders Peirce. 1994 ISBN 0-7923-2898-1
244. J. Srzednicki: *To Know or Not to Know.* Beyond Realism and Anti-Realism. 1994 ISBN 0-7923-2909-0
245. R. Egidi (ed.): *Wittgenstein: Mind and Language.* 1995 ISBN 0-7923-3171-0
246. A. Hyslop: *Other Minds.* 1995 ISBN 0-7923-3245-8

247. L. Pólos and M. Masuch (eds.): *Applied Logic: How, What and Why*. Logical Approaches to Natural Language. 1995 ISBN 0-7923-3432-9
248. M. Krynicki, M. Mostowski and L.M. Szczerba (eds.): *Quantifiers: Logics, Models and Computation*. Volume I: Surveys. 1995 ISBN 0-7923-3448-5
249. M. Krynicki, M. Mostowski and L.M. Szczerba (eds.): *Quantifiers: Logics, Models and Computation*. Volume II: Contributions. 1995 ISBN 0-7923-3449-3
Set ISBN (volume 248 + 249) 0-7923-3450-7

KLUWER ACADEMIC PUBLISHERS – DORDRECHT / BOSTON / LONDON